Understanding Probability

Chance events are commonplace in our daily lives. Every day, we encounter many situations in which the outcome is uncertain, and, perhaps without realizing it, we make a guess about the likelihood of one outcome or another. But mastering the concepts of probability can cast a new light on situations in which randomness and chance appear to rule.

Here, the reader can learn about the world of probability in an informal way. Lotteries and casino games provide a natural source of motivation, and these are carefully discussed with many worked examples to illustrate the key concepts and ideas from probability theory. The emphasis is on why probability works and how it can be applied.

The author introduces the reader to the law of large numbers, betting systems, random walks, the bootstrap, rare events, the central limit theorem, the multivariate normal distribution, the Bayesian approach, generating functions, and more.

Written with wit and clarity, this book can easily be read by anyone who is not put off by a few numbers and some high school algebra. It is also ideally suited to students of engineering or computer science undertaking a first course in probability.

HENK TIJMS is Professor in Operations Research at the Vrije University in Amsterdam. The author of several textbooks, including *A First Course in Stochastic Models*, he is intensely active in the popularization of applied mathematics and probability in Dutch high schools. He has also written numerous papers on applied probability and stochastic optimization for international journals, including *Applied Probability* and *Probability in the Engineering and Informational Sciences*.

Understanding Probability

Chance Rules in Everyday Life

HENK TIJMS
Vrije University

PUBLISHED BY THE PRESS SYNDICATE OF THE UNIVERSITY OF CAMBRIDGE
The Pitt Building, Trumpington Street, Cambridge, United Kingdom

CAMBRIDGE UNIVERSITY PRESS
The Edinburgh Building, Cambridge CB2 2RU, UK
40 West 20th Street, New York, NY 10011-4211, USA
477 Williamstown Road, Port Melbourne, VIC 3207, Australia
Ruiz de Alarcón 13, 28014 Madrid, Spain
Dock House, The Waterfront, Cape Town 8001, South Africa

http://www.cambridge.org

First published 2004

Printed in the United States of America

Typefaces Times Roman 10/13 pt., Helvetica and Courier *System* LaTeX 2_ε [TB]

A catalog record for this book is available from the British Library.

Library of Congress Cataloging in Publication Data

Tijms, H. C.

Understanding probability : chance rules in everyday life / Henk Tijms.

p. cm.

Includes bibliographical references and index.

ISBN 0-521-83329-9 – ISBN 0-521-54036-4 (pbk.)

1. Probabilities. 2. Mathematical analysis. 3. Chance. I. Title.

QA273.T48 2004
519.2 – dc22 2004040787

ISBN 0 521 83329 9 hardback
ISBN 0 521 54036 4 paperback

Contents

Preface

When I was a student, a class in topology made a great impression on me. The teacher asked me and my classmates not to take notes during the first hour of his lectures. In that hour, he explained ideas and concepts from topology in a nonrigorous, intuitive way. All we had to do was listen in order to grasp the concepts being introduced. In the second hour of the lecture, the material from the first hour was treated in a mathematically rigorous way, and the students were allowed to take notes. I learned a lot from this approach of interweaving intuition and formal mathematics.

This book, about probability as it applies to our daily lives, is written very much in the same spirit. It introduces the reader to the world of probability in an informal way. It is not written in a theorem-proof style. Instead, it aims to teach the novice the concepts of probability through the use of motivating and insightful examples. In the book, no mathematics are introduced without specific examples and applications to motivate the theory. Instruction is driven by the need to answer questions about probability problems that are drawn from real-world contexts. Most of the book can easily be read by anyone who is not put off by a few numbers and some high school algebra. The informal yet precise style of the book makes it suited for classroom use, particularly when more self-activation is required from students. The book is organized into chapters that may be understood if read in a nonlinear order. The concepts and the ideas are laid out in the first part of the book, while the second part covers the mathematical background. In the second part of the book, I have chosen to give a short account of the mathematics of the subject by highlighting the essentials in about 150 pages, which I believe better contributes to the understanding of the student than a diffuse account of many more pages. The book can be used for a one-quarter or one-semester course in a wide range of disciplines ranging from social sciences to engineering. Also, it is an ideal book to use as a supplementary text in more mathematical treatments of probability.

The book distinguishes itself from other introductory probability texts by its emphasis on why probability works and how to apply it. Simulation in interaction with theory is the perfect instrument to clarify and to enliven the basic concepts of probability. For this reason, computer simulation is used to give the reader insights into such key concepts as the law of large numbers, which come to life through the results of many simulation trials. The law of large numbers and the central limit theorem are at the center of the book, with numerous examples based on these main themes. Many of the examples deal with lotteries and casino games. The examples help the reader develop a "feel for probabilities." Good exercises are an essential part of each textbook. Much care has been paid to collecting exercises that appeal to the understanding and creativity of the reader rather than requiring the reader to plug numbers into formulas. Several of the examples and exercises in this book are inspired by material from the website "Chance News" (*www.dartmouth.edu/~chance*). This website contains a wealth of material on probability and statistics.

Acknowledgments

Many people helped me in one way or another during the writing of this book. My son Steven Tijms contributed to my Dutch book *Spelen met Kansen*, which formed the basis of this current edition. Sherrill Rose and Rob Luginbuhl are acknowledged for their invaluable help in translating parts of the Dutch text. The Dittmer Trust of the Vrije University and the Professor J. W. Cohen Foundation furnished financial support for the translation. My colleagues Ted Hill, Rein Nobel, Ad Ridder, Sheldon Ross, and Karl Sigman provided encouragement and helpful suggestions during the three years in which the book was written. Also, I owe a debt to my students Ton Dieker, Dion Heijnen, Sergio Panday, and Phi Hung Tran for their help in preparing the manuscript. Finally, the cartoons in the book were supplied by *www.cartoonstock.com*.

Amsterdam
November 2003

Introduction

It is difficult to say who had a greater impact on the mobility of goods in the preindustrial economy: the inventor of the wheel or the crafter of the first pair of dice. One thing, however, is certain: the genius that designed the first random number generator, like the inventor of the wheel, will very likely remain anonymous forever. We do know that the first dice-like exemplars were made a very long time ago. Excavations in the Middle East and in India reveal that dice were already in use at least fourteen centuries before Christ. Earlier still, around 3500 B.C., a board game existed in Egypt in which players tossed four-sided sheep bones. Known as the *astragalus*, this precursor to the modern-day die remained in use right up to the Middle Ages.

In the sixteenth century, the game of dice, or craps as we might call it today, was subjected for the first time to a formal mathematical study by the Italian mathematician and physician Gerolamo Cardano (1501–1576). An ardent gambler,

Cardano wrote a handbook for gamblers entitled *Liber de Ludo Aleae* (*The Book of Games of Chance*) about probabilities in games of chance. Cardano originated and introduced the concept of the set of outcomes of an experiment, and for cases in which all outcomes are equally probable, he defined the probability of any one event occurring as the ratio of the number of favorable outcomes and the total number of possible outcomes. This may seem obvious today, but in Cardano's day such an approach marked an enormous leap forward in the development of probability theory. This approach, along with a correct counting of the number of possible outcomes, gave the famous astronomer and physicist Galileo Galilei the tools he needed to explain to the Grand Duke of Tuscany, his benefactor, why it is that when you toss three dice, the chance of the sum being 10 is greater than the chance of the sum being 9 (the probabilities are $\frac{27}{216}$ and $\frac{25}{216}$, respectively).

By the end of the seventeenth century, the Dutch astronomer Christiaan Huygens (1625–1695) laid the foundation for current probability theory. His text *Van Rekeningh in Spelen van Geluck* (*On Reasoning in Games of Chance*), published in 1660, had enormous influence on later developments in probability theory (this text had already been translated into Latin under the title *De Ratiociniis de Ludo Aleae* in 1657). It was Huygens who originally introduced the concept of expected value, which plays such an important role in probability theory. His work unified various problems that had been solved earlier by the famous French mathematicians Pierre Fermat and Blaise Pascal. Among these was the interesting problem of how two players in a game of chance should divide the stakes if the game ends prematurely. Huygens' work led the field for many years until, in 1713, the Swiss mathematician Jakob Bernoulli (1654–1705) published *Ars Conjectandi* (*The Art of Conjecturing*) in which he presented the first general theory for calculating probabilities. Then, in 1812, the great French mathematician Pierre Simon Laplace (1749–1827) published his *Théorie Analytique des Probabilités*. This book unquestionably represents the greatest contribution in the history of probability theory.

Fermat and Pascal established the basic principles of probability in their brief correspondence during the summer of 1654, in which they considered some of the specific problems of odds calculation that had been posed to them by gambling acquaintances. One of the more well known of these problems is that of the Chevalier de Méré, who claimed to have discovered a contradiction in arithmetic. De Méré knew that it was advantageous to wager that a six would be rolled at least one time in four rolls of one die, but his experience as gambler taught him that it was not advantageous to wager on a double six being rolled at least one time in 24 rolls of a pair of dice. He argued that there were six possible outcomes for the toss of a single die and 36 possible outcomes for the toss of a pair of dice, and he claimed that this evidenced a contradiction to the arithmetic law of proportions, which says that the ratio of 4 to 6 should be the same as 24 to 36. De Méré turned to Pascal, who showed him with a few simple calculations that probability does not follow the law of proportions, as De Méré had mistakenly assumed (by De Méré's logic, the probability of at least one head in two tosses of a fair coin would be $2 \times 0.5 = 1$, which we know cannot be true). In any case, De Méré must have been an ardent player in order to have established empirically that the probability of rolling at least one double six in 24 rolls of a pair of dice lies just under one-half. The precise value of this probability is 0.4914. The probability of rolling at least one six in four rolls of a single die can be calculated as 0.5177. Incidentally, you may find it surprising that four rolls of a die are required, rather than three, in order to have about an equal chance of rolling at least one six.

Modern probability theory

Although probability theory was initially the product of questions posed by gamblers about their odds in the various games of chance, in its modern form, it has far outgrown any boundaries associated with the gaming room. These days, probability theory plays an increasingly greater roll in many fields. Countless problems in our daily lives call for a probabilistic approach. In many cases, better judicial and medical decisions result from an elementary knowledge of probability theory. It is essential to the field of insurance.[1] And, likewise, the stock market, "the largest casino in the world," cannot do without it. The telephone network, with its randomly fluctuating load, could not have been economically designed without the aid of probability theory. Call centers and airline companies apply probability theory to determine how many telephone lines and service desks will be needed based on expected demand. Probability theory is also essential in stock control to find a balance between the stock-out probability and the costs of holding inventories in an environment of uncertain demand. Engineers use probability theory when constructing dikes to calculate the probability of water levels exceeding their margins; this gives them the information they need to determine optimum dike elevation. These examples underline the extent to which the theory of probability has become an integral part of our lives. Laplace was right when he wrote almost 200 years ago in his *Théorie Analytique des Probabilités*:

[1] Actuarial scientists have been contributing to the development of probability theory since its early stages.

> The theory of probabilities is at bottom nothing but common sense reduced to calculus; it enables us to appreciate with exactness that which accurate minds feel with a sort of instinct for which ofttimes they are unable to account.... It teaches us to avoid the illusions which often mislead us; ... there is no science more worthy of our contemplations nor a more useful one for admission to our system of public education.

Probability theory and simulation

In terms of practical range, probability theory is comparable with geometry; both are branches of applied mathematics that are directly linked with the problems of daily life. But while pretty much anyone can call up a natural feel for geometry to some extent, many people clearly have trouble with the development of a good intuition for probability. Probability and intuition do not always agree. In no other branch of mathematics is it so easy to make mistakes as in probability theory. The development of the foundations of probability theory took a long time and went accompanied with ups and downs. The reader facing difficulties in grasping the concepts of probability theory might find comfort in the idea that even the genius Gottfried von Leibniz (1646–1716), the inventor of differential and integral calculus along with Newton, had difficulties in calculating the probability of throwing 11 with one throw of two dice. Probability theory is a difficult subject to get a good grasp of, especially in a formal framework. The computer offers excellent possibilities for acquiring a better understanding of the basic ideas of probability theory by means of simulation. With computer simulation, a concrete probability situation can be imitated on the computer. The simulated results can then be shown graphically on the screen. The graphic clarity offered by such a computer simulation makes it an especially suitable means to acquiring a better feel for probability. Not only a didactic aid, computer simulation is also a practical tool for tackling probability problems that are too complicated for scientific solution. Computer simulation, for example, has made it possible to develop winning strategies in the game of blackjack.

An outline

Part 1 of the book comprises Chapters 1 to 6. These chapters introduce the reader to the basic concepts of probability theory by using motivating examples to illustrate the concepts. A “feel for probabilities” is first developed through examples that endeavor to bring out the essence of probability in a compelling way. Simulation is a perfect aid in this undertaking of providing insight into the hows and whys of probability theory. We will use computer simulation, when needed, to illustrate subtle issues. The two pillars of probability theory – namely, the *law of large numbers* and the *central limit theorem* – receive in-depth treatment. The nature of these two laws is best illustrated through the coin-toss experiment. The law of large numbers says that the percentage of tosses to come out heads will be as close to 0.5 as you can imagine, provided that the coin is tossed often enough. How often the coin must be tossed in order to reach a prespecified precision for the percentage can be identified with the central limit theorem.

In Chapter 1, readers first encounter a series of intriguing problems to test their feel for probabilities. These problems will all be solved in the ensuing chapters. In Chapter 2, the law of large numbers provides the central theme. This law makes a connection between the probability of an event in an experiment and the relative frequency with which this event will occur when the experiment is repeated a very large number of times. Formulated by the aforementioned Jakob Bernoulli, the law of large numbers forms the theoretical foundation under the experimental determination of probability by means of computer simulation. The law of large numbers is clearly illuminated by the repeated coin-toss experiment, which is discussed in detail in Chapter 2. Astonishing results hold true in this simple experiment, and these results blow holes in many a mythical assumption, such as the "hot hand" in basketball. One remarkable application of the law of large numbers can be seen in the Kelly formula, a betting formula that can provide insight for the making of horse racing and investment decisions alike. The basic principles of computer simulation will also be discussed in Chapter 2, with emphasis on the subject of how random numbers can be generated on the computer.

In Chapter 3, we will tackle a number of realistic probability problems. Each problem will undergo two treatments, the first one being based on computer simulation and the second bearing the marks of a theoretical approach. Lotteries and casino games are sources of inspiration for some of the problems in Chapter 3.

The binomial distribution, the Poisson distribution, and the hypergeometric distribution are the subjects of Chapter 4. We will discuss which of these important probability distributions applies to which probability situations, and we will take a look into the practical importance of the distributions. Once again, we look to the lotteries to provide us with instructional and entertaining examples. We will see, in particular, how important the sometimes underestimated Poisson distribution, named after the French mathematician Siméon-Denis Poisson (1781–1840), really is.

In Chapter 5, two more fundamental principles of probability theory and statistics will be introduced: the central limit theorem and the normal distribution with its bell-shaped probability curve. The central limit theorem is by far the most important product of probability theory. The names of the mathematicians Abraham de Moivre and Pierre Simon Laplace are inseparably linked to this theorem and to the normal distribution. De Moivre discovered the normal distribution around 1730.[2] An explanation of the frequent occurrence of this distribution is provided by the central limit theorem. This theorem states that data influenced by many small and unrelated random effects are approximately normally distributed. It has been empirically observed that various natural phenomena – such as the heights of individuals, intelligence scores, the luminosity of stars, and daily returns of the S&P – follow approximately a normal distribution. The normal curve is also indispensable in quantum theory in physics. It describes the statistical behavior of huge numbers of atoms or electrons. A great many statistical methods are based on the central limit theorem. For one thing, this theorem makes it possible for us to evaluate how (im)probable certain deviations from the expected value are. For example, is the claim that heads came up 5,250 times in 10,000 tosses of a fair coin credible? What are the margins of errors in the predictions of election polls? The standard deviation concept plays a key roll in the answering of these questions. We devote considerable attention to this fundamental concept, particularly in the context of investment issues. At the same time, we also demonstrate in Chapter 5, with the help of the central limit theorem, how confidence intervals for the outcomes of simulation studies can be constructed. The standard deviation concept also comes into play here. The central limit theorem will also be used to link the random walk model with the Brownian motion model. These models, which are used to describe the behavior of a randomly moving object, are among the most useful probability models in science. Applications in finance will be discussed, including the Black–Scholes formula for the pricing of options.

The probability tree concept is discussed in Chapter 6. For situations in which the possibility of an uncertain outcome exists in successive phases, a probability tree can be made to systematically show what all of the possible paths are. Various applications of the probability tree concept will be considered, including the famous Monty Hall dilemma and the test paradox. In addition, we will also look at the Bayes formula in Chapter 6. This formula is a descriptive rule for revising probabilities in light of new information. Among other things, the Bayes rule is used in legal argumentation and in formulating medical diagnoses

[2]The French-born Abraham de Moivre (1667–1754) lived most of his life in England. The protestant de Moivre left France in 1688 to escape religious persecution. He was a good friend of Isaac Newton and supported himself by calculating odds for gamblers and insurers and by giving private lessons to students.

for specific illnesses. This eighteenth century formula, constructed by the English clergyman Thomas Bayes (1702–1761), laid the foundation for a separate branch of statistics, namely, Bayesian statistics.

Part 2 of the book comprises Chapters 7 to 14. These chapters are intended for the more mathematically oriented reader. Chapter 7 goes more deeply into the axioms and rules of probability theory. In Chapter 8, the concept of conditional probability and the nature of Bayesian analysis are delved into more deeply. Properties of the expected value are discussed in Chapter 9. Chapter 10 gives an explanation of continuous distributions, always a difficult concept for the beginner to absorb, and provides insight into the most important probability densities. Whereas Chapter 10 deals with the probability distribution of a single random variable, Chapter 11 discusses joint probability distributions for two or more dependent random variables. The multivariate normal distribution is the most important joint probability distribution and is the subject of Chapter 12. Chapter 13 deals with conditional distributions and discusses the law of conditional expectations. The final Chapter 14 covers the powerful tool of moment-generating functions.

ONE

Probability in action

1

Probability questions

In this chapter, we provide a number of probability problems that challenge the reader to test his or her feeling for probabilities. As stated in the introduction, it is possible to fall wide of the mark when using intuitive reasoning to calculate a probability or to estimate the order of magnitude of a probability. To find out how you fare in this regard, it may be useful to try one or more of these twelve problems. They are playful in nature, but are also illustrative of the surprises one can encounter in the solving of practical probability problems. Think carefully about each question before looking up its solution. All of the solutions to these problems can be found scattered throughout the ensuing chapters.

Question 1. A birthday problem (§3.1, §4.2.3)

You go with a friend to a football (soccer) game. The game involves 22 players of the two teams and one referee. Your friend wagers that, among these 23 persons on the field, at least two people will have birthdays on the same day. You will receive ten dollars from your friend if this is not the case. How much money should you, if the wager is to be a fair one, pay out to your friend if he is right?

Question 2. Probability of winning streaks (§2.1.3, §5.9.1)

A basketball player has a 50% success rate in free throw shots. Assuming that the outcomes of all free throws are independent from one another, what is the probability that, within a sequence of 20 shots, the player can score five baskets in a row?

Question 3. A scratch-and-win lottery (§4.2.3)

A scratch-and-win lottery dispenses 10,000 lottery tickets per week in Andorra and ten million in Spain. In both countries, demand exceeds supply. There are two numbers, composed of multiple digits, on every lottery ticket. One of these numbers is visible, and the other is covered by a layer of silver paint. The numbers on the 10,000 Andorran tickets are composed of four digits, and the numbers on the ten million Spanish tickets are composed of seven digits. These numbers are randomly distributed over the quantity of lottery tickets, but in such a way that no two tickets display the same open or the same hidden number. The ticket holder wins a large cash prize if the number under the silver paint is revealed to be the same as the unpainted number on the ticket. Do you think the probability of at least one winner in the Andorran Lottery is significantly different from the probability of at least one winner in Spain? What is your estimate of the probability of a win occurring in each of the lotteries?

Question 4. A lotto problem (§4.2.3)

In each drawing of Lotto 6/45, six distinct numbers are drawn from the numbers $1, \ldots, 45$. In an analysis of 30 such lotto drawings, it was apparent that some numbers were never drawn. This is surprising. In total, $30 \times 6 = 180$ numbers were drawn, and it was expected that each of the 45 numbers would be chosen about four times. The question arises as to whether the lotto numbers were drawn according to the rules, and whether there may be some cheating occurring. What is the probability that, in 30 drawings, at least one of the numbers $1, \ldots, 45$ will not be drawn?

Question 5. Hitting the jackpot (Appendix)

Is the probability of hitting the jackpot (getting all six numbers right) in a 6/45 Lottery greater or lesser than the probability of throwing heads only in 22 tosses of a fair coin?

Question 6. A legal problem (§4.2.2)

A robbery occurs in a big city. The perpetrator is an adult male, and it is certain that he has fled the city limits. Regarding his appearance, a number of identifying marks have been revealed. Coincidentally, the police encounter someone directly after the robbery, and this person fits the description of the perpetrator. In light of the resemblance, the man is arrested and further search is called off. There is no other evidence against the arrested man. Counsel for the prosecution argues that only one in a million adult males possesses the features that the perpetrator was observed to have had. On the basis of the extreme infrequency of these features occurring and the fact that the city is populated by only 150,000 adult males, the prosecutor jumps to the conclusion that the odds of the suspect not being the perpetrator are practically nil and calls for a tough sentence. What do you think of this conclusion?

Question 7. A coincidence problem (§4.3)

Two people, perfect strangers to one another, both living in the same city of one million inhabitants, meet each other. Each has approximately 500 acquaintances in the city. Assuming that for each of the two people, the acquaintances represent a random sampling of the city's various population sectors, what is the probability of the two people having an acquaintance in common?

Question 8. A sock problem (Appendix)

You have taken ten different pairs of socks to the laundromat, and during the washing, six socks are lost. In the best-case scenario, you will still have seven matching pairs left. In the worst-case scenario, you will have four matching pairs left. Do you think the probabilities of these two scenarios differ greatly?

Question 9. A statistical test problem (§3.6)

Using one die and rolling it 1,200 times, someone claims to have rolled the points 1, 2, 3, 4, 5, and 6 for a respective total of 196, 202, 199, 198, 202, and 203 times. Do you believe that these outcomes are, indeed, the result of coincidence or do you think they are fabricated?

Question 10. The best-choice problem (§2.3)

Your friend proposes the following wager: Twenty people are requested, independently of one another, to write a number on a piece of paper (the papers should be evenly sized). They may write any number they like, no matter how high. You fold up the twenty pieces of paper and place them randomly onto a tabletop. Your friend opens the papers one by one. Each time he opens one, he must decide whether to stop with that one or go on to open another one. Your friend's task is to single out the paper displaying the highest number. Once a paper is opened, your friend cannot go back to any of the previously opened papers. He pays you one dollar if he does not identify the paper with the highest number on it, otherwise you pay him five dollars. Do you take the wager? If your answer is no, what would you say to a similar wager where 100 people were asked to write a number on a piece of paper and the stakes were one dollar from your friend for an incorrect guess against ten dollars from you if he guesses correctly?

Question 11. The Monty Hall dilemma (§6.1)

A game-show climax draws nigh. A drum-roll sounds. The game show host leads you to a wall with three closed doors. Behind one of the doors is the automobile of your dreams, and behind each of the other two is a can of dog food. The three doors all have even chances of hiding the automobile. The host, a trustworthy person who knows precisely what is behind each of the three doors, explains how the game will work. First, you will choose a door without opening it, knowing that after you have done so, the host will open one of the two remaining doors to reveal a can of dog food. When this has been done, you will be given the opportunity to switch doors; you will win whatever is behind the door you choose at this stage of the game. Do you raise your chances of winning the automobile by switching doors?

Question 12. A daughter-son problem (§2.9, §6.1)

You are told that a family, completely unknown to you, has two children and that one of these children is a daughter. Is the chance of the other child also being a daughter equal to $\frac{1}{2}$ or $\frac{1}{3}$? Are the chances altered if, aware of the fact that the family has two children only, you ring their doorbell and a daughter opens the door?

The psychology of probability intuition is a main feature of some of these problems. Consider the birthday problem: how large must a group of randomly chosen people be such that the probability of two people having birthdays on the same day will be at least 50%? The answer to this question is 23. Almost no one guesses this answer; most people name much larger numbers. The number 183 is very commonly suggested on the grounds that it represents half the number of days in a year. A similar misconception can be seen in the words of a lottery official regarding his lottery, in which a four-digit number was drawn daily from the 10,000 number sequence 0000, 0001, . . . , 9999. On the second anniversary of the lottery, the official deemed it highly improbable that any of the 10,000 possible numbers had been drawn two or more times in the last 625 drawings. He added that this could only be expected after approximately half of the 10,000 possible numbers had been drawn. The lottery official was wildly off the mark: the probability that some number will not be drawn two or more times in 625 drawings is inconceivably small and is of the order of magnitude of 10^{-9}. This probability can be calculated by looking at the problem as a "birthday problem" with 10,000 possible birthdays and a group of 625 people; see Section 3.1 in Chapter 3. Canadian lottery officials, likewise, had no knowledge of the birthday problem and its treacherous variants when they put this idea into play: They purchased 500 automobiles from nonclaimed prize monies to be raffled off as bonus prizes among their 2.4 million registered subscribers. A computer chose the winners by selecting 500 subscriber numbers from a pool of 2.4 million registered numbers without regard for whether or not a given number had already appeared. The unsorted list of the 500 winning numbers was published and to the astonishment of lottery officials, one subscriber put in a claim for two automobiles. Unlike the probability of a given number being chosen two or more times, the probability of some number being chosen two or more times is not negligibly small in this case; it is in the neighborhood of 5%! The Monty Hall dilemma – which made it onto the front page of the New York Times in 1991 – is even more interesting in terms of the reactions it generates. Some people vehemently insist that it does not matter whether a player switches doors at the end of the game, whereas others confidently maintain that the player must switch. We will not give away the answer here, but suffice it to say that many a mathematics professor gets this one wrong. These types of examples demonstrate that, in situations of uncertainty, one needs rational methods in order to avoid mental pitfalls.[1] Probability theory provides us with these methods. In the chapters that follow, you will journey through the fascinating world of probability theory. This journey will not take you over familiar, well-trodden territory; it will provide you with interesting prospects.

[1]An interesting article on mistakes in reasoning in situations of uncertainty is K. McKean, 'Decisions, decisions, ...,' *Discover*, June 1985, 22–31. This article is inspired by the standard work of D. Kahneman, P. Slovic, and A. Tversky, *Judgment under Uncertainty: Heuristics and Biases*, Cambridge University Press, Cambridge, 1982.

2

The law of large numbers and simulation

In the midst of a coin-tossing game, after seeing a long run of tails, we are often tempted to think that the chances that the next toss will be heads must be getting larger. Or, if we have rolled a dice many times without seeing a six, we are sure that finally we will roll a six. These notions are known as the *gambler's fallacy*. Of course, it is a mistake to think that the previous tosses will influence the outcome of the next toss: a coin or die has no memory. With each new toss, each of the possible outcomes remains equally likely. Irregular patterns of heads and tails are even characteristic of tosses with a fair coin. Unexpectedly long runs of heads or tails can already occur with a relatively few number of tosses. To see five or six heads in a row in 20 tosses is not exceptional. It is the case, however, that as the number of tosses increases, the fractions of heads and tails should be about equal, but that is guaranteed only *in the long run*. In the theory of probability, this fact is known as *the law of large numbers*. Just as the name implies, this law only says something about the game after a large number of tosses. This law does not imply that the absolute difference between the numbers of heads and tails should oscillate close to zero. On the contrary. For games of chance, such as coin-tossing, it is even typical, as we shall see, that for long time periods, either heads or tails will remain constantly in the lead, with the absolute difference between the numbers of heads and tails tending to become larger and larger. The course of a game of chance, although eventually converging in an average sense, is a whimsical process. What else would you have expected?

In this chapter, *the law of large numbers* will play the central role. Together with the central limit theorem from Chapter 5, this law forms the fundamental basis for probability theory. With the use of some illustrative examples – especially coin-tossing – and the use of simulation of chance experiments on the computer, we hope to provide the reader with a better insight into the law of large numbers, and into what this law says, and does not say, about the properties of random processes. To clarify and illustrate probability concepts, the simulation approach has some advantages over the formal, purely theoretical approach: It allows us to almost *instantly* simulate chance experiments, and present the results in a clear and graphic form. A picture is worth a thousand words! In this chapter, our first goal is to help the reader develop "a feel for probabilities." Then, the theory will be gradually introduced to enable us to calculate probabilities in concrete situations, using a clear and systematic approach.

2.1 The law of large numbers for probabilities

Suppose that the weather statistics over the last 200 years show that, on average, it rained 7 of 30 days in June, with no definite pattern for which particular days it rained. Assuming things do not change, then the probability of rain on June 15 the following year has the numerical value $\frac{7}{30}$. In this case, the past relative frequency of rainy days in June is used to assign a numerical value to the probability of rain on a given day in June during the following year. Put another way, the so-called empirical law of large numbers suggests the choice of $\frac{7}{30}$ for the probability of rain on any given day. We can shed further light on this law by considering repeated tosses of a fair coin. If after each toss you observe the percentage of heads up to that point, then you will see that in the beginning this percentage can fluctuate considerably, but eventually it settles down near 50% as the number of tosses increases. In general, suppose that a certain chance experiment will be carried out a large number of times under exactly the same conditions and in a way so that the repetitions of the experiment are independent of each other. Let A be a given event in the experiment – for example A is the event that in a randomly selected group of 23 people, two or more people have the same birthday. The *relative frequency* of the event A in n repetitions of the experiment is defined as

$$f_n(A) = \frac{n(A)}{n},$$

where $n(A)$ is the number of times that event A occurred in the n repetitions of the experiment. The relative frequency is a number between 0 and 1. Intuitively, it is clear that

> **the relative frequency with which event A occurs will fluctuate less and less as time goes on, and will approach a limiting value as the number of repetitions increases without bound.**

This phenomenon is known as *the empirical law of large numbers*. Intuitively, we would like to define the probability of the occurrence of the event A in a single repetition of the experiment as the limiting number to which the relative frequency $f_n(A)$ converges as n increases. Introducing the notion of probability this way bypasses several rather serious obstacles. The most serious obstacle is that, for relative frequency, the standard meaning of the notion of a limit cannot be applied (because you cannot assume *a priori* that the limiting number will be the same each time). For the foundations of probability theory, a different approach is followed. The more formal treatise is based on the concepts of sample space and probability measure. A *sample space* of a chance experiment is a set of elements that is in a one-to-one correspondence with the set of all possible outcomes of the experiment. On the sample space, a so-called probability measure is defined that associates to each subset of the sample space a numerical probability. The probability measure must satisfy a number of basic principles (axioms), which we will go into in Section 2.2 and in Chapter 7. These principles are otherwise motivated by properties of relative frequency. After we accept that the relative frequency of an event gives a good approximation for the probability of the event, then it is reasonable to let probabilities satisfy the same relations as relative frequencies. From these basic principles, if theoretical results can be derived that agree with our experience in concrete probability situations, then we know that the basic principles chosen are reasonable. Indeed, the so-called *theoretical law of large numbers* can be derived from the basic principles of probability theory. This theoretical law makes mathematically precise what the empirical law of large numbers tries to express. The theoretical law of large numbers can best be understood in the context of a random process where a fair coin is tossed an unlimited number of times. An outcome of this random process can be described by an infinite sequence of H's and T's, recording whether a head or tail turns up with each toss. The symbol ω is used to designate an outcome of the random process. For each conceivable outcome ω, we define the number $K_n(\omega)$ as

$$K_n(\omega) = \text{number of heads in the first } n \text{ tosses in outcome } \omega.$$

For example, with the outcome

$$\omega = (H, T, T, H, H, H, T, H, H, \ldots .),$$

we have $K_5(\omega) = 3$ and $K_8(\omega) = 5$. Intuitively, we expect that "nature" will guarantee that ω will satisfy

$$\lim_{n\to\infty} K_n(\omega)/n = 1/2.$$

There are many conceivable sequences ω for which $K_n(\omega)/n$ does not converge to $\frac{1}{2}$ as $n \to \infty$, such as sequences containing only a finite number of H's. Nevertheless, "nature" chooses only sequences ω for which there is convergence to $\frac{1}{2}$. The theoretical law of large numbers says that the set of outcomes for which $K_n(\omega)/n$ does not converge to $\frac{1}{2}$ as $n \to \infty$ is "negligibly small" in a certain measure-theoretic sense. In probability theory, we say that the fraction of tosses that come up heads *converges with probability 1* to the constant $\frac{1}{2}$ (see also Chapter 7).

To give a mathematical formulation of *the theoretical law of large numbers*, advanced mathematics is needed. In words, we can formulate this law as follows:

> **If a certain chance experiment is repeated an unlimited number of times under exactly the same conditions, and if the repetitions are independent of each other, then the fraction of times that a given event A occurs will converge with probability 1 to a number that is equal to the probability that A occurs in a single repetition of the experiment.**

This strong law of large numbers corresponds directly to our world of experience. This result is also the mathematical basis for the widespread application of computer simulations to solve practical probability problems. In these applications, the (unknown) probability of a given event in a chance experiment is estimated by the relative frequency of occurrence of the event in a large number of computer simulations of the experiment. The application of simulations is based on the elementary principles of probability; it is a powerful tool with which extremely complicated probability problems can be solved.

The mathematical basis for the theoretical (strong) law of large numbers was given for the first time by the famous Russian mathematician A. N. Kolmogorov in the twentieth century.[1] A so-called weak version of the law of large numbers had already been formulated several centuries earlier by the Swiss mathematician Jakob Bernoulli in his masterpiece *Ars Conjectandi* that was published posthumously in 1713. In that book, which was partially based on Christiaan Huygens' work, Bernoulli was the first to make the mathematical connection between the probability of an event and the relative frequency with which the event occurs. It is important to bear in mind that the law of large numbers says nothing about the outcome of a single experiment. But what can be predicted with 100% certainty from this law is the average value of the system, because the experiment is carried out an unlimited number of times. Not only is the method of computer simulation based on this fact, but also the profit-earning capacities of insurance companies and casinos is based on the strong law of large numbers.

2.1.1 Coin-tossing

How can you better illustrate the law of large numbers than with the experiment of tossing a coin? We will do this experiment for both fair and unfair coins. We let p designate the probability that one toss of the coin shows "heads." For a

[1] Andrey Nikolayevich Kolmogorov (1903–1987) was active in many fields of mathematics and is considered one of the greatest mathematicians of the twentieth century. He is credited with the axiomatic foundation of probability theory.

Table 2.1 *Results of coin-toss simulations*

	Fair coin ($p = \frac{1}{2}$)		Unfair coin ($p = \frac{1}{6}$)	
n	$K_n - np$	f_n	$K_n - np$	f_n
10	1.0	0.6000	0.33	0.2000
25	1.5	0.5600	1.83	0.2400
50	2.0	0.5400	2.67	0.2200
100	2.0	0.5200	3.33	0.2040
250	1.0	0.5040	5.33	0.1880
500	−2.0	0.4960	4.67	0.1760
1,000	10.0	0.5100	−3.67	0.1630
2,500	12.0	0.5048	−15.67	0.1604
5,000	−9.0	0.4982	−5.33	0.1656
7,500	11.0	0.5015	21.00	0.1659
10,000	24.0	0.5024	−33.67	0.1633
15,000	40.0	0.5027	−85.00	0.1610
20,000	91.0	0.5045	−17.33	0.1658
25,000	64.0	0.5026	−30.67	0.1654
30,000	78.0	0.5026	−58.00	0.1647

fair coin, clearly $p = \frac{1}{2}$. Define the variables

$$K_n = \text{the total number of heads that will appear in the first } n \text{ tosses}$$

and

$$f_n = \text{the relative frequency with which heads will appear in the first } n \text{ tosses.}$$

Clearly, it follows that $f_n = K_n/n$. Even more interesting than K_n is the variable $K_n - np$, the difference between the actual number of heads and the expected number of heads. Table 2.1 gives the simulated values of $K_n - np$ for 30,000 tosses of a coin for a number of intermediate values of n. This is done for both a fair coin ($p = \frac{1}{2}$) and for an unfair coin ($p = \frac{1}{6}$). The numbers in Table 2.1 are the outcome of a particular simulation study. Any other simulation study will produce different numbers. It is worthwhile to take a close look at the results in Table 2.1. You see that the realizations of the relative frequency, f_n, approach the true value of the probability p in a rather irregular manner. This is a typical phenomenon (try it yourself with your own simulations!). You see the same sort of phenomenon in lists that lottery companies publish of the relative frequencies of the different numbers that have appeared in past drawings. Results like those in Table 2.1 make it clear that fluctuations in the relative frequencies of the numbers drawn are nothing other than "natural" turns of fortune. In Table 2.1, it also is striking that the relative frequency f_n converges more slowly to the true value of the probability p than most of us would expect intuitively. The smaller the value of p, the more simulation effort is needed to ensure that the empirical relative frequency is close to p. In Chapter 5, we will see that the simulation effort must be increased about a hundredfold in order to simulate an unknown probability with one extra decimal place of precision. Thus, in principle, you should be suspicious of simulation studies that consist of only a small number of simulation runs, especially if they deal with small probabilities!

2.1.2 Random walk

Let's go back to the experiment of the fair coin-toss. Many people mistakenly think that a number of tosses resulting in heads will be followed by a number of tosses resulting in tails, such that both heads and tails will that turn up approximately the same number of times. In the world of gambling, many gamblers make use of a system that is based on keeping track of the number of heads and tails that turn up as a game progresses. This is often described as the *gambler's fallacy*. Alas, it is absolute folly to think that a system of this kind will help. A coin simply does not have a memory and will therefore exhibit no compensatory behavior. In order to stimulate participation in lotteries, lottery sponsors publish lists of so-called "hot" and "cold" numbers, recording the number of wins for each number and the number of drawings that have taken place since each number was last drawn as a winning number. Such a list is often great fun to see, but will be of no practical use whatsoever in the choosing of a number for a future drawing. Lottery balls have no memory and exhibit no compensatory behavior.

For example, suppose a fair coin is tossed 100 times, resulting in heads 60 times. In the next 100 tosses, the absolute difference between the numbers of heads and tails can increase, whereas the relative difference declines. This would be the case, for example, if the next 100 tosses were to result in heads 51 times. In the long run, "local clusters" of heads or tails are *absorbed* by the average. It is certain that the relative frequencies of heads and tails will be the same over the long run. There is simply no law of averages for the absolute difference between the numbers of heads and tails. Indeed, the absolute difference between the numbers of heads and tails tends to become larger as the number of tosses increases. This surprising fact can be convincingly demonstrated using computer simulation. The graph in Figure 2.1 describes the path of the *actual*

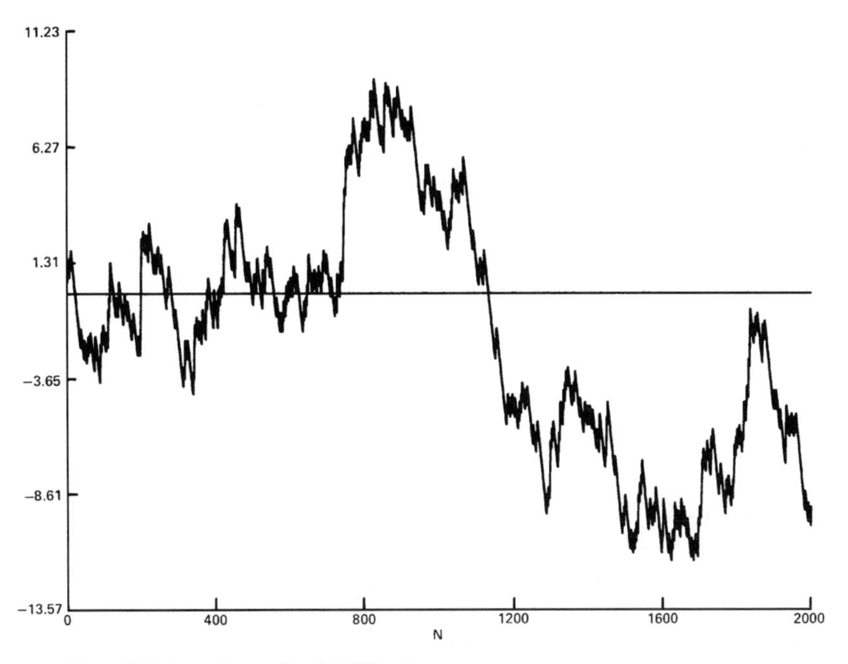

Figure 2.1 A random walk of 2,000 coin tosses.

number of heads turned up minus the *expected* number of heads when simulating 2,000 tosses of a fair coin. This process is called a *random walk*, based on the analogy of an indicator that moves one step higher if heads is thrown and one step lower, otherwise. A little bit of experimentation will show you that results such as those shown in Figure 2.1 are not exceptional. On the contrary, in fair coin-tossing experiments, it is typical to find that, as the number of tosses increases, the fluctuations in the random walk become larger and larger and a return to the zero-level becomes less and less likely. The appearance of these growing fluctuations can be clarified by looking at the central limit theorem, which will be discussed in Chapter 5. In that chapter, we demonstrate how the range of the difference between the actual number of heads and the expected number has a tendency to grow proportionally with $\sqrt{n}$ as n (= the number of tosses) increases. This result is otherwise not in conflict with the law of large numbers, which says that $\frac{1}{n}\times$(actual number of heads in n tosses $-\frac{1}{2}n$) goes to 0 when $n \to \infty$. It will be seen in Section 5.8.1 that the probability distribution of the proportion of heads in n tosses becomes more and more concentrated around the 50:50 ratio as n increases, where the deviations from the 50:50 ratio are on the order of $\frac{1}{\sqrt{n}}$.

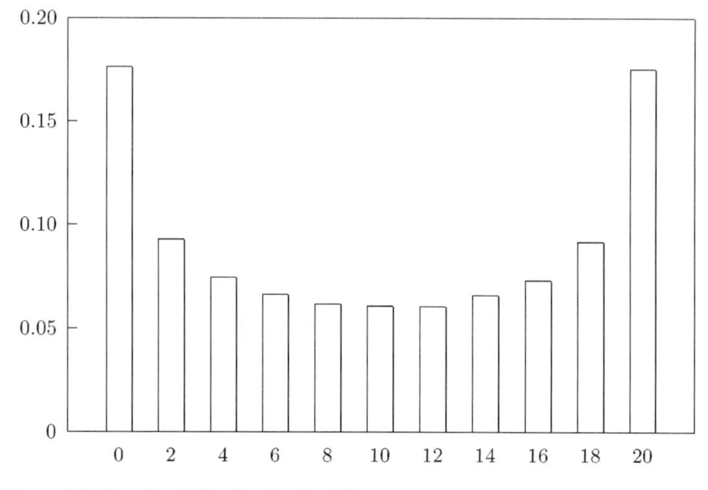

Figure 2.2 Simulated distribution for 20 tosses.

2.1.3 The arc-sine law[2]

The random walk resulting from the repeated tossing of a fair coin is filled with surprises that clash with intuitive thinking. We have seen that the random walk exhibits ever-growing fluctuations and that it returns to zero less and less frequently. Another characteristic of the fair coin toss that goes against intuition is that, in the vast majority of cases, the random walk tends to occur on one side of the axis line. Intuitively, one would expect that the most likely value of the percentage of total time the random walk occurs on the positive side of the axis will be somewhere near 50% when a *fixed* number of tosses is undertaken. But, quite the opposite is true, actually. This is illustrated by the simulation results in Figure 2.2. For this figure, we have simulated 100,000 repetitions of a match between two players A and B. The match consists of a series of 20 tosses of a fair coin, where player A scores a point each time heads comes up and player B scores a point each time tails comes up. Figure 2.2 gives the simulated distribution of the number of times that player A is in the lead during a series of 20 tosses. It is assumed that when the number of heads and tails is equal after 20 tosses, the winner will be the player who was in the lead after the penultimate toss (on the basis of this assumption, the number of times that player A is in the lead is always even).

Looking at the simulation results, it appears that player A has a probability of 17.5% of being in the lead during the whole match. Put differently, the player

[2]This specialized section may be omitted in a first reading of this book.

in the lead after the first toss has approximately a 35% probability of remaining in the lead throughout the 20-toss match. In contrast to this is the approximately 6% probability that player A will lead for half of the match and player B will lead for the other half. This specific result in the case of 20 matches can be more generally supported by the *arc-sine law*, given here without proofs. If the number of tosses in a match between players A and B is *fixed in advance*, and if this number is sufficiently large, then the following approximation formula holds true:

$$P\,(\text{player } A \text{ is at least } 100x\% \text{ of time in the lead}) \approx 1 - \frac{2}{\pi}\arcsin(\sqrt{x})$$

for each x satisfying $0 < x < 1$. From this approximation formula, it can be deduced that, for all α, β with $\frac{1}{2} \leq \alpha < \beta < 1$, it is true that

$$P(\text{one of the two players is in the lead for somewhere between } 100\alpha\% \text{ and } 100\beta\% \text{ of the time}) \approx \frac{4}{\pi}\{\arcsin(\sqrt{\beta}) - \arcsin(\sqrt{\alpha})\}.$$

Use $P(\alpha, \beta)$ to abbreviate $\frac{4}{\pi}\{\arcsin(\sqrt{\beta}) - \arcsin(\sqrt{\alpha})\}$. In Table 2.2, we give the value of $P(\alpha, \beta)$ for various values of α and β. The table shows that, in approximately 1 of 5 matches, one of the two players is in the lead for more than 97.5% of the time. It also shows that in 1 of 11 matches, one player is in the lead for more than 99.5% of the time. A fair coin, then, will regularly produce results that show no change in the lead for very long, continuous periods of time. Financial markets analysts would do well to keep these patterns in mind when analyzing financial markets. However controversial the assertion, some prominent economists claim that financial markets have no memory and behave according to a random walk. Their argument is quite simple: if share prices were predictable, then educated investors would buy low and sell high, but it would not be long before many others began to follow their lead, causing prices to adjust accordingly and to return to random behavior. This assertion is still

Table 2.2 *Probability $P(\alpha, \beta)$ in the arc-sine law*

(α, β)	$P(\alpha, \beta)$	(α, β)	$P(\alpha, \beta)$
(0.50, 0.505)	0.0064	(0.995, 1)	0.0901
(0.50, 0.510)	0.0127	(0.990, 1)	0.1275
(0.50, 0.525)	0.0318	(0.975, 1)	0.2022
(0.50, 0.550)	0.0638	(0.950, 1)	0.2871
(0.50, 0.600)	0.1007	(0.900, 1)	0.4097

extremely controversial because psychological factors (herd instinct) have a large influence on financial markets.[3]

Figure 2.2 and Table 2.2 demonstrate that the percentage of time of a random walk occurring on the positive side of the axis is much more likely to be near 0% or 100% than it is to be near the "expected" value of 50%. At first glance, few people cannot believe this to be the case. It is, however, true, and can be demonstrated with simulation experiments. These same simulations also demonstrate that the manner in which heads and tails switch off in a series of tosses with a fair coin is extremely irregular: unexpectedly, long series of heads or tails alone can occur. For example, in an experiment consisting of 20 tosses of a fair coin, simulation allows one to determine that the probability of a coin turning up heads five or more times in a row is approximately 25%, and that the probability of the coin landing on the same side, whether heads or tails, five or more times in a row is approximately 46%. On the grounds of this result, one need not be surprised if a basketball player with a free-throw success rate of 50% scores five or more baskets in a row in a series of 20 shots.

2.2 Events and random variables

This section deals with some of the fundamental theoretical concepts in probability theory. The *sample space* of an experiment has already been defined as a set of elements that is in a one-to-one correspondence with the set of all possible outcomes of the experiment. In many situations, you can make several choices for the sample space. Any subset of the sample space is called an *event*. That is, an event is a set consisting of possible outcomes of the experiment. If the outcome of the experiment is contained in the set E, it is said that the event E has occurred. A sample space in conjunction with a probability measure is called a *probability space*. A probability measure is simply a function P that assigns a numerical probability to each subset of the sample space. A probability measure must satisfy a number of consistency rules that will be discussed later.

Let's first illustrate a few things in light of an experiment that children sometimes use in their games to select one child out of the group. Three children simultaneously present their left or right fist to the group. If one of the children does not show the same fist as the other two, that child is "out." The sample space of this experiment can be described by the set $\{RRR, RRL, RLR, RLL, LLL, LLR, LRL, LRR\}$ consisting of eight elements, where $R(L)$ stands for a right

[3] See also Richard H. Thaler, *The Winner's Curse, Paradoxes and Anomalies in Economic Life*, Princeton University Press, Princeton, NJ, 1992.

(left) fist. The first letter of every element indicates which fist the first child shows, the second letter indicates the fist shown by the second child, and the third letter indicates the fist of the third child. If we assume that the children show the fists independently of one another, and each child chooses a fist randomly, then each of the outcomes is equally probable, and we can assign a probability of $\frac{1}{8}$ to each outcome. The outcome subset $\{RRL, RLR, RLL, LLR, LRL, LRR\}$ corresponds with the event that one of the children is declared "out." We assign a probability of $\frac{6}{8}$ to this event.

In many chance experiments, we are more interested in some function of the outcome of the experiment values than in the actual outcomes. A *random variable* is simply a function that is defined on the sample space of the experiment and assigns a numerical value to each possible outcome of the experiment. For example, in the experiment that consists of tossing a fair coin three times, the random variable X could be defined as the number of times the coin turns up heads. Or, in the experiment consisting of the simultaneous rolling of a pair of dice, the random variable X could be defined as the combined sum of the values rolled, or as the greater of the two values rolled. It is common to use uppercase letters such as X, Y, and Z to denote random variables, and lowercase letters x, y, and z to denote their possible numerical values. In many applications, the random variable X can take on only a finite number of possible values or values from a countably infinite set, such as the set of all nonnegative integers. In such a case, the random variable X is said to be a *discrete* random variable. In this book, we are mainly concerned with discrete random variables that take on a finite number of values. Let us assume that X can only take on values from the finite set $I = \{x_1, \ldots, x_M\}$. The event $\{X = x_j\}$ is defined as the set of those outcomes for which the random variable X takes on the value x_j. The probability of the event $\{X = x_j\}$ is thus defined as the sum of the probabilities of the individual outcomes for which X takes on the value x_j. This probability is denoted by $P(X = x_j)$. Abbreviating $P(X = x_j)$ as p_j for $j = 1, \ldots, M$, the collection of probabilities $(p_1, \ldots, p_M)$ is called the *probability distribution* of X.

Example 2.1 Consider the experiment involving the rolling of a fair pair of dice. What is the distribution of the random variable X when X is defined as the greater of the two point values rolled?

This random variable X has $I = \{1, \ldots, 6\}$ as its set of possible values. To find the distribution of X, you will need the sample space of the experiment. A logical choice is the set

$$\{(1, 1), \ldots, (1, 6), (2, 1), \ldots, (6, 1), \ldots, (6, 6)\},$$

where the outcome (i, j) corresponds with the event that a roll of the first ("blue") die gives i points and a roll of the second ("red") die gives j points. Each of the 36 possible outcomes is equally probable with fair dice. One translates this fact by assigning an equal probability of $\frac{1}{36}$ to each outcome. The random variable X assumes the value $\max(i, j)$ for outcome (i, j). For example, X assumes the value 3 for each of the five outcomes $(1, 3)$, $(3, 1)$, $(2, 3)$, $(3, 2)$, and $(3, 3)$. Consequently, $P(X = 3) = \frac{5}{36}$. In this way, one finds

$$P(X = 1) = \frac{1}{36}, \quad P(X = 2) = \frac{3}{36}, \quad P(X = 3) = \frac{5}{36},$$

$$P(X = 4) = \frac{7}{36}, \quad P(X = 5) = \frac{9}{36}, \quad P(X = 6) = \frac{11}{36}.$$

Example 2.2 You are participating in a game involving three tosses of a fair coin. The entry stake is two dollars. The payoff is three dollars plus your entry stake if heads comes up three times and one dollar plus entry stake if heads turns up twice. All other outcomes result in no payoff and a forfeiture of the entry stake. What is the distribution of the random variable X when X is defined as your net win?

To answer this question, we will take the set

$$\{HHH, HHT, HTH, THH, TTH, THT, HTT, TTT\},$$

as the sample space for the experiment, wherein an H (T) indicates that the toss in question turned up heads (tails). Each of the eight possible outcomes is equally probable and is assigned a probability of $\frac{1}{8}$. The random variable X takes on a value of 3 for the outcome HHH, a value of 1 for each of the three outcomes HHT, HTH, and THH, and a value of -2 for each of the other four outcomes. Thus, the random variable X has the set of possible values $I = \{3, 1, -2\}$ with the corresponding probability distribution

$$P(X = 3) = \frac{1}{8}, \quad P(X = 1) = \frac{3}{8}, \quad P(X = -2) = \frac{4}{8}.$$

In the following two examples we will discuss experiments in which not every element of the sample space is equally probable. These two examples involve so-called *compound* experiments. A compound experiment is one that is based on a sequence of elementary experiments. When the outcomes of the elementary experiments are independent of one another, then the probabilities assigned in the compound experiment are based on the multiplication of the probabilities of the outcomes in the individual elementary experiments. The theoretical construct for this *product rule* is discussed in Chapter 7.

Example 2.3 You participate in the following game. A fair coin is tossed into the air. If it lands heads, it will be tossed one more time, otherwise it will be tossed two more times. You will win ten dollars if heads does not come up at all; you must pay one dollar each time heads does turn up. What is the distribution of the random variable X when X is defined as your net win?

To answer this question, we will use the set

$$\{HH, HT, THH, THT, TTH, TTT\}$$

as our sample space for the experiment. A logical assignment of probabilities to the elements of this sample space is to assign the probability $\frac{1}{2} \times \frac{1}{2} = \frac{1}{4}$ to each of the outcomes HH and HT, and to assign the probability $\frac{1}{2} \times \frac{1}{4} = \frac{1}{8}$ to each of the outcomes THH, THT, TTH, and TTT. The random variable X takes on a value of 10 for the outcome TTT; the value -1 for outcomes HT, THT, and TTH; and the value -2 for the outcomes HH and THH. In this way, one finds the distribution

$$P(X = 10) = \frac{1}{8}, \quad P(X = -1) = \frac{1}{4} + \frac{1}{8} + \frac{1}{8} = \frac{1}{2},$$
$$P(X = -2) = \frac{1}{4} + \frac{1}{8} = \frac{3}{8}.$$

Example 2.4 Two desperados A and B are playing Russian roulette, and they have agreed that they will take turns pulling the trigger of a revolver with six cylinders and one bullet. This dangerous game ends when the trigger has been pulled six times without a fatal shot occurring (after each attempt the magazine is spun to a random position). Desperado A begins. What is the distribution of the number of times desperado A pulls the trigger?

The sample space for this experiment is taken as

$$\{F, GF, GGF, GGGF, GGGGF, GGGGGF, GGGGGG\},$$

where an F stands for an attempt resulting in a fatal shot, and G stands for an attempt that has a good ending. The results of the consecutive attempts are independent from one another. On these grounds, we will assign the probabilities

$$\frac{1}{6}, \left(\frac{5}{6}\right)\frac{1}{6}, \left(\frac{5}{6}\right)^2\frac{1}{6}, \left(\frac{5}{6}\right)^3\frac{1}{6}, \left(\frac{5}{6}\right)^4\frac{1}{6}, \left(\frac{5}{6}\right)^5\frac{1}{6} \text{ and } \left(\frac{5}{6}\right)^6$$

to the consecutive elements of the sample space. The random variable X will be defined as the number of times that desperado A pulls the trigger. The random variable X takes on the value 1 for outcomes F and GF, the value 2 for outcomes

GGF and $GGGF$ and the value 3 for all other outcomes. This gives

$$P(X = 1) = \frac{1}{6} + \left(\frac{5}{6}\right)\frac{1}{6} = 0.30556,$$

$$P(X = 2) = \left(\frac{5}{6}\right)^2 \frac{1}{6} + \left(\frac{5}{6}\right)^3 \frac{1}{6} = 0.21219,$$

$$P(X = 3) = \left(\frac{5}{6}\right)^4 \frac{1}{6} + \left(\frac{5}{6}\right)^5 \frac{1}{6} + \left(\frac{5}{6}\right)^6 = 0.48225.$$

We constructed a probability model for the various situations occurring in the above examples. The ingredients necessary for the making of a model are a sample space and the probabilities assigned to the elements of the sample space.[4] These ingredients are part of a translation process from a physical context into a mathematical framework. The probabilities assigned to the outcomes of the chance experiment do not just appear out of nowhere, we must choose them. Naturally, this must be done in such a way that the model is in agreement with reality. In most cases, when the experiment can be repeated infinitely under stable conditions, we have the empirical relative frequencies of the outcomes in mind along with the assignment of probabilities to the possible outcomes. For the case of a chance experiment with a *finite* sample space, it suffices to assign a probability to each individual element of the sample space. These elementary probabilities must naturally meet the requirement of being greater than or equal to 0 and adding up to 1. An event in the experiment corresponds with a subset in the sample space. It is said that an *event* A occurs when the outcome of the experiment belongs to the *subset* A of the sample space. A numerical value $P(A)$ is assigned to each subset A of the sample space. This numerical value $P(A)$ tells us how likely the event A is to occur. The probability function $P(A)$ is logically defined as:

$P(A)$ **is the sum of the probabilities of the individual outcomes in the set** A.

For the special case, when all outcomes are equally probable, $P(A)$ is found by dividing the number of outcomes in set A by the total number of outcomes. The model with equally probable outcomes is often called the *Laplace model.* It has numerous applications.

[4]To gain an understanding of this concept, the beginner would do well to undertake solving some of the problems at the end of this chapter that deal with specifying the sample space and determining the probabilities of the elements of the sample space in various probability situations. A knowledge of these principles forms the basis for a good command of probability theory. Once you have acquired this basis, you will often perform probability calculations without explicitly specifying a sample space.

The function P that assigns a numerical probability $P(A)$ to each subset A of the sample space is called a *probability measure*. A sample space in conjunction with a probability measure is called a *probability space*. The probability measure P also determines the probability distribution of every random variable X that is defined on the sample space. In fact, the probability $P(X = x)$ is none other than the probability of event A occurring, where set A consists of the outcomes for which the random variable X takes on the value x. The probability measure P satisfies the axioms of modern probability theory:

Axiom 1. $P(A) \geq 0$ *for every event* A.

Axiom 2. $P(A) = 1$, *when* A *is equal to the sample space.*

Axiom 3. $P(A \cup B) = P(A) + P(B)$ *for disjoint events* A *and* B.

Events A and B are said to be *disjoint* when the subsets A and B have no common elements. It is important to keep in mind that these axioms only provide us with the conditions that the probabilities must satisfy; they do not tell us how to assign probabilities in concrete cases. They are either assigned on the basis of relative frequencies (as in a dice game) or on the grounds of subjective consideration (as in a horse race). In both of these cases, the axioms are natural conditions that must be satisfied. The third axiom says that the probability of event A or event B occurring is equal to the sum of the probability of event A and the probability of event B, when these two events cannot occur simultaneously.[5] In the case of a nonfinite sample space, the *addition rule* from the third axiom must be modified accordingly. Rather than going into the details of such a modification here, we would direct interested readers to Chapter 7. The beauty of mathematics can be seen in the fact that these simple axioms suffice to derive such profound results as the theoretical law of large numbers. Compare this with a similar situation in geometry, where simple axioms about points and lines are all it takes to establish some very handsome results.

To illustrate, take another look at the above Example 2.4. Define A as the event that desperado A dies with his boots on and B as the event that B dies with his boots on. Event A is given by $A = \{F, GGF, GGGGF\}$. This gives

$$P(A) = \frac{1}{6} + \left(\frac{5}{6}\right)^2 \frac{1}{6} + \left(\frac{5}{6}\right)^4 \frac{1}{6} = 0.3628.$$

[5] The choice of the third axiom can be reasoned by the fact that relative frequency has the property $f_n(A \cup B) = f_n(A) + f_n(B)$ for disjoint events A and B, as one can directly see from the definition of relative frequency in Section 2.1 ($\frac{n(A)+n(B)}{n} = \frac{n(A)}{n} + \frac{n(B)}{n}$).

Likewise, one also finds that $P(B) = 0.3023$. The probability $P(A \cup B)$ represents the probability that one of the two desperados will end up shooting himself. Events A and B are disjoint and so

$$P(A \cup B) = P(A) + P(B) = 0.6651.$$

2.3 Expected value and the law of large numbers

The concept of expected value was first introduced into probability theory by Christiaan Huygens in the seventeenth century. Huygens established this important concept in the context of a game of chance, and to gain a good understanding of precisely what the concept is, it helps to retrace Huygens' footsteps. Consider a casino game where the player has a 0.70 probability of losing 1 dollar and probabilities of 0.25 and 0.05 of winning 2 and 3 dollars, respectively. A player who plays this game a large number of times reasons intuitively as follows in order to determine the average win per game in n games. In approximately $0.70n$ repetitions of the game, the player loses 1 dollar per game and in approximately $0.25n$ and $0.05n$ repetitions of the game, the player wins 2 and 3 dollars, respectively. This means that the total win in dollars is approximately equal to

$$(0.70n) \times (-1) + (0.25n) \times 2 + (0.05n) \times 3 = -(0.05)n,$$

or the average win per game is approximately -0.05 dollars (meaning that the average "win" is actually a loss). If we define the random variable X as the win achieved in just a single repetition of the game, then the number -0.05 is said to be the expected value of X. The expected value of X is written as $E(X)$. In the casino game $E(X)$ is given by

$$E(X) = (-1) \times P(X = -1) + 2 \times P(X = 2) + 3 \times P(X = 3).$$

The general definition of expected value is reasoned out in the example above. Assume that X is a random variable with a discrete probability distribution $p_j = P(X = x_j)$ for $j = 1, \ldots, M$. The *expected value* or *expectation* of the random variable X is then defined by

$$E(X) = x_1 p_1 + x_2 p_2 + \cdots + x_M p_M.$$

Abbreviating this formula to include the commonly used summation sign $\sum$, we get:

$$E(X) = \sum_{j=1}^{M} x_j p_j.$$

Stating this formula in words, $E(X)$ is a weighted average of the possible values that X could assume, where each value is weighted with the probability that X would assume the value in question. The term "expected value" can be misleading. It must not be confused with the "most probable value." An insurance agent who tells a 40-year-old person that he/she can expect to live another 37 years naturally means that you come up with 37 more years when you multiply the possible values of the person's future years with the corresponding probabilities and then add the products together. The expected value $E(X)$ is not restricted to values that the random variable X could possibly assume. For example, let X be the number of points accrued in one roll of a fair die. Then

$$E(X) = 1 \times \frac{1}{6} + 2 \times \frac{1}{6} + 3 \times \frac{1}{6} + 4 \times \frac{1}{6} + 5 \times \frac{1}{6} + 6 \times \frac{1}{6} = 3\frac{1}{2}.$$

The value $3\frac{1}{2}$ can never be the outcome of a single roll with the die. When we are taking a very large number of rolls of the die, however, it does appear that the average value of the points will be close to $3\frac{1}{2}$. One can look into this empirical result intuitively with the law of large numbers for probabilities. This law teaches us that, when you have a very large number of rolls with a die, the fraction of rolls with j points is closely equal to $\frac{1}{6}$ for every $j = 1, \ldots, 6$. From here, it follows that the average number of points per roll is close to $\frac{1}{6}(1 + 2 + \cdots + 6) = 3\frac{1}{2}$.

The empirical finding that the average value of points accrued in the rolls of a fair die gets ever closer to $3\frac{1}{2}$ as the number of rolls increases can be placed in a more general framework. Consider, therefore, a chance experiment that can be repeatedly performed under exactly the same conditions. Let X be a random variable that is defined on the probability space of the experiment. In order to keep the train of thought running smoothly, it is helpful to suppose that the experiment is a certain (casino) game and that X is the random payoff of the game. Suppose the game is carried out a large number of times under exactly the same conditions, and in a way such that the repetitions of the game are independent of each other. It would appear, then, that:

the average payment per game will fluctuate less and less as time goes on, and will approach a limiting value as the number of repetitions of the game increases without bound.

This empirical result has a mathematical counterpart that stems from probability theory axioms. If we define the random variable X_k as the payoff in the kth repetition of the game, then the *theoretical law of large numbers for expected value* can be stated as:

> **the average payment $\frac{1}{n}(X_1 + X_2 + \cdots + X_n)$ over the first n repetitions of the game will converge with probability 1 to a constant as $n \to \infty$ and this constant is equal to the expected value $E(X)$.**

In an effort to keep things uncomplicated, we will allow the precise meaning of the term "with probability 1" to remain somewhat vague at this point. Suffice it to say that, intuitively, it means that "nature," apparently, always takes care of realizations for which the long-term average payment per game is equal to the theoretical expected value $E(X)$ (see also Chapter 7). In many practical problems, it is helpful to interpret the expected value of a random variable as a long-term average.

Example 2.5 In the game "Unders and Overs," two dice are rolled and you can bet whether the total of the two dice will be under 7, over 7, or equal to 7.[6] The gambling table is divided into three sections marked as "Under 7," "7," and "Over 7." The payoff odds for a bet on "Under 7" are 1 to 1, for a bet on "Over 7" are 1 to 1, and for a bet on "7" are 4 to 1 (payoffs of r to 1 mean that you get $r + 1$ dollars back for each dollar bet if you win; otherwise, you lose your stake). Each player can put chips on one or more sections of the gambling table. Your strategy is to bet one chip on "Under 7" and one chip on "7" each time. What is your average win or loss per round if you play the game over and over?

Let the random variable X denote the number of chips you get back in any given round. The possible values of X are 0, 2, and 5. The random variable X is defined on the sample space consisting of the 36 equiprobable outcomes $(1, 1), (1, 2), \ldots, (6, 6)$. Outcome (i, j) means that i points turn up on the first die and j points on the second die. The total of the two dice is 7 for the six outcomes $(1, 6), (6, 1), (2, 5), (5, 2), (3, 4)$, and $(4, 3)$. Thus, $P(X = 5) = \frac{6}{36}$. Similarly, $P(X = 0) = \frac{15}{36}$ and $P(X = 2) = \frac{15}{36}$. This gives

$$E(X) = 0 \times \frac{15}{36} + 2 \times \frac{15}{36} + 5 \times \frac{6}{36} = 1\frac{2}{3}.$$

You bet two chips each round. Thus, your average loss is $2 - 1\frac{2}{3} = \frac{1}{3}$ chip per round when you play the game over and over.

[6] In the old days, the game was often played at local schools in order to raise money for the school.

2.3.1 Expected value and risk

In the case that the random variable X is the random payoff in a game that can be repeated many times under identical conditions, the expected value of X is an informative measure on the grounds of the law of large numbers. However, the information provided by $E(X)$ is usually not sufficient when X is the random payoff in a nonrepeatable game. Suppose your investment has yielded a profit of \$3,000 and you must choose between the following two options: the first option is to take the sure profit of \$3,000, and the second option is to reinvest the profit of \$3,000 under the scenario that this profit increases to \$4,000 with probability 0.8 and is lost with probability 0.2. The expected profit of the second option is $0.8 \times$ \$4,000 $+ 0.2 \times$ \$0 $=$ \$3,200 and is larger than the \$3,000 from the first option. Nevertheless, most people would prefer the first option. The downside risk is too big for them. A measure that takes into account the aspect of risk is the variance of a random variable. This concept will be discussed in detail in Chapter 5.

2.3.2 Best-choice problem

In order to answer Question 10 from Chapter 1, you must know which strategy your friend is using to correctly identify the piece of paper with the largest number. Suppose your friend allows the first half of the papers to pass through his hands, but keeps a mental note of the highest number that appears. As he opens and discards the papers in the subsequent group, he stops at the appearance of the first paper showing a number higher than the one he took note of earlier. Of course, this paper will only appear if the ultimate highest number was not among the first 10 papers opened. Let p represent the (unknown) probability that your friend will win the contest using this simple strategy. Imagine that you will have to pay five dollars to your friend if he wins and that otherwise you receive one dollar. The expected value of your net win in a given contest is then:

$$(1-p) \times 1 - p \times 5 = 1 - 6p.$$

The contest is unfavorable to you if $p > \frac{1}{6}$. With a simple model not only can you show that this is the case, but also that p is actually greater than $\frac{1}{4}$. Now, try to visualize that the paper with the highest number has a 1 stamped on it in invisible ink, that the paper with the next highest number has a 2 stamped on it, etc. Then imagine that the 20 pieces of paper are randomly lined up. The relative ranking of the numbers on the 20 papers corresponds to a permutation (ordered sequence) of the numbers $1, \ldots, 20$. This suggests a sample space consisting of all the possible permutations $(i_1, i_2, \ldots, i_{20})$ of the numbers $1, \ldots, 20$. The outcome $(i_1, i_2, \ldots, i_{20})$ corresponds to the situation in which i_1 is stamped in invisible ink on the outside of the first paper your friend chooses, i_2 on the second paper your friend chooses, etc. The total number of permutations of the integers $1, \ldots, 20$ is $20 \times 19 \times \cdots \times 1$. The notation $n!$ is used for the product $1 \times 2 \times \cdots \times n$ (see the Appendix). Thus, the sample space consists of 20! different elements. Each element is assigned the same probability $\frac{1}{20!}$. Let A represent the event that the second highest number is among the first 10 papers, but that the highest number is not. In any case, your friend will win the contest if event A occurs. In order to find $P(A)$, one must count the number of elements $(i_1, i_2, \ldots, i_{20})$ where one of the numbers $i_1, \ldots, i_{10}$ is equal to 2 and one of the numbers $i_{11}, \ldots, i_{20}$ is equal to 1. This number is equal to $10 \times 10 \times 18!$. Thus,

$$P(A) = \frac{10 \times 10 \times 18!}{20!} = \frac{100}{19 \times 20} = 0.263.$$

The probability p that your friend will win the contest is greater than $P(A)$ and is then, indeed, greater than 25%. Using this reasoning, you will also come to the same conclusion if 100 people or even one million people write down a random number on a piece of paper. Using computer simulation, it can be verified that the simple strategy described above gives your friend the probabilities 0.359 and 0.349 of winning when the number of people participating is 20 and 100, respectively. On the computer, the contest can be played out a great many times. You would take the fraction of contests won by your friend as an estimate for the probability p of your friend winning. In order to simulate the model on the computer, you need a procedure for generating a random permutation of the numbers $1, \ldots, n$ for a given value of n. Such a procedure is discussed in Section 2.9.

In Problem 3.25 of Chapter 3, we come back to the best-choice problem, and you may be surprised by the solution here. When we speak of n papers with n being high (say, $n \geq 100$), then the maximum probability of winning is approximately equal to $\frac{1}{e} = 0.368$, irrespective of the value of n. The optimal strategy is to open the first $\frac{n}{e}$ papers and then to choose the next paper to appear with a number higher than those contained in all of the previous papers. This strategy might guide you when you are looking for a restaurant in a city you visit for the first time!

2.4 The drunkard's walk

The drunkard's walk is named for the drunkard exiting a pub who takes a step to the right with a probability of $\frac{1}{2}$ or a step to the left with a probability of $\frac{1}{2}$. Each successive step is executed independently of the others. The following questions arise: what is the probability that the drunkard will ever return to his point of origin, and what is the expected distance back to the point of origin after the drunkard has taken many steps? These questions seemingly fall into the category of pure entertainment; but, in actuality, nothing could be further from the truth. The drunkard's walk has many important applications in physics, chemistry, astronomy, and biology. These applications usually consider two- or three-dimensional representations of the drunkard's walk. The biologist looks at the transporting of molecules through cell walls. The physicist looks at the electrical resistance of a fixed particle. The chemist looks for explanations for the speed of chemical reactions. The climate specialist looks for evidence of global warming, etc. The model of the drunkard's walk is extremely useful for this type of research.[7]

We first look at the model of the drunkard walking along a straight line. Plotting the path of the drunkard's walk along a straight line is much the same as tracing the random walk of the fair-coin toss. Imagine a drunkard at his point of origin. His steps are of unit length, and there is a probability of $\frac{1}{2}$ that in any given step he will go to the right and a probability of $\frac{1}{2}$ that he will go to the left. The drunkard has no memory, i.e., the directions of the man's successive steps are independent of one another. Define the random variable D_m as

$$D_m = \text{the drunkard's distance from his point of origin after } m \text{ steps.}$$

It holds that the expected value of the *quadratic* distance of the drunkard from his point of origin after m steps is given by

$$E\left(D_m^2\right) = m$$

for every value of m. This result will be proved below. For now, it is worth noting that the result does *not* allow us to conclude that $E(D_m)$ is equal to $\sqrt{m}$, although this erroneous conclusion is often cited as true. Rather, the actual answer for $E(D_m)$ is that $\sqrt{m}$ must be amended by a factor of less than 1. For m large, this correction factor is approximately equal to 0.798. The following can then be said:

$$E(D_m) \approx \sqrt{\frac{2}{\pi}m},$$

where the symbol $\approx$ stands for "is approximately equal to." This result will be explained in Section 5.8, with the help of the central limit theorem.

[7]See G. H. Weiss, "Random walks and their applications," *American Scientist*, January–February 1983, 71, 65–70.

2.4.1 Derivation of quadratic distance

The proof that $E(D_m^2) = m$ is based on "first principles." As the sample space for the chance experiment of the drunkard's walk, choose the set of all m-tuples $(x_1, \ldots, x_m)$, where the number x_i is equal to 1 if the drunkard steps to the right in his ith step and is otherwise equal to -1. The sample space has 2^m elements. Every element is assigned with the same probability $\frac{1}{2^m} = (\frac{1}{2})^m$. For a given realization $(x_1, \ldots, x_m)$ of the drunkard's walk, the distance of the drunkard from his point of origin is equal to the absolute value $|x_1 + \cdots + x_m|$. If D_m is a random variable, then D_m^2 is also a random variable. The possible values of the random variable D_m^2 are $0, 1, 4, \ldots, m^2$. We first prove that $E(D_m^2) = m$ for the case that m is even, say $m = 2r$. In this case, the random variable D_m can only take on even values (verify!). The random variable D_m is equal to $2k$ for a certain k with $0 \leq k \leq r$ only if the drunkard takes $2k + \frac{1}{2}(m - 2k)$ steps to the right and $\frac{1}{2}(m - 2k)$ steps to the left or if the drunkard takes $2k + \frac{1}{2}(m - 2k)$ steps to the left and $\frac{1}{2}(m - 2k)$ steps to the right. The probability of the drunkard taking $2k + \frac{1}{2}(m - 2k) = k + r$ steps to the right and $\frac{1}{2}(m - 2k) = r - k$ steps to the left is equal to

$$\binom{2r}{k+r}\left(\frac{1}{2}\right)^{2r}.$$

Here, $\binom{p}{\ell}$ gives the total number of ways to choose ℓ positions for a one and $p - \ell$ positions for a -1 out of p positions. A basic result in combinatorial analysis is $\binom{p}{\ell} = \frac{p!}{\ell!(p-\ell)!}$ (see the Appendix). The probability that the drunkard will take $k + r$ steps to the left and $r - k$ steps to the right is the same as the probability of $k + r$ steps to the right and $r - k$ steps to the left. This gives

$$P(D_m = 2k) = 2 \times \binom{2r}{k+r}\left(\frac{1}{2}\right)^{2r} = \binom{2r}{k+r}\left(\frac{1}{2}\right)^{2r-1}$$

for $k = 0, 1, \ldots, r$. Since $P(D_m^2 = 4k^2) = P(D_m = 2k)$ for all k, it follows from the definition of expected value that

$$E(D_m^2) = 0 \times \binom{2r}{r}\left(\frac{1}{2}\right)^{2r-1} + 4 \times \binom{2r}{r+1}\left(\frac{1}{2}\right)^{2r-1} + \cdots$$
$$+ 4r^2 \times \binom{2r}{2r}\left(\frac{1}{2}\right)^{2r-1}.$$

From combinatorial mathematics, the identity

$$\sum_{k=1}^{r} k^2 \binom{2r}{k+r} = \frac{1}{2} r 2^{2r-1}$$

is known. This gives $E(D_m^2) = 2r = m$ for m even. For m odd, an analogy can show that $E(D_m^2) = m$. A shorter way to prove that $E(D_m^2) = m$ is to make use of the rules of calculation for expected values that are discussed in Chapter 9. In order to apply these rules of calculation, $E(D_m^2)$ is written as $E\left[(X_1 + \cdots + X_m)^2\right]$, where the random variable X_i is equal to 1 if the drunkard goes to the right in the ith step and is equal to -1 otherwise. This approach is worked out in Section 9.4 and allows us to prove that the result $E(D_m^2) = m$ also holds for the drunkard's walk in higher dimensions.

2.4.2 The drunkard's walk in higher dimensions

For the drunkard's walk on the two-dimensional plane, the expected value of the distance of the drunkard from his point of origin after taking m steps is approximately given by

$$E(D_m) \approx \frac{1}{2}\sqrt{\pi m}.$$

This approximation formula is applicable both in the case where the drunkard leaves from point (x, y) with equal probability $\frac{1}{4}$ toward each of the four bordering grid points $(x + 1, y)$, $(x - 1, y)$, $(x, y + 1)$, and $(x, y - 1)$ and in the case where the drunkard takes steps of unit length each time in a randomly chosen direction between 0 and 2π. The approximation formula for the drunkard's walk in three-dimensional space is:

$$E(D_m) \approx \sqrt{\frac{8}{3\pi}m}.$$

We delve into these approximations further on in Chapter 12. The approximation for $E(D_m)$ has many applications. How long does it take a photon to travel from the sun's core to its surface? The answer is that it takes approximately 10 million years, and it is found by using the model of the drunkard's walk. A photon has a countless number of collisions on its way to the sun's surface. The distance traveled by a photon between two collisions can be measured as 6×10^{-6} mm. The sun's radius measures 70,000 km. A photon travels at a speed of 300,000 km per second. Taking into consideration that 70,000 km is equal to 7×10^{10} mm, the equality

$$\sqrt{\frac{8}{3\pi}m} = \frac{7 \times 10^{10}}{6 \times 10^{-6}}$$

shows that the average number of collisions that a photon undergoes before reaching the sun's surface is approximately equal to $m = 1.604 \times 10^{32}$. The speed of light is 300,000 km per second, meaning that the travel time of a photon between two collisions is equal to $(6 \times 10^{-6})/(3 \times 10^{11}) = 2 \times 10^{-17}$ seconds. The average travel time of a photon from the sun's core to its surface is thus approximately equal to 3.208×10^{15} seconds. If you divide this by $365 \times 24 \times 3{,}600$, then you find that the average travel time is approximately 10 million years. It takes a photon only 8 minutes to travel from the surface of the sun to the earth (the distance from the sun to the earth is 149,600,000 km).

2.4.3 The probability of returning to the point of origin

The drunkard's walk provides surprising results with regard to the probability of the drunkard returning to his point of origin if he keeps at it long enough. This probability is equal to 1 both for the drunkard's walk on the line and the drunkard's walk in two dimensions, but it is less than 1 for the drunkard's walk in the third dimension, assuming that the drunkard travels over a discrete grid of points. In the third dimension, the probability of ever returning to the point of origin is 0.3405.[8] To make it even more surprising, the drunkard will eventually visit every grid point with probability 1 in the dimensions 1 and 2, but the expected value of the number of necessary steps back to his point of origin is infinitely large. An advanced knowledge of probability theory is needed to verify the validity of these results.

2.5 The St. Petersburg paradox

This paradox comes down to us from Nicolaus Bernoulli (1695–1726), one of the many mathematicians of the well-known Bernoulli family and a distinguished professor in St. Petersburg. What is the essential problem underlying the paradox? In a certain casino game, a fair coin is tossed successively until the moment that heads appears for the first time. The casino payoff is two dollars if heads turns up in the first toss, four dollars if heads turns up for the first time in the second toss, etc. In general, the payoff is 2^n if heads turns up for the first time in the nth toss. What amount must the casino require the player to stake such that, over the long term, the game will not be a losing endeavor for the casino? To answer this question, we need to calculate the expected value of the casino payoff for a single repetition of the game. The probability of getting heads in the first toss is $\frac{1}{2}$, the probability of getting tails in the first toss and heads in the second toss is $\frac{1}{2} \times \frac{1}{2}$, etc., and the probability of getting tails in the first $n - 1$ tosses and heads in the nth toss is $(\frac{1}{2})^n$. The expected value of the

[8] On earth, all roads lead to Rome, but not in space!

casino payoff for a single repetition of the game is thus equal to

$$\frac{1}{2} \times 2 + \frac{1}{4} \times 4 + \cdots + \frac{1}{2^n} \times 2^n + \cdots$$

dollars. In this infinite series, a figure equal to 1 is added to the sum each time. In this way, the sum exceeds every conceivable large value and mathematicians would say that the sum of the infinite series is infinitely large. The expected value of the casino payoff for a single repetition of the game is thus an infinitely large dollar amount. This means that casino owners should not allow this game to be played, whatever amount a player is willing to stake. However, no player in his right mind would be prepared to stake, say, 10 million dollars for the opportunity to play this game. The reality of the situation is that the game is simply not worth that much. In Bernoulli's day, a heated discussion grew up around this problem. Some of those involved even began to question whether there was not a problem with the mathematics. But no, the math was good, and the mathematical model for the game is calculated correctly. The trouble is that the model being used simply does not provide a good reflection of the actual situation in this case! For one thing, the model implicitly suggests that the casino is always in a position to pay out, whatever happens, even in the case of a great number of tosses being executed before the first heads appears, which adds up to a dazzlingly high payoff. The practical reality is that the casino is only in possession of a limited amount of capital and cannot pay out more than a limited amount. The paradox can be explained thus: if the mathematical model does not provide a good reflection of reality, the conclusion it forms will have no practical relevance.

The problem does become more realistic when the following modification is made to the game. The casino can only pay out up to a limited amount. To simplify the matter, let's assume that the maximum payoff is a given multiple of 2. Let the maximum casino payoff per game be equal to 2^M dollars for some given integer M (e.g., $M = 15$ would correspond with a maximum payoff of \$32,768). In every repetition of the game, a fair coin is tossed until either heads appears for the first time or M tosses are executed without heads appearing. The casino pays the player 2^k dollars when heads appears for the first time in the kth toss and pays nothing if tails is tossed M times in a row. What must the player's minimum stake be such that the game will not be a loss for the casino over the long term? The same reasoning we used before says that the expected value of the casino payoff for a single execution of the game is equal to

$$\frac{1}{2} \times 2 + \frac{1}{4} \times 4 + \cdots + \frac{1}{2^M} \times 2^M = M \text{ dollars.}$$

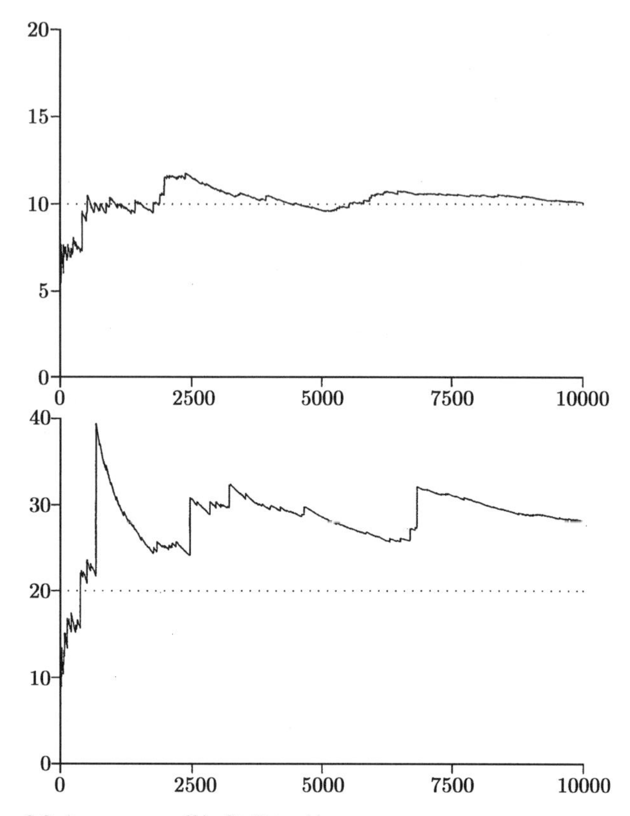

Figure 2.3 Average payoff in St. Petersburg game.

This means that the modified game is profitable for the casino if the player's stake is above M dollars. It is instructive to look at how the average payoff per game converges to the theoretical expected value M if the game is executed a great number of times. In Figure 2.3, we give the simulated results for 10,000 repetitions of the game both for $M = 10$ and $M = 20$. From these results, it appears that as the value of M increases, many more plays are necessary before the average payoff per play converges to the theoretical expected value. The explanation for the slow convergence when M is large lies in the fact that very large payoffs occurring with a very small probability contribute a nonnegligible amount to the expected value. The simulation confirms this. In situations where a very small probability plays a nonnegligible role, very long simulations are required in order to get reliable estimates. The lesson to be gained here is that, in situations of this kind, it is especially dangerous to conclude results from simulations that are "too short." In addition, this underscores the importance of evaluating the reliability of results gained through the process of simulation.

Such evaluation can only be achieved with help from a concept called the confidence interval, which will be discussed in Chapter 5.

2.6 Roulette and the law of large numbers

The law of large numbers is the basis for casino profits. If a sufficient number of players stake money at the gaming tables (and stake amounts are limited to a given maximum), then the casino will not operate at a loss and in fact will be ensured of steadily growing profits.[9]

Roulette is one of the oldest casino games. The basis for this game was established at the beginning of the seventeenth century by French mathematician Blaise Pascal. The most common version of roulette uses the numbers $0, 1, \ldots, 36$, where the number 0 is always reserved as the winning number for the house (European roulette). Players bet against the house on a number to emerge when the roulette wheel stops spinning. Bets may be placed on either single numbers or combinations of numbers. A winning bet placed on a combination of k numbers earns $\frac{36}{k} - 1$ times the amount staked plus the initial stake itself in winnings. For each separate bet, the expected value of the casino take for each dollar staked is equal to

$$1 \times \left(1 - \frac{k}{37}\right) - \left(\frac{36}{k} - 1\right) \times \frac{k}{37} = \frac{1}{37} \text{ dollars.}$$

In other words, casinos get 2.7 cents for every dollar staked over the long run, and this translates into a *house percentage* of 2.70%. In American roulette, which differs from the European version in that the roulette wheel has a "house double-zero" (00) in addition to the single (house) zero (0), the house percentage for each bet is 5.26%, except for five-number combination bets; these offer an even higher house percentage of 7.89%. It is impossible to win this game over the long run. No matter what betting system you use, you can count on giving away 2.7 cents for every dollar you stake. It is impossible to make a winning combination of bets when every individual bet is a losing proposition. Betting systems are only of interest for their entertainment and excitement value. Betting systems that are much in use are the Big–Martingale system and the D'Alembert system.

[9]Blackjack (or twenty-one) is the only casino game in which the player has a theoretical advantage over the casino. Around 1960, computer-simulated winning blackjack strategies were developed. Casinos can be glad that these strategies are not only difficult to put into practice, but also provide only a small advantage to players. Players with large bankrolls attempting to use this system usually find either that small changes in game rules thwart their attempt or that they are simply escorted from the premises.

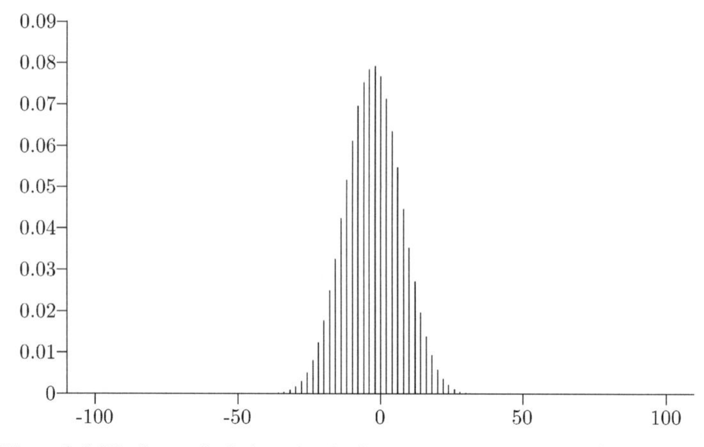

Figure 2.4 Win/loss calculations for the flat system.

In both systems, the game is played according to simple probability patterns with 18 numbers (always betting on red, for example), where the payoff equals twice the amount staked. The Big–Martingale system works thus: the amount of your initial stake is one chip. If you lose, your next stake will be twice your previous stake plus one chip. If you win, your next stake will be one chip. Should you score your first win after four attempts, your first four stake amounts will have been 1, 3, 7, and 15 chips, and after four turns you will have gained $30 - (1 + 3 + 7 + 15) = 4$ chips. In the D'Alembert system, the amount of your initial stake will also be one chip. After a loss, you raise your stake with one chip, and after a win you decrease your stake by one chip. Engaging as these systems may be, they, too, will result in a loss over the long run of 2.7 cents for every dollar staked. Attempting to influence your average loss in roulette by using a betting system is as nonsensical as it was, long ago, for a despot to try and influence the ratio of newborn boys to girls by prohibiting women from bearing any more children as soon as they gave birth to a boy. The latter merely the folly of the gambler dressed up in different clothes!

Betting systems for roulette that claim to be winners, whether in book form or on the Internet, represent nothing more than charlatanism. To underline the fact that one betting system is no better than another in roulette, we chart the results of a simulation study that compares the Big–Martingale system with the flat system, which calls for a stake of one chip on each round. The study was composed of one million simulated repetitions of the game for both systems. For each repetition, the initial playing capital consisted of 100 chips, and a maximum of 100 bets were made, always on red. Under the flat system, one chip was staked on each spin of the wheel. Under the Big–Martingale system, 100 bets or less were made, depending on how long the chips lasted. The

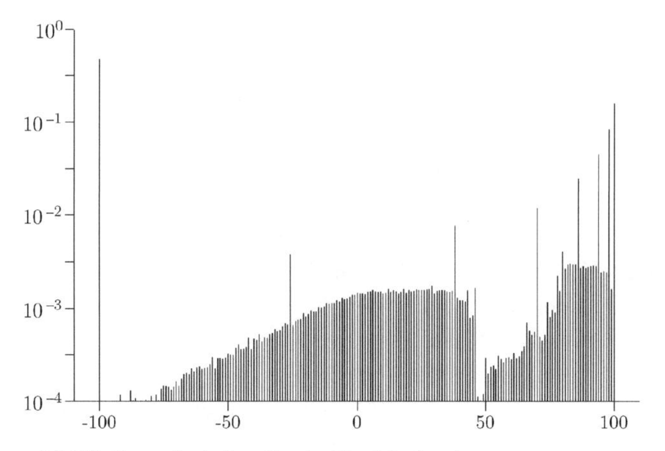

Figure 2.5 Win/loss calculations for the Big–Martingale system.

following total scores were found for the one million simulation runs:

Flat system

total amount staked	=	100,000,000
total loss	=	2,706,348
loss per unit staked	=	0.0271

Big–Martingale system

total amount staked	=	384,718,672
total loss	=	10,333,828
loss per unit staked	=	0.0269

As you can see, the quotient of the total loss and the total amount staked, in both cases, lies near the house advantage of 0.027. It is also interesting to note the probability distribution of the number of chips that are won or lost at the end of one repetition of the game. We give the simulated probability distribution for the flat system in Figure 2.4 and for the Big–Martingale system in Figure 2.5. In Figure 2.5, a logarithmic scale is used. As might have been expected, the probability distribution for the Big–Martingale system is much more strongly concentrated at the outer ends than the distribution for the flat system.

2.7 The Kelly betting system[10]

You are playing a game where you have an edge. How should you bet to manage your money in a good way? The idea is always to bet a *fixed proportion* of your

[10]This paragraph can be skipped at first reading.

present bankroll. When your bankroll decreases you bet less, as it increases you bet more. This strategy is called the Kelly system, after the American mathematician J. F. Kelly, Jr., who published this system in 1956. The objective of Kelly betting is to maximize the long-run rate of growth of your bankroll. The optimal value of the fraction to bet can be found by simple arguments based on the law of large numbers.

Suppose you are offered a sequence of bets, each bet being a losing proposition with probability 0.6 and paying out three times your stake with probability 0.4. How to gamble if you must? Note that each bet is favorable, because the expected net payoff is positive ($0.4 \times 3 - 1 > 0$). However, it is not wise to bet your whole bankroll each time; if you do, you will certainly go bankrupt after a while. Indeed, it is better to bet 10% of your current bankroll each time. This strategy maximizes the long-run rate of growth of your bankroll and achieves an effective rate of return of 0.98% over the long run. To derive this result, it is helpful to use a general notation. Let's assume that the payoff odds are $f - 1$ to 1 for a given $f > 1$. That is, in case of a win, you get a payoff of f times the amount bet; otherwise, you lose the amount bet. Letting p denote the probability of the player winning the bet, it is assumed that $0 < p < 1$ and $pf > 1$ (favorable bet).

Assuming that your starting bankroll is V_0, define the random variable V_n as

$$V_n = \text{the size of your bankroll after } n \text{ bets},$$

when you bet a fixed fraction α ($0 < \alpha < 1$) of your current bankroll each time. Here, it is supposed that winnings are reinvested and that your bankroll is infinitely divisible. It is not difficult to show that

$$V_n = (1 - \alpha + \alpha R_1) \times \cdots \times (1 - \alpha + \alpha R_n)\, V_0 \quad \text{for} \quad n = 1, 2, \ldots,$$

where the random variable R_k is equal to the payoff factor f if the kth bet is won and is otherwise equal to 0. Evidence of this relationship appears at the end of this section. In mathematics, a growth process is most often described with the help of an exponential function. This motivates us to define the exponential growth factor G_n via the relationship

$$V_n = V_0 e^{nG_n},$$

where $e = 2.718\ldots$ is the base of the natural logarithm. If you take the logarithm of both sides of this equation, you see that the definition of G_n is equivalent to

$$G_n = \frac{1}{n} \ln\left(\frac{V_n}{V_0}\right).$$

If you apply the above product formula for V_n and use the fact that $\ln(ab) = \ln(a) + \ln(b)$, then you find

$$G_n = \frac{1}{n}\left[\ln(1 - \alpha + \alpha R_1) + \cdots + \ln(1 - \alpha + \alpha R_n)\right].$$

The law of large numbers applies to the growth rate G_n if n (= the number of bets) is very large. Indeed, the random variables $X_i = \ln(1 - \alpha + \alpha R_i)$, $i = 1, 2, \ldots$, form a sequence of independent random variables having a common distribution. If you apply the law of large numbers, you find that

$$\lim_{n\to\infty} G_n = E\left[\ln(1 - \alpha + \alpha R)\right],$$

where the random variable R is equal to f with probability p and is equal to 0 with probability $1 - p$. This leads to

$$\lim_{n\to\infty} G_n = p\ln(1 - \alpha + \alpha f) + (1 - p)\ln(1 - \alpha).$$

Under a strategy that has a fixed betting fraction α, the long-run growth factor of your bankroll is thus given by

$$g(\alpha) = p\ln(1 - \alpha + \alpha f) + (1 - p)\ln(1 - \alpha).$$

It is not difficult to verify that an α_0 with $0 < \alpha_0 < 1$ exists such that the long-run growth factor $g(\alpha)$ is positive for all α with $0 < \alpha < \alpha_0$ and negative for all α with $\alpha_0 < \alpha < 1$. Choose a betting fraction between 0 and α_0 and your bankroll will ultimately exceed every large level if you simply keep playing for a long enough period of time. It is quite easy to find the value of α for which the long-run growth factor of your bankroll is maximal. Toward that end, set the derivative of the function $g(\alpha)$ equal to 0. This leads to $p(f - 1)/(1 - \alpha + f\alpha) - (1 - p)/(1 - \alpha) = 0$. Hence, the optimal value of α is given by

$$\alpha^* = \frac{pf - 1}{f - 1}.$$

This is the famous formula for the *Kelly betting fraction*. This fraction can be interpreted as the ratio of the expected payoff for a one-dollar bet and the payoff odds. The Kelly system is of little use for casino games, but may be useful for the situation of investment opportunities with positive expected payoff. In such situations, it may be more appropriate to use a modification of the Kelly formula that takes into account the interest accrued on the noninvested part of your bankroll. In Problem 2.13, the reader is asked to modify the Kelly formula for a situation where a fixed interest is attached to a player's nonstaked capital.

2.7.1 Long-run rate of return

For the strategy under which you bet the same fraction α of your bankroll each time, define the return factor γ_n by

$$V_n = (1 + \gamma_n)^n V_0.$$

The random variable γ_n gives the rate of return on your bankroll over the first n bets. It follows from the relationship $V_n = e^{nG(n)} V_0$ that

$$\gamma_n = e^{G(n)} - 1.$$

Earlier, we saw that the random variable $G(n)$ converges to the constant $g(\alpha) = p \ln(1 - \alpha + \alpha f) + (1 - p) \ln(1 - \alpha)$ as $n \to \infty$. This means that γ_n converges to the constant $\gamma_{eff} = e^{g(\alpha)} - 1$ as $n \to \infty$. The constant γ_{eff} gives the effective rate of return for the long run if you bet the same fraction α of your bankroll each time. Substituting the expression for $g(\alpha)$ and using $e^{b \ln(a)} = a^b$, you find that the long-run rate of return is given by

$$\gamma_{eff} = (1 - \alpha + \alpha f)^p (1 - \alpha)^{1-p} - 1.$$

2.7.2 Simulation of Kelly's strategy

Consider these data:

$$p(=\text{win probability}) = 0.4, \qquad f(=\text{payoff factor}) = 3,$$
$$n(=\text{number of bets}) = 100, \qquad V_0(=\text{starting capital}) = 1.$$

Under the Kelly system, a fraction $\alpha^* = 0.1$ of your current bankroll is bet each time. Let us compare this strategy with the alternative strategy under which the fixed fraction $\alpha = 0.25$ of your bankroll is bet each time. The comparison is done by executing a simulation experiment where both strategies are exposed to the same experimental conditions. The results of this simulation are given in Figure 2.6. The simulation outcomes confirm that, in the long run, the Kelly strategy is superior to other strategies. From the formula for γ_{eff}, it follows that the Kelly betting strategy with $\alpha = 0.1$ has an effective rate of return of 0.98% over the long run, whereas the betting strategy with $\alpha = 0.25$ has an effective rate of return of -1.04% over the long run. In Chapter 5, we come back to another property of the Kelly strategy: it minimizes the expected time needed to reach a specified, but large value for your bankroll.

At the height of the Internet frenzy, one dot-com company went public on the stock exchange every week. Particularly in their first week of being listed on the exchange, the stock of such companies demonstrated a lot of movement

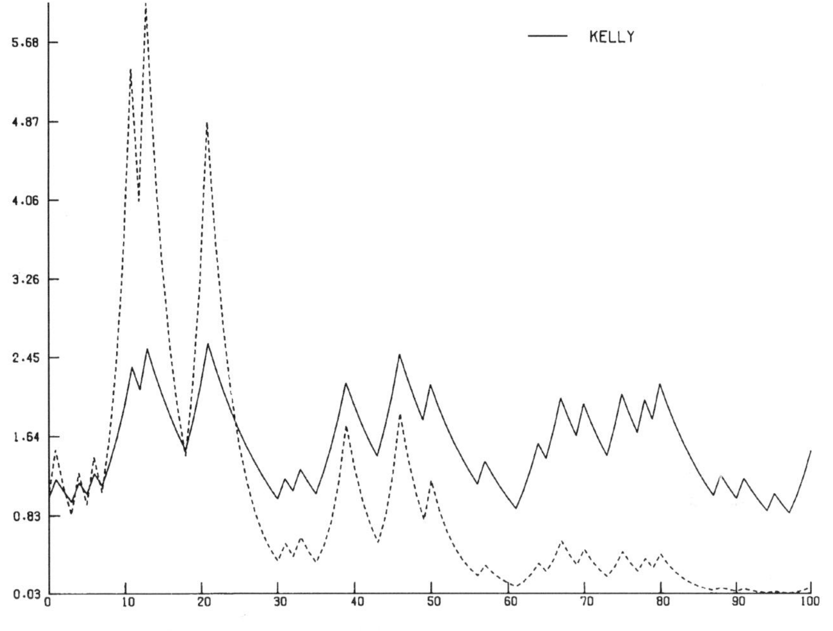

Figure 2.6 Kelly strategy and an alternative strategy.

in the market. Suppose the following hypothetical situation occurred: in the first week of being listed, the stock of a new dot-com company had a 50% probability of doubling in value and a 50% probability of falling three-fifths in value. You buy stock from a newly listed dot-com company at the beginning of every week, and at the end of the week you sell it again. In one week, your expected gain is positive and equals 20% at your investment. Now, in one particular case-in-point, the worst-case scenario is that your investment is not lost altogether but becomes 60% less valuable. Getting 40% of a doubling of one dollar is less than one dollar. Are you headed for long-term financial disaster if you continue to invest all of your money in a newly listed dot-com company each week? The answer is yes: in the long run, the effective rate return on your bankroll is equal to -10.56% if you invest all of your money each time. Here, as above, you would show yourself wise to spread the risk according to a Kelly strategy. If you invest a fixed fraction $\alpha^* = \frac{1}{3}$ of your bankroll every week in a new dot-com company, and keep the other $\frac{2}{3}$ of your bankroll in your pocket, then the long-term rate of return on your bankroll is maximized and equal to 3.28%. These results follow from a small modification of the formulas for the Kelly fraction and the effective rate of return (see Problem 2.14).

2.7.3 Derivation of the growth rate

Proof of the relationship

$$V_n = (1 - \alpha + \alpha R_1) \times \cdots \times (1 - \alpha + \alpha R_n) V_0 \quad \text{for} \quad n = 1, 2, \ldots$$

is as follows. If you invest a fraction α of the capital you possess every time, then

$$V_k = (1 - \alpha) V_{k-1} + \alpha V_{k-1} R_k \quad \text{for} \quad k = 1, 2, \ldots .$$

At this point, we apply the mathematical principle of induction to prove the product formula for V_n. This formula is correct for $n = 1$ as follows directly from $V_1 = (1 - \alpha) V_0 + \alpha V_0 R_1$. Suppose the formula is proven for $n = j$. It would then be true for $n = j + 1$ that

$$\begin{aligned} V_{j+1} &= (1 - \alpha) V_j + \alpha V_j R_{j+1} = (1 - \alpha + \alpha R_{j+1}) V_j \\ &= (1 - \alpha + \alpha R_{j+1})(1 - \alpha + \alpha R_1) \times \cdots \times (1 - \alpha + \alpha R_j) V_0 \\ &= (1 - \alpha + \alpha R_1) \times \cdots \times (1 - \alpha + \alpha R_{j+1}) V_0, \end{aligned}$$

which proves the assertion for $n = j + 1$ and so the induction step is complete.

2.8 Random-number generator

Suppose you are asked to write a long sequence of H's and T's that would be representative of the tossing of a fair coin, where H stands for heads and T for tails. You may not realize just how incredibly difficult a task this is. Virtually no one is capable of writing down a sequence of H's and T's such that they would be statistically indistinguishable from a randomly formed sequence of H's and T's. Anyone endeavoring to accomplish this feat will likely avoid clusters of H's and T's. But such clusters do appear with regularity in truly random sequences. For example, as we saw in Section 2.1, the probability of tossing heads five successive times in 20 tosses of a fair coin is not only nonnegligible, but also it actually amounts to 25%. A sequence of H's and T's that does not occasionally exhibit a long run of H's or a long run of T's cannot have been randomly generated. In probability theory, access to random numbers is of critical importance. In the simulation of probability models, a random-number generator, as it is called, is simply indispensable.

A random-number generator produces a sequence of numbers that are picked at random between 0 and 1 (excluding the values 0 and 1). It is as if fate falls on a number between 0 and 1 by pure coincidence. When we speak of generating a random number between 0 and 1, we assume that the probability of the generated number falling in any given subinterval of the unit interval (0,1) equals the length of that subinterval. In other words, the probability distribution of a random number between 0 and 1 is a *uniform* distribution on (0,1). This is a continuous distribution, which means that it only makes sense to speak of the probability of a randomly chosen number falling in a *given interval*. It makes no sense to speak of the probability of an individual value. For example, a randomly chosen number between 0 and 1 will fall in the interval (0.7315, 0.7325) with a probability of 0.001. The probability that a randomly chosen number will take on a *prespecified* value, say 0.732, is equal to 0. A random-number generator immediately gives us the power to simulate the outcome of a fair-coin toss without actually having to toss the coin. The outcome is heads if the random number lies between 0 and $\frac{1}{2}$ (the probability of this is 0.5), and otherwise the outcome is tails.

Producing random numbers is not as easily accomplished as it seems, especially when they must be generated quickly, efficiently, and in massive amounts.[11] For occasional purposes, the use of a watch might be suitable if it were equipped with a stopwatch that could precisely measure time in tenths of seconds. Around 1920, crime syndicates in New York City's Harlem used the last five digits of the daily published U.S. treasure balance to generate the winning number for their illegal "Treasury Lottery." But this sort of method is of course not really practical. Even for simple simulation experiments, the required amount of random numbers runs quickly into the tens of thousands or higher. Generating a very large amount of random numbers on a one-time only basis, and storing them up in a computer memory, is also practically infeasible. But there is a solution to this kind of practical hurdle that is as handsome as it is practical. Instead of generating *truly* random numbers, a computer can generate *pseudo-random* numbers, as they are known, and it achieves this with the help of a nonrandom procedure. This procedure is iterative by nature and is determined by a suitably chosen function f. Starting with an arbitrary number z_0, the numbers $z_1, z_2, \ldots$ are successively generated by

$$z_1 = f(z_0), z_2 = f(z_1), \ldots, z_n = f(z_{n-1}), \ldots .$$

We refer to the function f as a pseudo-random generator, and it must be chosen such that the series $\{z_i\}$ is indistinguishable from a series of truly random numbers. In other words, the output of function f must be able to stand up to a great many statistical tests for "randomness." When this is the case, we are

[11]An interesting account of the history of producing random numbers can be found in D. J. Bennett's *Randomness*. Cambridge, MA: Harvard University Press, 1999.

in command of a simple and efficient procedure to produce random numbers. An added advantage is that a series of numbers generated by a pseudo-random generator is *reproducible* by beginning the procedure over again with the same seed number z_0. This can come in very handy when you want to make a simulation that compares alternative system designs: the comparison of alternative systems is purest when it can be achieved (to the extent that it is possible) under identical experimental conditions.

In practice, most random-number generators in use can be referred to as a *multiplicative* generator:

$$z_n = az_{n-1} \text{ (modulo } m),$$

where a and m are fixed positive integers. For the seed number z_0, a positive integer must always be chosen. The notation $z_n = az_{n-1}$ (modulo m) means that z_n represents the whole remainder of az_{n-1} after division by m; for example, 17 (modulo 5) = 2. This scheme produces one of the numbers $0, 1, \ldots, m-1$ each time. It takes no more than m steps until some number repeats itself. Whenever z_n takes on a value it has had previously, exactly the same sequence of values is generated again, and this cycle repeats itself endlessly. When the parameters a and m are suitably chosen, the number 0 is not generated and each of the numbers $1, \ldots, m-1$ appears exactly once in each cycle. In this case, the parameter m gives the length of the cycle. This explains why a very large integer should be chosen for m. The number z_n determines the random number u_n between 0 and 1 by $u_n = \frac{z_n}{m}$. The quality of the generator is strongly dependent on the choice of parameters a and m (a much used generator is characterized as $a = 630{,}360{,}016$ and $m = 2^{31} - 1$). We will not delve into the theory behind this. An understanding of the theory is not necessary in order to use this random generator on your computer. Today, most computers come equipped with a good random-number generator (this was not the case in days of yore). The simulation programs listed at the end of this chapter show very clearly how to use the random-number generator.

2.8.1 Pitfalls encountered in randomizing

The development of a good pseudo-random number generator must not be taken lightly. It is foolish, when using a multiplicative generator, to choose parameters for a and m oneself, or to piece together a patchwork algorithm by combining fragments from a number of existing methods, for example. That something is wild or complicated does not automatically mean that it is also random. The task of mixing objects together (lotto balls, for example) through physical means, such that we can say that the result is a random mix, is even more difficult than making a good random generator. A useful illustration of the difficulties involved in this undertaking can be seen in the example of the drafting of soldiers into the U.S. Armed Forces during the period of the Vietnam War. In 1970, widely varying drafting programs that had been run by individual states were scrapped in favor of a national lottery. The framework of the lottery was built on a plan to use birthdays as a means of choosing the young men to be drafted. Preparations for the drawing were made as follows. First, the 31 days of January were recorded on pieces of paper that were placed into capsules, and these, in turn, were placed into a large receptacle. After that, the 29 days of February (including February 29) were recorded, placed into capsules, and added to the receptacle. At this point, the January and February capsules were mixed. Next, the 31 days of March were recorded, encapsulated, and mixed through the January/February mixture, and the days of all the other months were treated similarly. When it was time for the drawing, the first capsule to be drawn was assigned the number 1, the second capsule drawn was assigned a number 2, etc., until all capsules had been drawn and assigned a number between 1 and 366. The men whose birth dates were contained in capsules receiving low-end numbers were called up first. Doubts about the integrity of the lottery were raised immediately following the drawing. Statistical tests demonstrated, indeed, that the lottery was far from random (see also Section 3.5). The failure of the lottery is easily traced to the preparatory procedures that occurred prior to the drawing. The mixing of the capsules was inadequately performed: the January capsules were mixed through the others 11 times, whereas the December capsules were mixed only once. In addition, it appeared that, during the public drawing, most capsules were chosen from the top of the receptacle. Preparations for the 1971 drawing were made with a great deal more care, partly because statisticians were called in to help. This time, two receptacles were used: one with 366 capsules for the days of the year, and another with 366 capsules for the numbers 1 through 366. One capsule was chosen from each receptacle in order to couple the days and numbers. The biggest improvement was that the *order* in which the capsules with the 366 days and the 366 numbers went into their respective receptacles was determined *beforehand* by letting a computer generate two random permutations of the integers $1, \ldots, 366$. The random permutations governed the order in which the capsules containing the days of the year and the lottery numbers were put into the receptacles. Next, a physical hand-mixing of the capsules took place. In fact, the physical mixing was not necessary but served as a public display of what people think of as random. The actual mixing took place through the random permutations. In Section 2.9, it is shown how the computer generates a random permutation of the integers $1, \ldots, 366$.

2.8.2 The card shuffle

Another example of how informal procedures will not lead to a random mix can be seen in the shuffling of a deck of cards. Most people will shuffle a pack of 52 cards three or four times. This is completely inadequate to achieve a random mix of the cards. A card mix is called random when it can be said that each card in the deck is as likely to turn up in any one given position as in any other. For a pack of 52 cards, it is reasonable to say that *seven* "riffle shuffles" are needed to get a mix of cards that, for all practical purposes, is sufficiently random. In a riffle shuffle, the deck of cards is divided into two more or less equal stacks that are then intermingled (riffled) to form one integrated stack. It is assumed that the riffle shuffle is imperfect and thus contains a nonnegligible element of chance (in the perfect riffle shuffle, the deck is exactly halved and every single card is interwoven back and forth; it can be mathematically demonstrated that, after eight such perfect riffle shuffles, a new deck of 52 cards will have returned to its original order). Some experienced bridge players are capable of taking advantage of situations in which the cards are shuffled such that their resulting distribution is not random. In order to raise their chances of winning, some regular casino gamblers make use of the knowledge that cards are usually not shuffled to a random mix (one deck should be shuffled seven times, two decks should be shuffled nine times, and six packs should be shuffled twelve times). In professional bridge tournaments and in casinos, computers are being used more and more to ensure a random mix of cards. It took advanced mathematics to explain the fact that only after seven or more imperfect riffle shuffles could one expect to find a more or less random mix of a deck of 52 cards (see D. J. Aldous and P. Diaconis, "Shuffling cards and stopping times," *The American Mathematical Monthly* 93, 333–348, 1986). The mix of cards resulting from seven riffle shuffles is sufficiently random for common card games such as bridge, but it is not really random in the mathematical sense. This can be seen in Peter Doyle's fascinating card game called "Yin Yang Solitaire." To play this game, begin with a new deck of cards. In the United States, a new deck of cards comes in the order specified here: with the deck laying face down, you will have ace, two, . . . , king of hearts, ace, two, . . . , king of clubs, king, . . . , two, ace of diamonds, and king, . . . , two, ace of spades. Hearts and clubs are yin suits, and diamonds and spades are yang suits. The deck is shuffled seven times, cut, and placed face down on the table. The player's goal is to make a stack of cards for each suit. This is achieved by removing each card from the top of the deck and turning it over to reveal its face value. A stack for a suit is started as soon as the ace of that suit appears. Cards of the same suit are added to the stack according to the rule that they must be added in the order ace, two, . . . , king. A single pass through the deck is normally not enough to complete the stack for all four suits. Having made a pass, the remaining deck is turned back over and another pass is made. The game is over as soon as either the two yin-suit stacks or the two yang-suit stacks are complete. Yin wins if the two yin suits are completed first. If the deck has been thoroughly permuted (by being put through a clothes dryer cycle, say), the yins and yangs will be equally likely to be completed first. But it turns out that, after seven ordinary riffle shuffles and a cut, it is significantly more likely that the yins will be completed before the yangs. The probability of yin winning is about 81% in this case. This demonstrates the difficulty of getting a fully random mix of the cards by hand. Finally, it is interesting to note that the probability of yin winning is approximately equal to 67%, 59%, and 54%, respectively, after eight, nine, and ten riffle shuffles. Only after fifteen riffle shuffles can we speak of a nearly 50% probability of yin winning.

2.9 Simulating from probability distributions

A random-number generator for random numbers between 0 and 1 suffices for the simulation of random samples from an arbitrary probability distribution. A handsome theory with all kinds of efficient methods has been developed for this purpose; however, we will confine ourselves to mentioning just the few basic methods that serve our immediate purposes.

2.9.1 Simulating from an interval

You want to surprise some friends by arriving at their party at a completely random moment in time between 2:30 and 5:00. How can you determine that moment? You must generate a random number between $2\frac{1}{2}$ and 5. How do you blindly choose a number between two given real numbers a and b when $a < b$? First, you have your computer generate a random number u between 0 and 1. Then you find a random number between a and b by

$$a + (b - a)u.$$

2.9.2 Simulating from integers

How can you designate one fair prize-winner among the 2,725 people who correctly answered a contest question? You achieve this by numbering the correct entries as $1, 2, \ldots, 2{,}725$ and generating randomly an integer out of the integers $1, 2, \ldots, 2{,}725$. How can you blindly choose an integer out of the integers $1, \ldots, M$? First, have your computer generate a random number u between 0 and 1. Then the integer

$$1 + \lfloor Mu \rfloor$$

can be considered as a random integer sampled from the integers $1, \ldots, M$. The common notation for the integer that arises when the real number x is rounded down is $\lfloor x \rfloor$. In other terms,

$$\lfloor x \rfloor = k \quad \text{if} \quad k \leq x < k + 1$$

for some integer k. One application is the simulation of the outcome of a roll of a fair die ($M = 6$). For example, the random number $u = 0.428\ldots$ leads to the outcome $3 = (1 + \lfloor 6u \rfloor)$ of the roll of the die.

Verify for yourself that a simulation from the integers $a, a + 1, \ldots, b$ is given by $a + \lfloor (b - a + 1)u \rfloor$ when u is a random number between 0 and 1.

2.9.3 Simulating from a discrete distribution

For a football pool, how can you come up with a replacement outcome for a cancelled football match for which a group of experts has declared a homewin with 50% probability, a visitor's win with 15% probability, and a tie game with 35% probability? You can do this by simulating from a distribution that has assigned probabilities 0.50, 0.15, and 0.35 to the numbers 1, 2, and 3, respectively. How do you simulate from a discrete distribution of a random variable X that assumes a finite number of values $x_1, \ldots, x_M$ with corresponding probabilities $p_1, \ldots, p_M$? This is very simple for the special case of a two-point distribution in which the random variable X can only assume the values x_1 and x_2. First, you have your computer generate a random number u between 0 and 1. Next, for the random variable X, you find the simulated value x_1 if $u \leq p_1$ and the value x_2 otherwise. You generate in this way the value x_1 with probability p_1 and the value x_2 with probability $p_2 = 1 - p_1$ (why?). In particular, the outcome of "heads or tails" in the toss of a fair coin can be simulated in this way. Generate a random number u between 0 and 1. If $u \leq \frac{1}{2}$, then the outcome is "heads" and otherwise the outcome is "tails." The inversion method for simulating from a two-point distribution can also be extended to that of a general discrete distribution, but this leads to an inefficient approach for the general case of $M > 2$. A direct search for the index l satisfying $p_1 + \cdots + p_{l-1} < u \leq p_1 + \cdots + p_l$ is too time-consuming for simulation purposes when M is not small. However, if each of the probabilities $p_j = P(X = x_j)$ is given only in a few decimals, then there is a simple but useful method called the *array method.* As a means of understanding this method, consider the case in which each probability p_j is given in precisely two decimals. That is, p_j can be represented by $k_j/100$ for some integer k_j with $0 \leq k_j \leq 100$ for $j = 1, \ldots, M$. You then form an array `A[i]`, $i = 1, \ldots, 100$, by setting the first k_1 elements equal to x_1, the next k_2 elements equal to x_2, etc., and the last k_M elements equal to x_M. To illustrate, suppose that $M = 4$ and that the distribution $p_j = P(X = x_j)$ for $j = 1, \ldots, 4$ is given by

$$p_1 = 0.20, \quad p_2 = 0.30, \quad p_3 = 0.15, \quad p_4 = 0.35.$$

You find then

$$\mathtt{A[1]} = \cdots = \mathtt{A[20]} = x_1, \qquad \mathtt{A[21]} = \cdots = \mathtt{A[50]} = x_2,$$

$$\mathtt{A[51]} = \cdots = \mathtt{A[65]} = x_3, \qquad \mathtt{A[66]} = \cdots = \mathtt{A[100]} = x_4.$$

Now have your computer generate a random number u between 0 and 1. Calculate the integer $m = 1 + \lfloor 100u \rfloor$. This simulated integer m is a randomly chosen integer from the integers $1, \ldots, 100$. Next, take `A[`m`]` as the simulated value of the random variable X. For example, suppose that the random number $u = 0.373\ldots$ has been generated. This gives $m = 38$ and thus the simulated value x_2 for the random variable X. It will be clear that the array method applies with an array of thousand elements when each probability p_j is given to exactly three decimal places.

2.9.4 Random permutation

How can you randomly assign numbers from the integers $1, \ldots, 10$ to ten people such that each person gets a different number? This can be done by making a random permutation of the integers $1, \ldots, 10$. A random permutation of the integers $1, \ldots, 10$ is a sequence in which the integers $1, \ldots, 10$ are put in random order. The following algorithm generates a random permutation of $1, \ldots, n$ for a given positive integer n:

Algorithm for random permutation

1. Initialize $t := n$ and $a[j] := j$ for $j = 1, \ldots, n$.
2. Generate a random number u between 0 and 1.
3. Set $k := 1 + \lfloor tu \rfloor$ (random integer from the indices $1, \ldots, t$). Interchange the current values of $a[k]$ and $a[t]$.
4. Let $t := t - 1$. If $t > 1$, return to step 2; otherwise, stop and the desired random permutation $(a[1], \ldots, a[n])$ is obtained.

The idea of the algorithm is first to randomly choose one of the integers $1, \ldots, n$ and to place that integer in position n. Next, you randomly choose one of the remaining $n - 1$ integers and place it in position $n - 1$, etc. For the simulation of many probability problems, this is a very useful algorithm.

2.9.5 Simulating a random subset of integers

How does a computer generate the Lotto 6/45 "Quick Pick," that is, six different integers from the integers $1, \ldots, 45$? More generally, how does the computer generate randomly r different integers from the integers $1, \ldots, n$? This is accomplished by following the first r iteration steps of the above algorithm for a random permutation until the positions $n, n-1, \ldots, n-r+1$ are filled. The elements $a[n], \ldots, a[n-r+1]$ in these positions constitute the desired random subset.

2.9.6 Simulation programs

In the field of physics, it is quite common to determine the values of certain constants in an experimental way. Computer simulation makes this kind of approach possible in the field of mathematics, too. For example, the value of π can be estimated with the help of some basic principles of simulation (of course, this is not the simplest method for the calculation of π). This is the general idea: take a unit circle (radius $= 1$) with the origin $(0, 0)$ as middle point. In order to generate random points inside the circle, position the unit circle in a square that is described by the four corner points $(-1, 1)$, $(1, 1)$, $(1, -1)$, and $(-1, -1)$. The area of the unit circle is π and the area of the square is equal to 4. Now, generate a large number of points that are randomly spread out over the surface of the square. Next, count the number of points that fall within the surface of the unit circle. If you divide this number of points by the total number of generated points, you get an estimate for $\frac{\pi}{4}$. You can identify a blindly chosen point (x, y) in the square by generating two random numbers u_1 and u_2 between 0 and 1 and then taking $x = -1 + 2u_1$ and $y = -1 + 2u_2$. Point (x, y), then, only belongs to the unit circle if $x^2 + y^2 \leq 1$. The *hit-or-miss method* used to generate random points inside the circle can also be used to generate random points in any given bounded region in the plane or in other higher dimensional spaces.

Below, for the interested reader, we have listed a Pascal program for simulating the value of π. Pascal was chosen as the programming language because of its clarity and readability. In Pascal, all information occurring between brackets { } is commentary included for instructional purposes and is not read by the computer.

```
PROGRAM SimulatePi(Input, Output);

CONST
  n = 10000; { number of runs }
```

```
VAR
  k : Integer; { current run }
  DartsInCircle : Integer; { total number of hits }
  Fraction : Real; { proportion of hits }
  Pi : Real; { estimate of pi }

PROCEDURE ThrowDart;
  VAR
    x, y : Real; {(x,y) coordinates of the position of
       the dart }
  BEGIN { x and  y are randomly sampled from [-1,1] }
    x := -1+2*Random;
    y := -1+2*Random;
    { Test whether (x,y) falls within the unit circle }
    IF (Sqr(x) + Sqr(y) <= 1) THEN
       DartsInCircle := DartsInCircle + 1;
  END;

BEGIN { main program }
  Randomize; { initializes the random-number generator
               on your computer}
  DartsInCirkel := 0;
  { execute the simulation }
  for k := 1 to n do
    ThrowDart;
  { compute the desired results}
  Fraction := DartsInCircle / n;
  Pi := 4*Fraction;
  WriteLn('the simulated value of pi is:',Pi);

END.
```

This simulation program requires only a minor adjustment to verify experimentally that the volume of a sphere with radius r is equal to $4\pi r^3/3$ (try it!). In general, computer simulation can be used for the numerical evaluation of (complicated) integrals. It suffices to have a random-number generator. Also, the reader might find computer simulation a quick and useful approach to solve geometric probability problems that are otherwise not easily amenable to an analytical approach. Several examples of such problems are given in Problem 2.23.

We close this chapter with a simulation program for the daughter-son problem from Chapter 1. This problem can lead to heated discussions over the right answer. The right answer to the first question posed in this problem is $\frac{1}{3}$, but the probability in question changes to $\frac{1}{2}$ in the second situation discussed in the problem. In Chapter 6, we give a probabilistic derivation of these answers based on the assumption that the probability of a newborn infant being a girl is the same as the probability of its being a boy. In answering the second question, we have also made an assumption that, randomly, one of the two children will open the door. If it is assumed that when a family has one boy and one girl, the girl always opens the door, the situation changes dramatically. We give, here, a simulation program that is bound to convince mathematicians and non-mathematicians alike of the correctness of the answer. Writing a simulation program forces you to make implicit assumptions explicit and, in every step, to indicate precisely what you mean, unambiguously. In this way, you fall into fewer mental traps than would be the case if you were giving a purely verbal account.

```
PROGRAM Family(Input, Output);

CONST
  number_runs = 10000; { total number of simulation runs}

VAR
  number_d : Integer; { total number of runs with 1 or
    2 daughters }
  number_dd : Integer; { total number of runs with 2 daughters }
  number_d_opens : Integer;
    { total number of runs in which a daughter opens the door }
  k : integer; { index of the current run }

PROCEDURE init;
  BEGIN
     number_d := 0;
     number_dd := 0;
     number_d_opens := 0;
  END;

PROCEDURE run;
  VAR
    composition: (DD, DS, SD, SS);
        { composition of the family, D=daughter, S=son }
    opens_door : (D, S};
        { daughter or son opens the door}
    drawing : Integer; { random drawing from 1,2,3,4 }
  BEGIN
    { determine the composition of the family: draw from 1,2,3,4 }
     drawing := 1 + trunc(4*Random);
     CASE drawing OF
       1 : composition := DD;
       2 : composition := DS;
       3 : composition := SD;
       4 : composition := SS;
     END;

     { determine whether daughter or son opens the door }
     IF ( composition = DD) THEN
       opens_door := D
     ELSE IF ( composition = SS) THEN
       opens_door := S
```

```
    ELSE BEGIN { composition is DS of SD: additional random
           number is needed to determine who opens the door,
           D or S; each possibility has a probability of 0.5 }
      IF (Random < 0.5) THEN
         opens_door := D
      ELSE
         opens_door := S;
    END;

    { update counters }
    IF (composition <> SS) THEN
      number_d := number_d + 1;
    IF (composition = DD) THEN
      number_dd := number_dd + 1;
    IF (opens_door = D) then
      number_d_opens := number_d_opens + 1;

 END;

BEGIN { main program }
   Randomize; { initialize the  random-number generator
               of your computer }
   { execute the simulation }
   init;
   for k := 1 to number_runs do
      run;
   { results }
   Writeln('probability of two daughters given one daughter :',
      number_dd/number_d);
   Writeln('probability of two daughters given a daughter opens:',
      number_dd/number_d_opens);
END.
```

2.10 Problems

2.1 On a modern die, the face value 6 is opposite to the face value 1, the face value 5 to the face value 2, and the face value 4 to the face value 3. In other words, by turning a die upside down, the face value k is changed into $7 - k$. This fact may be used to explain why when rolling three dice, the totals 9 and 12 ($= 3 \times 7 - 9$) are equally likely. Old Etruscan dice show 1 and 2, 3 and 4, and 5 and 6 on opposite sides. Would the totals 9 and 12 remain equally likely when rolling three Etruscan dice?

2.2 In the television program "Big Sisters," 12 candidates remain. The public chooses four candidates for the final round. Each candidate has an equal probability of being chosen. The Gotham Echo reckons that the local heroine, Stella Stone, has a probability of 38.5% of getting through to the final: they give her a $\frac{1}{12}$ probability of being chosen first, a $\frac{1}{11}$

probability of being chosen second, a $\frac{1}{10}$ probability of being chosen third, and a $\frac{1}{9}$ probability of being chosen fourth. Is this calculation correct?

2.3 Three friends go to the cinema together on a weekly basis. Before buying their tickets, all three friends toss a fair coin into the air once. If one of the three gets a different outcome than the other two, that one pays for all three tickets; otherwise, everyone pays his own way. Set up a probability model to calculate the probability that one of the three friends will have to pay for all three tickets. What is the long-run average frequency with which one of the three friends pays for all the tickets?

2.4 Answer each of the following five questions by choosing an appropriate sample space and assigning probabilities to the various elements of the sample space.

(a) In Leakwater township, there are two plumbers. On a particular day, three Leakwater residents call village plumbers independently of each other. Each resident randomly chooses one of the two plumbers. What is the probability that all three residents will choose the same plumber?

(b) You roll a fair die three times in a row. What is the probability that the second roll will deliver a higher point count than the first roll and the third roll a higher count than the second?

(c) Independently of each other, two people think of a number between 1 and 10. What is the probability that five or more numbers will separate the two numbers chosen at random by the two people?

(d) In the "Reynard the Fox" café, it normally costs \$3.50 to buy a pint of beer. On Thursday nights, however, customers pay \$0.25, \$1.00, or \$2.50 for the first pint. In order to determine how much they will pay, customers must throw a dart at a dartboard that is divided into eight segments of equal size. Two of the segments read \$0.25, four of the segments read \$1.00, and two more of the segments read \$2.50. You pay whatever you hit. Two friends, Lou and Henry, each throw a dart at the board and hope for the best. What is the probability that the two friends will have to pay more than \$2.00 between them for their first pint?

(e) Two players A and B each roll one die. The absolute difference of the outcomes is computed. Player A wins if the difference is 0, 1, or 2; and otherwise, player B wins. Is this a fair game?

2.5 A dog has a litter of four puppies. Set up a probability model to answer the following question. Can we correctly say that the litter more likely consists of three puppies of one gender and one of the other than that it consists of two puppies of each gender?

2.6 Sixteen bridge teams, including the teams Johnson and Smith, participate in a tournament. The tournament is organized as a knock-out tournament and has four rounds. The sixteen teams are evenly matched. In each round, the remaining teams are paired by drawing lots.

(a) What is the probability that the teams Johnson and Smith will meet in the first round?

(b) What is the probability that these two teams will meet in the final? *Hint*: use different sample spaces for the questions (a) and (b).

2.7 Use an appropriate sample space with equiprobable elements to answer the following question. You enter a grand-prize lottery along with nine other people. Ten numbered lots, including the winning lots, go into a box. One at a time, participants draw a lot out of the box. Does it make a difference to your probability of winning whether you are the first or the last to draw a lot?

2.8 In the daily lottery game "Guess 3," three different numbers are picked randomly from the numbers $0, 1, \ldots, 9$. The numbers are picked in order. To play this game, you must choose between "Exact order" and "Any order" on the entry form. In either case, the game costs \$1 to play. Should you choose to play "Exact order," you must tick three different numbers in the order you think they will be picked. If those numbers are picked in that order, you win a \$360 payoff. Should you opt to play "Any order," you tick three numbers without regard for their order of arrangement. You win a \$160 payoff if those three numbers are picked. Set up a probability model to calculate the expected value of the payoff amount for both options.

2.9 In the dice game known as "seven," two fair dice are rolled and sum of scores is counted. You bet on "manque" (that a sum of 2, 3, 4, 5, or 6 will result) or on "passe" (that a sum of 8, 9, 10, 11, or 12 will result). The sum of 7 is a fixed winner for the house. A winner receives a payoff that is double the amount staked on the game. Nonwinners forfeit the amount staked. Define an appropriate probability space for this experiment. Then calculate the expected value of the payoff per dollar staked.

2.10 Sic Bo is an ancient Chinese dice game that is played with three dice. There are many possibilities for betting on this game. Two of these are "big" and "small." When you bet "big," you win if the total points rolled equals 11, 12, 13, 14, 15, 16, or 17, except when three 4s or three 5s are rolled. When you bet "small," you win if the total points rolled equals 4, 5, 6, 7, 8, 9, or 10, except when three twos or three threes are rolled. Winners of "big" and "small" alike receive double the amount staked on the game. Calculate the house percentage for each of these betting formats.

2.11 You are playing a game in which four fair dice are rolled. A \$1 stake is required. The payoff is \$100 if all four dice show the same number and \$10 if two dice show the same even or odd number. What is an appropriate probability space for this experiment, and what is the expected value of the payoff?

2.12 Calculate the expected value of the greater of two numbers when two different numbers are picked at random from the numbers $1, \ldots, n$. What is the expected value of the difference between the two numbers?

2.13 Consider the Kelly betting model from Section 2.7. In addition to the possibility of investing in a risky project over a large number of successive periods, you can get a fixed-interest rate at the bank for the portion of your capital that you do not invest. You can reinvest your money at the end of each period. Let the interest rate be r, i.e., every dollar you do not invest in a certain period will be worth $1 + r$ dollars at the end of the period. Assume that the expected value of the payoff of the risky project satisfies $pf > 1 + r$.

(a) For the growth factor G_n in the representation $V_n = e^{nG_n} V_0$, show that it holds true that

$$\lim_{n\to\infty} G_n = p \ln[(1-\alpha)(1+r) + \alpha f] + (1-p)\ln[(1-\alpha)(1+r)].$$

Verify that this expression is maximal for $\alpha^* = \frac{pf-(1+r)}{f-(1+r)}$.

(b) Suppose you are faced with a 100% safe investment returning 5% and a 90% safe investment returning 25%. Calculate how to invest your money using the Kelly strategy. Calculate also the effective rate of return on your investment over the long-term.

2.14 A particular game pays f_1 times the amount staked with a probability of p and f_2 times the amount staked with a probability of $1 - p$, where $f_1 > 1, 0 \leq f_2 < 1$, and $pf_1 + (1-p)f_2 > 1$. You play this game a large number of times and each time you stake the same fraction α of your bankroll. Verify that the Kelly fraction is given by

$$\alpha^* = \min\left(\frac{pf_1 + (1-p)f_2 - 1}{(f_1 - 1)(1 - f_2)}, 1\right)$$

with $(1 - \alpha^* + \alpha^* f_1)^p (1 - \alpha^* + \alpha^* f_2)^{1-p} - 1$ as the corresponding effective rate of return over the long-term.

2.15 Someone proposes the following wager. You and the person in question, independently of one another, are each to draw a random number from the numbers $1, \ldots, 100$. If the sum of the two numbers drawn is either in the range 2, ..., 50 or in the range 101, ..., 150, then you will receive ten dollars; otherwise you must pay ten dollars. Is this a fair wager?

2.16 Three players, A, B, and C each put ten dollars into a pot with a list on which they have, independently of one another, predicted the outcome of three successive tosses of a fair coin. The fair coin is then tossed three times. The player having most correctly predicted the three outcomes gets the contents of the pot. The contents are to be divided if multiple players guess the same number of correct outcomes.

(a) Calculate the expected value of the amount that player A will get.

(b) Suppose that players A and B decide to collaborate, unbeknownst to player C. The collaboration consists of the two players agreeing that player B's list will always be a mirror image of player A's list (should player A predict an outcome of HTT, for example, then player B would predict TTH). Calculate the expected value of the amount that player A receives.

2.17 The following game is played in a particular carnival tent. The carnival master has two covered beakers, each containing one die. He shakes the beakers thoroughly, removes the lids, and peers inside. You have agreed that whenever at least one of the two dice shows an even number of points, you will bet with even odds that the other die will also show an even number of points. Is this a fair bet?

2.18 Three players enter a room and are given a red or a blue hat to wear. The color of each hat is determined by a fair coin toss. Players cannot see the color of their own hats, but do see the color of the other two players' hats. The game is won when at least one of the players correctly guesses the color of his own hat, and no player gives an incorrect answer. In addition to having the opportunity to guess a color, players may also pass. Communication of any kind between players is not permissible after they have been given hats, however, they may agree on a group strategy beforehand. Verify that there is a group strategy that results in a $\frac{3}{4}$ probability of winning. (This puzzle was discussed in the *New York Times* of April 10, 2001. The hat problem with many players is related to problems in coding theory. The strategy gets far more complicated for larger numbers of players. In the game with $2^m - 1$ players, there is a strategy for which the group is victorious with a probability of $(2^m - 1)/2^m$.)

2.19 At a completely random moment between 6:30 and 7:30 A.M., the morning newspaper is delivered to Mr. Johnson's residence. Mr. Johnson leaves for work at a completely random moment between 7:00 and 8:00 A.M., regardless of whether the newspaper has been delivered. What is the probability that Mr. Johnson can take the newspaper with him to work? Use computer simulation to find the probability.

2.20 Two people have agreed to meet at the train station between 12:00 and 1:00 P.M. Independently of one another, each person is to appear at a completely random moment between the hours of 12:00 and 1:00. The agreement is that each will wait no longer than ten minutes for the other. What is the probability that the two people will actually meet up? Use computer simulation to find the probability.

2.21 Three numbers a, b, and c are chosen at random from the interval $(-1, 1)$, independently of each other. What is the probability that the quadratic equation $ax^2 + bx + c = 0$ has two real roots? Use computer simulation to find the probability.

2.22 A stick is broken at random into two pieces. You bet on the ratio of the length of the longer piece to the length of the smaller piece. You receive $\$\, k$ if the ratio is between k and $k + 1$ for some $1 \leq k \leq m - 1$, while you receive $\$\, m$ if the ratio is larger than m. Here, m is a given positive integer. Using computer simulation, verify that your expected payoff is approximately equal to $2[\ln(m + 1) - 0.4228 + 2/(m + 1)]$. Do you see a resemblance with the St. Petersburg paradox?

2.23 Use computer simulation to find

(a) the expected value of the distance between two points that are chosen at random inside the interval (0, 1).

(b) the expected value of the distance between two points that are chosen at random inside the unit square

(c) the expected value of the distance between two points that are chosen at random inside the unit circle

(d) the expected value of the distance between two points that are chosen at random inside an equilateral triangle with sides of unit length.

2.24 A millionaire plays European roulette every evening for pleasure. He begins every time with $A = 100$ chips of the same value and plays on until he has gambled away all 100 chips. When he has lost his 100 chips for that evening's entertainment, he quits. Use computer simulation to find the average number of times the millionaire will play per round, for the Big-Martingale betting system and for the D'Alembert betting system. Also determine the probability that on a given evening the millionaire will acquire $B = 150$ chips before he is finished playing. Do the same for $A = 50$ and $B = 75$. Can you give an intuitive explanation for why the average value of the total number of chips the millionaire stakes per evening is equal to $37A$ over the long-term, regardless of the betting system he uses?

2.25 You decide to bet on ten spins of the roulette wheel in European roulette and to use the double-up strategy. Under this strategy, you bet on red each time and you double your bet if red does not come up. If red comes up, you go back to your initial bet of 1 euro. Use computer simulation to find the expected value of your loss after a round of ten bets and to find the expected value of the total amount bet during a round. Can you explain why the ratio of these two expected values is equal to $\frac{1}{37}$?

2.26 Seated at a round table, five friends are playing the following game. One of the five players opens the game by passing a cup to the player seated either to his left or right. That player, in turn, passes the cup to a player on his left or right and so on until the cup has progressed all the way around the table. As soon as one complete round has been achieved, the player left holding the cup pays for a round of drinks. A coin-toss is performed before each turn in order to determine whether the cup will go to the left or right. Use computer simulation to find, for each player, the probability that the player will have to buy a round of drinks.

2.27 What is the probability of two consecutive numbers in a particular lotto drawing of six numbers from the numbers $1, \ldots, 45$? Use computer simulation to find this probability.

2.28 You have been asked to determine a policy for accepting reservation for an airline flight. This particular flight uses an aircraft with 15 first-class seats and 75 economy-class seats. First-class tickets on the flight cost \$500 and economy-class tickets cost \$250. The number of individuals who seek to reserve seats takes on the equally likely values of $10, 11, \ldots, 20$ in first class and the equally likely values of $40, 41, \ldots, 120$ in economy class. The demand for first-class seats and economy-class seats is independent. The airline allows itself to sell somewhat more tickets than it has seats. This is a common practice called overbooking. You are asked to analyze the four possible booking polices which permit the overbooking of either up to 0, or up to 3 first-class seats and the overbooking of either up to 5, or up to 10 economy seats. Each passenger who buys a first-class ticket has a 10% probability of not showing up for the flight. The probability of not showing-up is 5% for economy-class passengers. Passengers decide whether to show up independently of each other. First-class passengers who do not show up can return their unused tickets for a full refund. No-shows in the economy class are not entitled to any refund. Any first-class passengers who show up for the flight but cannot be seated in the first class are entitled to a full refund plus \$400 compensation. If there are free seats in first class and economy class is full, economy-class passengers can be seated in first class.

If an economy-class passenger shows up and is denied a seat, however, they get a full refund plus \$200 compensation. Use computer simulation to find for each of the four possible booking polices both the expected value of the net profit for the flight and the probabilities of the net profit falling in each of the ranges [\$15,000, \$16,000], ..., [\$27,000, \$28,000].

2.29 The card game called Ace-Jack-Two is played between one player and the bank. It goes this way: a deck of 52 cards is shuffled thoroughly, after which the bank repeatedly reveals three cards next to each other on a table. If an ace, jack or two is among the three cards revealed, the bank gets a point. Otherwise, the player gets a point. The points are tallied after 17 rounds are played. The one with the most points is the winner. Use computer simulation to determine the probability of the bank winning and the average number of points that the bank will collect per game.

2.30 Consider the best-choice problem from Section 2.3 with 100 pieces of paper and the strategy of discarding the numbers on the first 25 pieces of paper. Use computer simulation to find the probability of obtaining either the largest or the second largest possible number. *Hint*: use a random permutation of the integers $1, \ldots, 100$.

2.31 Two candidates A and B remain in the finale of a television game show. At this point, each candidate must spin a wheel of fortune. The twenty numbers $5, 10, \ldots, 95, 100$ are listed on the wheel and when the wheel has stopped spinning, a pointer randomly stops on one of the numbers. Each candidate has a choice of spinning the wheel one or two times, whereby a second spin must immediately follow the first. The goal is to reach a total closest to but not exceeding 100 points. The winner is the candidate who gets the highest score. Should there be a tie, then the candidate to spin the wheel first is the winner. The candidate who spins second has the advantage of knowing what the score of the first candidate was. Lots are drawn to determine which player begins. Suppose that candidate A spins first and scores G points in a first spin. Use computer simulation to determine the probability that candidate A will lose by stopping after the first spin and the probability that candidate A will lose by spinning again. Determine these probabilities for $G = 50$ and $G = 55$.

2.32 Reconsider the wheel of fortune from Problem 2.31. This time, let us assume that the two players are competing against each other thus: player A spins the wheel either once or twice as desired, after which player B is obliged to spin the wheel two times. A player's total score equals the total number of points acquired in the spins of the wheel, with the stipulation that spins tallying up to more than 100 points will result in a score of 0. Player A uses the following strategy: should the first spin give more than L points, stop; otherwise, spin again. Using computer simulation, determine the value of L that gives player A a maximum probability of having a higher total score than player B.

2.33 Using five dice, you are playing a game consisting of accumulating as many points as possible in five rounds. After each round, you may "freeze" one or more of the dice, i.e., a frozen die will not be rolled again in successive rounds, but the amount of points showing will be recounted in successive rounds. You apply the following strategy: if there are still i rounds to go, you freeze a die only when it displays more than α_i points, where $\alpha_4 = 5$, $\alpha_3 = 4$, $\alpha_2 = 4$, $\alpha_1 = 3$, and $\alpha_0 = 0$. A grand total of s points results in a payoff of $s - 25$ dollars if $s \geq 25$, and a forfeiture of $25 - s$ dollars if $s < 25$. Using computer simulation, determine the expected value of the payoff. Also determine the probability that your grand total will be 25 or more points.

2.34 A fair coin is tossed repeatedly. Answer the following questions.

(a) What is the probability that n tosses are needed before the coin lands on the same side twice in a row? What is the expected value of the number of tosses required? *Hint*: use the sample space of the coin-toss experiment with a fixed number of n tosses.

(b) Use computer simulation to find the probability that n tosses will be required before the coin lands heads twice in a row. Verify experimentally that this probability is equal to $f_{n-1}/2^n$, where f_k is the kth Fibonacci number ($f_1 = f_2 = 1,\ f_k = f_{k-1} + f_{k-2}$ for $k > 2$). Also verify that the expected value of the number of required tosses is 6.

(c) Use computer simulation to find the probability that the number of heads ever exceeds twice the number of tails if the coin is tossed 5 times. What is the probability if the coin is tossed 25 times. What is the probability if the coin is tossed 50 times? Verify experimentally that the probability approaches the value $\frac{1}{2}(\sqrt{5} - 1)$ if the number of tosses increases.

(d) A fair coin is tossed no more than n times, where n is fixed in advance. After each toss, you can decide to stop the coin-toss experiment. Your payoff is 1,000 dollars multiplied by the proportion of heads at the moment the experiment is stopped. Your strategy is to stop as soon as the proportion of heads exceeds $\frac{1}{2}$ or as soon as n tosses are done, whichever occurs first. Use computer simulation to find your expected payoff for $n = 5, 10$, and 25. Verify experimentally that your expected payoff approaches the value $\frac{1}{4}\pi$ times 1,000 if n becomes large. Can you devise a better strategy than the one proposed?

2.35 In a TV program, the contestant can win one of three prizes. The prizes consist of a first prize and two lesser prizes. The dollar value of the first prize is a five-digit number and begins with 1, whereas the dollar values of the lesser prizes are three-digit numbers. There are initially four unexposed digits in the value of first prize and three in each of the values of the other two prizes. The game involves trying to guess the digits in the dollar value of the first prize before guessing the digits in either of the dollar values of the other two prizes. Each of the digits 0-9 is used only once among the three prizes. The contestant chooses one digit at a time until all of the digits in the dollar value of one of the three prizes have been completed. What is the probability that the contestant will win the first price? Use computer simulation to find this probability.

2.36 A random sequence of 0's and 1's is generated by tossing a fair coin N times. A 0 corresponds to the outcome heads and a 1 to the outcome tails. A run is an uninterrupted sequence of 0's or 1's only. Use computer simulation to verify experimentally that the length of the longest run exhibits little variation and has its probability mass concentrated around the value $\log_2(N) - \frac{2}{3}$ when N is sufficiently large.

2.37 You are playing the following game: a fair coin is tossed until it lands heads three times in a row. You get 12 dollars when this occurs, but you must pay one dollar for each toss. Using computer simulation, make a determination as to whether this is a fair contest.

2.38 A drunkard is standing in the middle of a very large town square. He begins to walk. Each step he takes is a unit distance in a randomly chosen direction. The direction for each step taken is chosen independently of the direction of the others. Suppose that the drunkard takes a total of n steps for a given value of n. Using computer simulation, verify that the expected value of the quadratic distance between the starting and ending points is equal to n, whereas the expected value of the distance between starting and ending points is approximately equal to $0.886\sqrt{n}$ if n is sufficiently large. Also use computer simulation to find, for both $n = 25$ and $n = 100$, the probability that the maximal distance of the drunkard to his starting point during the n steps will be greater than $1.18\sqrt{n}$.

2.39 A particle moves over the flat surface of a grid such that an equal unit of distance is measured with every step. The particle begins at the origin (0,0). The first step may be to the left, right, up, or down, with equal probability $\frac{1}{4}$. The particle cannot move back in the direction that the previous step originated from. Each of the remaining three directions has an equal probability of $\frac{1}{3}$. Suppose that the particle makes a total of n steps for a given value of n. Using computer simulation, verify that the expected value of the distance between the particle's starting and ending points is approximately equal to $1.25\sqrt{n}$ if n is sufficiently large. Also use computer simulation to find, for both $n = 25$ and $n = 100$, the probability that the maximal distance of the particle to its starting point during the n steps will be greater than $1.65\sqrt{n}$.

2.40 You have received a reliable tip that in the local casino the roulette wheel is not exactly fair. The probability of the ball landing on the number 13 is twice what it should be. The roulette table in question will be in use that evening. In that casino, European roulette is played. You go with 1,000 euros and intend to make 100 bets. Your betting strategy is as follows: each time you stake a multiple of five euros on the number 13 and you choose that multiple that is closest to 2.5% of your bankroll. Your will receive a payoff of 36 times the amount staked if the ball lands on 13. Use computer simulation to determine the probability distribution of your bankroll at the end of the night. Specifically, determine the probability of your leaving the casino with more than 2,000 euros.

2.41 Sixteen teams remain in a soccer tournament. A drawing of lots will determine which eight matches will be played. Before the drawing takes place, it is possible to place bets with bookmakers over the outcome of the drawing. Use computer simulation to find the probability of correctly predicting i matches for $i = 0$, 1, 2, and 3. *Hint*: simulate drawing lots by generating a random permutation of the integers $1, \ldots, 16$.

2.42 Two teams A and B are engaged in a series of matches that will continue until one of the two teams wins three matches in a row. Every match ends in a win for one of the teams (a draw is not possible). The probability of team A winning any given match is equal to p. The results of the consecutive matches are independent of each other. Use computer simulation to determine the expected value of the number of required matches and the probability that team A will take the series when $p = 0.45$. Do the same for $p = 0.5$.

2.43 Independently of each other, ten numbers are randomly drawn from the interval $(0, 1)$. You may view the numbers one by one in the order in which they are drawn. After viewing each individual number, you are given the opportunity to take it or let it pass. You are not allowed to go back to numbers you have passed by. Your task is to pick out the highest number. Your strategy is as follows. If, after you have viewed a number, there are still k numbers left to view, you will take the number if it is the highest number to appear up to that point and if it is higher than critical level a_k, where $a_0 = 0$, $a_1 = 0.500$, $a_2 = 0.690$, $a_3 = 0.776$, $a_4 = 0.825$, $a_5 = 0.856$, $a_6 = 0.878$, $a_7 = 0.894$, $a_8 = 0.906$, and $a_9 = 0.916$. Use computer simulation to determine the probability that you will pick out the highest number.

2.44 In a certain betting contest, you may choose between two games A and B at the start of every turn. In game A, you always toss the same coin, while in game B you toss either coin 1 or coin 2 depending on your bankroll. In game B, you must toss coin 1 if your bankroll is a multiple of three; otherwise, you must toss coin 2. A toss of the coin from game A will land heads with a probability of $\frac{1}{2} - \epsilon$ and tails with a probability of $\frac{1}{2} + \epsilon$, where $\epsilon = 0.005$. Coin 1 in game B will land heads with probability $\frac{1}{10} - \epsilon$ and tails with probability $\frac{9}{10} + \epsilon$; coin 2 in game B will land heads with probability $\frac{3}{4} - \epsilon$ and tails with probability $\frac{1}{4} + \epsilon$. In each of the games A and B, you win one dollar if heads is thrown and you lose one dollar if tails is thrown. An unlimited sequence of bets is made in which you may continue to play even if your bankroll is negative (a negative bankroll corresponds to debt). Following the strategy $A, A, \ldots$, you win an average of 49.5% of the bets over the long term. Use computer simulation to verify that using strategy $B, B, \ldots$, you will win an average of 49.6% of the bets over the long-term, but that using strategy $A, A, B, B, A, A, B, B, \ldots$ you will win 50.7% of the bets over the long term (the paradoxical phenomenon, that in special betting situations winning combinations can be made up of individually losing bets, is called *Parrondo's paradox* after the Spanish physicist Juan Parrondo see also G. P. Harmer and D. Abbott, "Losing strategies can win by Parrondo's paradox," *Nature*, Vol. 402, 23/30 December 1999. An explanation of the paradox lies in the dependency between the betting outcomes. Unfortunately, such a dependency is absent in casino games).

2.45 Center court at Wimbledon is buzzing with excitement. The dream finale between Alassi and Bicker is about to begin. The weather is fine, and both players are in top condition. In the past, these two players have competed multiple times under similar conditions. On the basis of past outcomes, you know that 0.631 and 0.659 give the respective probabilities that Alassi and Bicker will win their own service points when playing against each other. Use computer simulation to determine the probability of Alassi winning the finale. Now assume that the first set has been played and won by Alassi, and that the second set is about to begin. Bookmakers are still accepting bets. Determine, for this scenario, what Alassi's probability of winning the finale is.

3
Probabilities in everyday life

Computer simulation can be extremely useful to those who are trying to develop an understanding of the basic concepts of probability theory. The previous chapter recommended simulation as a means of explaining such phenomena as chance fluctuations and the law of large numbers. Fast computers allow us to simulate models swiftly and to achieve a graphic rendering of our outcomes. This naturally enhances our understanding of the laws of probability theory.

Monte Carlo simulation is the name given to the type of simulation used to solve problems that contain a random element. In such simulations, the computer's random-number generator functions as a sort of roulette wheel. Monte

Carlo simulation is widely applicable. Often, it is the only possible method of solving probability problems. This is not to say, however, that it does not have its limitations. It is not a "quick fix" to be applied haphazardly. Before beginning, one must think carefully about the model to be programmed. The development of simulation models for complex problems can require a lot of valuable time. Monte Carlo simulation gives numerical results, but the vast amounts of numerical data resulting can make it difficult to draw insightful conclusions, and insight is often more important than the numbers themselves. In general, the mathematical solution of a model will render both numbers and insight. In practice, then, a purely mathematical model that is limited to the essentials of a complex problem can be more useful than a detailed simulation model. Sometimes a combination of the two methods is the most useful, as in the use of simulation to test the practical usefulness of results gained from a simplified mathematical equation.

In this chapter, we will discuss a number of interesting probability problems. We will solve each of these problems twice: first by means of simulation and subsequently by means of a theoretical model that can be solved mathematically. The first problem we will tackle is the birthday problem, one of the most surprising problems in the field of probability theory. Thereafter, we will look at a few of the casino problems encountered in the games of craps and roulette, and a scratch-lottery problem. The common element in all of these problems is that they playfully demonstrate some of the important concepts and solution methods used in the field of probability theory.

3.1 The birthday problem

The birthday problem is very well known in the field of probability theory. It raises the following interesting questions: What is the probability that, in a group of randomly chosen people, at least two of them will have been born on the same day of the year? How many people are needed to ensure a probability greater than 0.5? Excluding February 29 from our calculations and assuming that the remaining 365 possible birthdays are all equally probable, we may be surprised to realize that, in a group of only 23 people, the probability of two people having the same birthday is greater than 0.5 (the exact probability is 0.5073). Then, again, perhaps this result is not so very surprising: think back to your school days and consider how often two or more classmates celebrated birthdays on the same day. In Section 4.2.3 of Chapter 4, further insight will be given to the fact that a group of only 23 people is large enough to have about a 50-50 chance of at least one coincidental birthday. What about the assumption that birthdays are uniformly distributed throughout the year? In reality, birthdays are not uniformly distributed. The answer is that the probability of a match only becomes larger for any deviation from the uniform distribution. This result can be mathematically proved. Intuitively, you might better understand the result by thinking of a group of people coming from a planet on which all people are born on the same day.

3.1.1 Simulation approach

A simulation model is easily constructed. Imagine that you want to calculate the probability of two people out of a group of 23 randomly chosen people having their birthdays on the same day. In each simulation experiment, 23 random drawings will be made out of the numbers $1, \ldots, 365$. A random drawing from these numbers is given by $1 + \lfloor 365u \rfloor$, when u is a random number between 0 and 1 (see Section 2.9). A simulation experiment is said to be successful when the same number is drawn at least twice. After a sufficiently large number of experiments, the probability of at least two persons having the same birthday can be estimated by

$$\frac{\text{number of successful simulation experiments}}{\text{total number simulation experiments}}.$$

For the interested reader, a Pascal program follows:

```
PROGRAM birthdayproblem(Input, Output);

  CONST
    size = 23;                    { number of pupils in the classroom }
    days= 365;                     { number of days in the year }
    simulation_length = 5000; { number of simulation runs }

  VAR
    birthday : ARRAY[1..days] OF Boolean;
    { birthday[i] = true  if somebody has a birthday on day i
      and false otherwise }
    number_successes : Integer; { total number of simulation runs
        with two or more birthdays on a same day}
    run : Integer;     { index of the current simulation run }

  PROCEDURE init;
    VAR i : Integer;
    BEGIN
      FOR i := 1 TO days DO
        birthday[i] := False;
    END;

  PROCEDURE classroom;
    VAR
      ready : Boolean;    { if true, the stop the simulation run }
```

```
    pupil : Integer; { index of the next pupil }
    birthday_next_pupil : Integer;

  BEGIN
    { pupils call in turn their birthdays until all pupils have
      been asked or a pupil calls a birthday mentioned before}
    ready := False;
    pupil := 0;
    WHILE NOT hit DO
      BEGIN
        pupil := pupil+1;
        birthday_next_pupil := 1 + Trunc(days*Random);
           { a random drawing for the birthday of
             the next pupil }
        IF birthday[birthday_next_pipil] THEN
          { an aforementioned birthday appears ! }
          BEGIN
            number_successes := number_successes+1;
            ready := True;
          END
        ELSE birthday[birtday_next_pupil] := True;
        IF pupil = size THEN { all pupils have been asked }
          ready := True;
      END;
  END;

BEGIN
  Randomize;  { initializes the random-number generator
                on your computer}
   number_successes := 0;
  FOR run := 1 TO simulation_length DO
    BEGIN
      init;
      classroom;
    END;
  WriteLn('  the simuled value of the probability is : ',
             number_successes/simulation_length);
 END.
```

3.1.2 Theoretical approach

In order to calculate the probability of two people in a randomly chosen group of n people having birthdays on the same day, the following approach is applicable. First, calculate the *complementary probability* (i.e., the probability of no two birthdays falling on the same day). This probability is simpler to calculate.[1]

[1] The simple technique of working with complementary probabilities is also handy in the solution of the De Méré problem described in the introduction to this book: the probability of rolling a double six in n rolls of a fair pair of dice is equal to 1 minus the complementary probability of rolling no double sixes at all in n rolls ($= 1 - \frac{35^n}{36^n}$).

Table 3.1 *Probabilities for the birthday problem*

n	15	20	23	25	30	40	50	75
p_n	0.2529	0.4114	0.5073	0.5687	0.7063	0.8912	0.9704	0.9997

Imagine that the n people are numbered in order from $1, \ldots, n$. There are 365^n outcomes for the possible birth dates of the n ordered people. Each of these outcomes is equally probable. The number of outcomes showing no common birthdays is equal to $365 \times 364 \times \cdots \times (365 - n + 1)$. The probability then, of no two of the n people having a common birthday, is equal to $365 \times 364 \times \cdots \times (365 - n + 1)$ divided by 365^n. From this, it follows that, in a group of n people, the probability of two people having the same birthday can be given by

$$p_n = 1 - \frac{365 \times 364 \times \cdots \times (365 - n + 1)}{365^n}.$$

In Table 3.1, the probability p_n is given for various values of n. It is surprising to see how quickly this probability approaches 1 as n grows larger. In a group of 75 people, it is practically certain that at least two people will have the same birthday. An approximation formula for probability p_n shows how quickly p_n increases as n grows larger. In Problem 3.12, you are asked to derive the approximation formula

$$p_n \approx 1 - e^{-\frac{1}{2}n(n-1)/365}.$$

We come back to this approximation formula in Section 4.2.3 of Chapter 4.

John Allen Paulos' *Innumeracy* contains a wonderful example of the misinterpretation of probabilities in everyday life. On late-night television's "The Tonight Show with Johnny Carson," Carson was discussing the birthday problem in one of his famous monologues. At a certain point, he remarked to his audience of approximately 100 people: "Great! There must be someone here who was born on my birthday!" He was off by a long shot. Carson had confused two distinctly different probability problems: 1) the probability of one person out of a group of 100 people having the same birth date as Carson himself, and 2) the probability of any two or more people out of a group of 101 people having birthdays on the same day. How can we calculate the first of these two probabilities? First, we must recalculate the complementary probability of no one person in a group of 100 people having the same birth date as Carson. A random person in the group will have a probability of $\frac{364}{365}$ of having a different birth date than Carson. The probability of no one having the same birthday as Carson is equal to $(\frac{364}{365}) \times \cdots \times (\frac{364}{365}) = (\frac{364}{365})^{100}$. Now, we can calculate the probability of at least one audience member having the same birthday as Carson to be equal to $1 - (\frac{364}{365})^{100} = 0.240$ (and not 0.9999998). Verify for yourself that the audience would have had to consist of 253 people in order to get about a 50-50 chance of someone having the same birthday as Carson.

3.1.3 Another birthday surprise

On Wednesday, June 21, 1995, a remarkable thing occurred in the German Lotto 6/49, in which six different numbers are drawn from the numbers $1, \ldots, 49$. On the day in question, the mid-week drawing produced this six-number result: 15-25-27-30-42-48. These were the same numbers as had been drawn previously on Saturday, December 20, 1986, and it was for the first time in the 3,016 drawings of the German Lotto that the same sequence had been drawn twice. Is this an incredible occurrence, given that in German Lotto there are nearly 14 million possible combinations of the six numbers in question? Actually, no, and this is easily demonstrated if we set the problem up as a birthday problem. In this birthday problem, there are 3,016 people and 13,983,816 possible birthdays. The 3,016 people correspond with the 3,016 drawings, while the binomial coefficient $\binom{49}{6} = 13{,}983{,}816$ gives the total number of possible combinations of six numbers drawn from the numbers $1, \ldots, 49$. The same reasoning used in the classic birthday problem leads to the conclusion that there is a probability of

$$\frac{13{,}983{,}816 \times (13{,}983{,}816 - 1) \times \cdots (13{,}983{,}816 - 3{,}015)}{(13{,}983{,}816)^{3016}} = 0.7224$$

that no combination of the six numbers will be drawn multiple times in 3,016 drawings of the German Lotto. In other words, there is a probability of 0.2776 that a same combination of six numbers will be drawn two or more times in 3,016 drawings. And this probability is not negligibly small!

3.1.4 The almost-birthday problem

In the almost-birthday problem, we undertake the task of determining the probability of two or more people in a randomly assembled group of n people having their birthdays within r days of each other. Denoting this probability by $p_n(r)$, it is given by

$$p_n(r) = 1 - \frac{(365 - 1 - nr)!}{365^{n-1}(365 - (r + 1)n)!}.$$

The proof for this formula being quite difficult, we have omitted it here, referring the interested reader to P. Diaconis and F. Mosteller's "Methods for Studying Coincidences." *Journal of the American Statistical Association* 84 (1989): 853–861.

Although the almost-birthday problem is far more complicated than the ordinary birthday problem when it comes to theoretical analysis, this is not the case when it comes to computer simulation. Just a slight adjustment to the simulation program for the birthday problem makes it suitable for the almost-birthday problem. This is one of the advantages of simulation. For several values of n, Table 3.2 gives the value of the probability $p_n(1)$ that in a randomly assembled group of n people at least two people will have birthdays within one day of each other ($r = 1$). A group of 14 people is large enough to end up with a probability of more than 50% that at least two people will have birthdays within one day of each other. Taking $r = 7$, one calculates that if seven students are renting a house together, there is a probability of more than 50% that at least two of them will have birthdays within one week of each other.

3.1.5 Coincidences

The birthday and almost-birthday problems handsomely illustrate the fact that concurrent circumstances are often less coincidental than we tend to think. It pays to be aware of a world full of apparently coincidental events that, on closer examination, are less improbable than intuition alone might lead one to suppose.

The following example represents another case of coincidence turning out to be something less than coincidental. You answer the telephone and find yourself in conversation with a certain friend whose name had come up earlier that day in a conversation with others. How coincidental is this? A few rough calculations on a piece of scrap paper will show that, over a period of time, this is less coincidental than you might think. Making a rough calculation on scrap paper means simplifying without detracting from the essence of the problem. Let's begin by roughly estimating that over the years, you have discussed your friend with others one hundred or so times and that the probability of this friend

Table 3.2 *Probabilities for the almost-birthday problem* ($r = 1$)

n	10	14	20	25	30	35	40
$p_n(1)$	0.3147	0.5375	0.8045	0.9263	0.9782	0.9950	0.9991

telephoning you on any given day is equal to $p = \frac{1}{100}$. Instead of calculating the probability of your friend calling on a day when you have previously mentioned his name, let's calculate the complementary probability of your friend not telephoning you on any of the $n = 100$ days when he has been the subject of a conversation. This complementary probability is equal to $(1 - p)^n$. The probability then, of your being telephoned at least one time by your friend on a day when you had previously mentioned him, is given by $1 - (1 - p)^n$. For every value of $p > 0$, this probability comes arbitrarily close to 1 if n is large enough. In particular, the probability $1 - (1 - p)^n$ has a value of 0.634 for $n = 100$ and $p = \frac{1}{100}$. Over a period of time then, it is not particularly exceptional to have been telephoned by someone whom you had spoken of earlier that same day. This argumentation is applicable to many comparable situations, for example, a newspaper story reporting the collision of two Mercedes at a particular intersection, and that both drivers were called John Smith. This seems like an exceptional occurrence, but if you think about it, there must be quite a few men called John Smith who drive Mercedes and pass one another every day at intersections. Of course, the newspaper never mentions these noncollisions. We only receive the filtered information about the collision, and it therefore appears to be exceptional.

On August 18, 1913, a memorable event occurred in a Monte Carlo casino: the roulette wheel stopped no less than twenty-six times in a row on black. How exceptional can we consider a streak of this kind to be? In 1913, the Monte Carlo casino had been in operation for approximately 100 years. We can roughly estimate that over all of those 100 years, the roulette table had completed between three and five million runs. The probability of the wheel stopping on either red or black twenty-six times in a row in n rounds has the value of 0.022 for $n = 3{,}000{,}000$ and the value 0.037 for $n = 5{,}000{,}000$ (see Problem 5.34 in Chapter 5). Thus, it can be said to be exceptional that, in the first 100 years of the existence of the world's first casino, the roulette wheel stopped twenty-six times in a row on one and the same color. Today, well-trafficked casinos are to be found far and wide, and each is likely to have quite a number of roulette tables. On these grounds, one could hardly call it risky to predict that somewhere in the world during the coming twenty-five years, a roulette ball will stop on either red or black twenty-six or more times in a row. Any event with a nonzero probability will eventually occur when it is given enough opportunity to occur. This principle can be seen most clearly in the Lotto. Each participant has a probability almost equal to zero of winning the jackpot. Nevertheless, there is a large probability of the jackpot being won when the number of participants is sufficiently large.

3.2 The coupon collector's problem

In order to introduce a new kind of chips, the producer has introduced a campaign offering a "flippo" in each bag of chips purchased. There are ten different flippos. How many bags of chips do you expect to buy in order to get all ten flippos? In probability theory, this problem is known as the coupon collector's problem. The problem comes in many variations.

3.2.1 Simulation approach

In the Monte Carlo simulation, each simulation experiment consists of generating random drawings from the numbers $1, \ldots, 10$ until each of the ten numbers has been drawn at least one time. The number of drawings necessary is seen as the result of this experiment. After a sufficiently large number of experiments, the expected value we are looking for can be estimated by

$$\frac{\text{the sum of the outcomes of the experiments}}{\text{the total number of experiments}}.$$

The Monte Carlo study has to be redone when the number of flippos involved changes. This is not the case for the theoretical approach. This approach gives a better qualitative insight than the simulation approach.

3.2.2 Theoretical approach

Let's assume that there are n different flippos. Define the random variable X as the number of bags of chips that must be purchased in order to get a complete set of flippos. The random variable X can, in principle, take on any of the values $1, 2, \ldots$ and has thus a discrete distribution with infinitely many possible values. The expected value of X is defined by

$$E(X) = 1 \times P(X = 1) + 2 \times P(X = 2) + 3 \times P(X = 3) + \cdots .$$

A straightforward calculation of $E(X)$ is far from simple. Nevertheless, $E(X)$ is fairly easy to find indirectly by defining the random variable Y_i as

$$Y_i = \text{the number of bags of chips needed in order to go from } i-1 \text{ to } i \text{ different flippos.}$$

Now, we can write X as

$$X = Y_1 + Y_2 + \cdots + Y_n.$$

The trick of representing a random variable by a sum of simpler random variables is a very useful one in probability theory. The expected value of the original random variable follows by taking the sum of the expected values of the simpler random variables. In Chapter 9, it will be shown that the expected value of a finite sum of random variables is always equal to the sum of the expected values. In order to calculate $E(Y_i)$, the so-called geometric probability model is used. Consider an experiment having two possible outcomes. Call these outcomes "success" and "failure" and notate the probability of a "success" as p. In the geometric model, independent trials of an experiment are done until the first "success" occurs. Since the outcomes of the trials are independent of each other, it is reasonable to assign the probability $(1-p)^{k-1}p$ to the event of the first $k-1$ trials of the experiment delivering no success, and the kth delivering a success. It is obvious that the geometric probability model is applicable in the case of the Y_i variables. Let p_i represent the probability that the next bag of chips purchased will contain a new flippo when as many as $i-1$ differing flippos have already been collected. The probability p_i is equal to $\frac{n-(i-1)}{n}$ and the distribution of Y_i is given by

$$P(Y_i = k) = (1-p_i)^{k-1}\, p_i \quad \text{for} \quad k = 1, 2, \ldots .$$

For each $i = 1, \ldots, n$, the expected value of Y_i is given by

$$\begin{aligned} E(Y_i) &= p_i + 2(1-p_i)\,p_i + 3(1-p_i)^2\,p_i + \cdots \\ &= p_i[1 + 2(1-p_i) + 3(1-p_i)^2 + \cdots] \\ &= \frac{p_i}{[1-(1-p_i)]^2} = \frac{n}{n-i+1}, \end{aligned}$$

using the fact that the infinite series $1 + 2a + 3a^2 + \cdots$ has the sum $\frac{1}{(1-a)^2}$ for each a with $0 < a < 1$ (see the Appendix). The sought-after value of $E(X)$ now follows from

$$E(X) = E(Y_1) + E(Y_2) + \cdots + E(Y_n).$$

Filling in the expression for $E(Y_i)$ leads to

$$E(X) = n\left[\frac{1}{n} + \frac{1}{n-1} + \cdots + 1\right].$$

For $n = 10$, we find then that the expected number of bags of chips needed in order to get a complete set of flippos is equal to 29.3.

The formula given above for $E(X)$ can be rewritten in a form that gives more insight into the way that $E(X)$ increases as a function of n. A well-known mathematical approximation formula is

$$1 + \frac{1}{2} + \cdots + \frac{1}{n} \approx \ln(n) + \gamma + \frac{1}{2n},$$

where $\gamma = 0.57722\ldots$ is the Euler constant. This leads to the insightful approximation

$$E(X) \approx n\ln(n) + \gamma n + \tfrac{1}{2}.$$

The coupon's collector problem appears in many forms. For example, how many rolls of a fair die are needed on average before each of the point surfaces has turned up at least one time? This problem is identical to the flippo problem with $n = 6$ flippos. Taking $n = 365$ flippos, the flippo problem also gives us the expected value of the number of people needed before we can assemble a group of people in which all of the possible 365 birthdays are represented.

3.3 Craps

The wildly popular game of craps, first played in the United States in the twentieth century, is based on the old English game of Hazard. Craps is an extremely simple game in its most basic form; however, casinos have added on twists and turns enough to make most players' heads spin. The basic rules are as follows. A player rolls a pair of dice and the sum of the points is tallied. The player has won if the sum of the points is equal to seven or eleven, and has lost if the sum is equal to two, three, or twelve. In the case of all other point combinations, the player continues to roll until the sum of the first roll is repeated, in which case the player wins, or until rolling a total of seven, in which case the player loses. What is the probability of the player winning in craps?

3.3.1 Simulation approach

In a simulated craps experiment the rolls of a pair of dice are perpetuated until the game is ended. We simulate a roll of the dice by drawing a random number twice out of the numbers $1, \ldots, 6$, and adding up the sum of the two numbers. A key variable in the simulation is the total obtained in the first roll. Let's call this number the chance point. The experiment ends immediately if the chance point turns out to be seven or eleven (a win), or if it turns out to be a two, three, or twelve (a loss). If none of these totals occurs, the simulation continues to "roll" until the chance point turns up again (a win), or until a total of seven appears (a loss). The probability of the player winning is estimated by dividing the number of simulated experiments leading to wins by the total number of experiments.

3.3.2 Theoretical approach

A simulation approach first requires looking at the number of points received in the first roll of the game. Depending on that, the next step of the simulation program is determined. In the theoretical model, we work along the same lines. In this case, we make use of the concept of *conditional probability*. Conditional probabilities have a bearing on a situation in which partial information over the outcome of the experiment is available. Probabilities alter when the available information alters. The notation $P(A \mid B)$ refers to the conditional probability that event A will occur *given* that event B has occurred.[2] In most concrete situations, the meaning of conditional probability and how it is calculated are obvious.

The law of conditional probabilities is an extremely useful result of applied probability theory. Let A be an event that can only occur after one of the events $B_1, \ldots, B_n$ has occurred. It is essential that the events $B_1, \ldots, B_n$ are disjoint, that is, only one of the $B_1, \ldots, B_n$ events can occur at a time. Under these conditions, the *law of conditional probabilities* says that:

$$P(A) = P(A \mid B_1)P(B_1) + P(A \mid B_2)P(B_2) + \cdots + P(A \mid B_n)P(B_n)$$

or, in abbreviated form,

$$P(A) = \sum_{i=1}^{n} P(A \mid B_i)P(B_i).$$

We find the (unconditional) probability $P(A)$, then, by averaging the conditional probabilities $P(A \mid B_i)$ over the probabilities $P(B_i)$ for $i = 1, \ldots, n$. It is insightful to represent schematically the law of conditional probabilities by the tree diagram shown in Figure 3.1. A mathematical proof of this law will be given in Chapter 8.

Usually conditional probabilities are easy to calculate when the disjoint events $B_1, \ldots, B_n$ are suitably chosen. In determining the choice of these events, it may helpful to think of what you would do when writing a simulation program. In the craps example, we choose B_i as the event in which the first roll of the dice delivers i points for $i = 2, \ldots, 12$. Denote by $P(win)$ the probability of

[2]The formal definition is that $P(A \mid B) = P(AB)/P(B)$ assuming that $P(B) \neq 0$, where $P(AB)$ represents the probability that both event A and event B will occur (see Chapter 8).

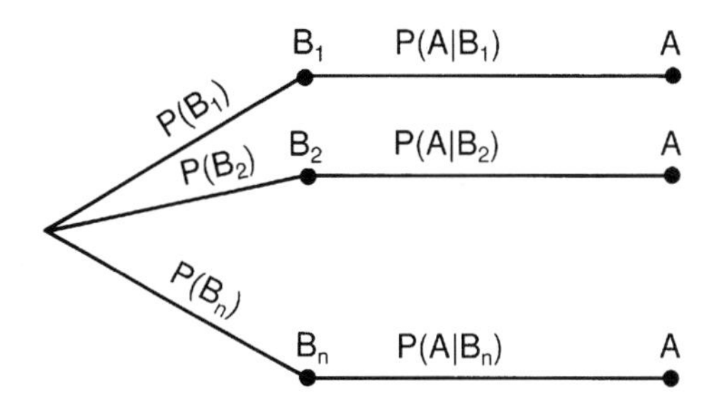

Figure 3.1 Tree diagram for the law of conditional probabilities.

the player winning in craps and let $P(win \mid B_i)$ denote the revised value of this probability given the information of the occurrence of the event B_i. Then

$$P(win) = \sum_{i=2}^{12} P(win \mid B_i)P(B_i).$$

The conditional win probabilities are easy to calculate. Naturally,

$$P(win \mid B_i) = \begin{cases} 1 & \text{for} \quad i = 7, 11, \\ 0 & \text{for} \quad i = 2, 3, 12. \end{cases}$$

Prior to calculating $P(win \mid B_i)$ for the other values of i, we first determine the probabilities $P(B_i)$. The sample space for the experiment of rolling a pair of dice consists of the 36 outcomes (j, k), where $j, k = 1, 2, \ldots, 6$. The outcome (j, k) occurs if j points turn up on the first (red) die and k points turn up on the second (blue) die. The dice are fair, so the same probability $\frac{1}{36}$ is assigned to each of the 36 possible outcomes. The outcome (j, k) results in the value $i = j + k$ for the total of the points. Using the shorthand p_i for the probability $P(B_i)$, it is readily verified that

$$p_2 = \frac{1}{36}, \quad p_3 = \frac{2}{36}, \quad p_4 = \frac{3}{36}, \quad p_5 = \frac{4}{36}, \quad p_6 = \frac{5}{36}, \quad p_7 = \frac{6}{36},$$

$$p_8 = \frac{5}{36}, \quad p_9 = \frac{4}{36}, \quad p_{10} = \frac{3}{36}, \quad p_{11} = \frac{2}{36}, \quad p_{12} = \frac{1}{36}.$$

Then, we calculate the conditional probabilities $P(win \mid B_i)$. In order to do this, we first give the meaning of these probabilities in the concrete situation of the craps game. For example, the conditional probability $P(win \mid B_4)$ is no other than the unconditional probability that the total of 4 will appear before the total of 7 does in the (compound) experiment of repetitive dice rolling. The total of 4 will appear before the total of 7 only if one of the disjoint events $A_1, A_2, \ldots$ occurs, where A_k is the event that the first consecutive $k - 1$ rolls give neither the total of 4 nor the total of 7 and the kth consecutive roll gives a total of 4. Since the events $A_1, A_2, \ldots$ are mutually disjoint, $P(A_1 \cup A_2 \cup \cdots)$ is obtained by adding the probabilities $P(A_k)$ for $k = 1, 2, \ldots$. This gives

$$P(4 \text{ before } 7) = P(A_1 \cup A_2 \cup \cdots) = P(A_1) + P(A_2) + \cdots.$$

The event A_k is generated by physically independent subexperiments and thus the probabilities of the individual outcomes in the subexperiments are multiplied by each other in order to obtain

$$P(A_k) = (1 - p_4 - p_7)^{k-1} p_4 \quad \text{for} \quad k = 1, 2, \ldots.$$

This leads to the formula

$$\begin{aligned} P(4 \text{ before } 7) &= p_4 + (1 - p_4 - p_7)p_4 + (1 - p_4 - p_7)^2 p_4 + \cdots \\ &= \frac{p_4}{p_4 + p_7}, \end{aligned}$$

using the fact that the geometric series $1 + a + a^2 + \cdots$ has a sum of $\frac{1}{1-a}$ for a with $0 < a < 1$ (see the Appendix). In this way, we find that

$$P(win \mid B_i) = \frac{p_i}{p_i + p_7} \quad \text{for} \quad i = 4, 5, 6, 8, 9, 10.$$

If we fill in the p_i values, we get

$$\begin{aligned} &P(win \mid B_4) = \frac{3}{9}, \quad P(win \mid B_5) = \frac{4}{10}, \quad P(win \mid B_6) = \frac{5}{11}, \\ &P(win \mid B_8) = \frac{5}{11}, \quad P(win \mid B_9) = \frac{4}{10}, \quad P(win \mid B_{10}) = \frac{3}{9}. \end{aligned}$$

Putting it all together, we get

$$\begin{aligned} P(win) = \; & 0 \times \frac{1}{36} + 0 \times \frac{2}{36} + \frac{3}{9} \times \frac{3}{36} + \frac{4}{10} \times \frac{4}{36} + \frac{5}{11} \times \frac{5}{36} \\ & + 1 \times \frac{6}{36} + \frac{5}{11} \times \frac{5}{36} + \frac{4}{10} \times \frac{4}{36} + \frac{3}{9} \times \frac{3}{36} + 1 \times \frac{2}{36} \\ & + 0 \times \frac{1}{36} = 0.4929. \end{aligned}$$

In other words, the probability of the player losing is 0.5071. The casino payout is 1:1, so that you would lose on average $(0.5071 - 0.4929) \times 100 = 1.42$ cents per dollar staked. The fact that the house percentage is lower with the game of craps than with other casino games partly explains the popularity of the game. In the most basic version of craps, the players are passive during follow-up rolls of the dice, when the first roll has not been decisive. Players like action, and casinos like to keep players active. For this reason, casinos have added quite a few options onto the basic formula such that during the passive rounds, players can make seemingly attractive bets (which actually only raise the house advantage).

3.4 Gambling systems for roulette

The origins of probability theory lie in the gambling world. The best-known casino game is roulette. The oldest form of roulette is European roulette, which was developed in France around the year 1800. Players bet on the outcome of a turning wheel, which is outfitted with 37 spokes numbering from 0 to 36. Of the spokes numbered from 1 to 36, eighteen are red and eighteen are black. The 0, neither red nor black, represents a win for the casino. Players can bet on individual numbers, on combinations of numbers or on colors. The casino payout depends on the type of bet made. For example, a bet on the color red has a 1 to 1 payout. This means that if you stake one dollar on red and the wheel falls on a red number, you win back your dollar plus one more dollar. If the wheel does not stop on a red number, you lose your bet and forfeit the dollar you staked. For a bet on red, the probability of the player winning is $\frac{18}{37}$ and so the expected value of the casino payout is $2 \times \frac{18}{37} = 0.973$ dollars. In the long run, the casino keeps \$0.0270 of a one dollar bet on red, or rather a house percentage of 2.70% for a bet on red. This house percentage of 2.70% remains constant for every type of bet in European roulette, as shown at the end of Section 2.6. The term *house percentage* (or *house advantage*) is much used by casinos and lotteries. The house percentage is defined as 100% times the casino's long-run average win per dollar staked.

3.4.1 Doubling strategy

A seemingly attractive strategy is known as the doubling strategy for a bet on red. This system works as follows. The player begins by staking one dollar on red. If he loses, he doubles his stake, and continues doubling until red wins. Theoretically, this system guarantees the player of an eventual one dollar win. But, in practice, a player cannot continue to double unlimitedly. At a certain point, he will either cross over the high stake limit or simply run out of money. Whatever the high stake limit is, over the long run a player loses 2.70 cents on every dollar staked. We will illustrate this by manner of a stake limit of \$1,000. Players reach this limit after losing ten times in a row. And by the tenth bet, players are staking an amount of $2^9 = 512$ dollars. A doubling round then, consists of eleven bets at the most, of which the maximum stake of \$1,000 is made in the eleventh bet. We will assume that the starting capital is sufficiently large to play out the entire round.

3.4.2 Simulation approach

A simulation model for this problem is simply constructed. In each simulation experiment, a doubling round is replicated. This consists of taking random drawings from the numbers $0, 1, \ldots, 37$. The experiment ends as soon as a number corresponding to red is drawn, or when the eleventh drawing has been completed. The outcome of the experiment is 1 if you end a winner; otherwise, it is the negative of the total amount staked in the experiment. After a sufficiently large number of experiments, you can estimate the percentage of your win or loss per dollar staked by

$$\frac{\text{the sum of the outcomes in the experiments}}{\text{the sum of the total amounts staked in the experiments}} \times 100\%.$$

In running the simulation study, you will end up with a loss percentage estimated somewhere in the neighborhood of 2.70%.

3.4.3 Theoretical approach

Using a theoretical approach to calculate the average loss per doubling round, we must first determine the distribution of the random variable X that represents the number of bets in a single doubling round. The probability of the wheel stopping on red in the first bet is $\frac{18}{37}$, so we can say that $P(X = 1) = \frac{18}{37}$. The random variable X takes on the value k with $2 \leq k \leq 10$ if red does not result in the first $k - 1$ bets and then does result in the kth bet. The random variable X takes on the value 11 when red has not resulted in the first 10 bets. This leads to

$$P(X = k) = \begin{cases} \left(\frac{19}{37}\right)^{k-1} \frac{18}{37} & \text{for} \quad k = 1, \ldots, 10, \\ \\ \left(\frac{19}{37}\right)^{10} & \text{for} \quad k = 11. \end{cases}$$

Denote by a_k, the total amount staked when the doubling round ends after k bets. Then,

$$a_k = \begin{cases} 1 & \text{for} \quad k = 1, \\ 1 + 2 + \ldots + 2^{k-1} & \text{for} \quad k = 2, \ldots, 10, \\ 1 + 2 + \ldots + 2^9 + 1{,}000 & \text{for} \quad k = 11. \end{cases}$$

If we fill in the values for $P(X = k)$ and a_k, we find that

$$E(\text{amount staked in a doubling round}) = \sum_{k=1}^{11} a_k P(X = k) = 12.583 \text{ dollars.}$$

In a doubling round, a player's win is equal to one dollar if the round lasts for fewer than eleven bets. The player's loss is \$23 if the round goes to eleven bets and the eleventh bet is won; the loss is \$2,023 if the round goes to eleven bets and the eleventh bet is lost. A doubling round goes to eleven bets with a probability of $(\frac{19}{37})^{10}$, this being the probability of losing ten bets in a row. This gives

$$E(\text{win in a doubling round}) = 1 \times \left[1 - \left(\frac{19}{37}\right)^{10}\right] - 23 \times \left(\frac{19}{37}\right)^{10} \times \frac{18}{37} - 2{,}023 \times \left(\frac{19}{37}\right)^{10} \times \frac{19}{37} = -0.3401 \text{ dollars.}$$

The stake amounts and losses (wins) will vary from round to round. The law of large numbers guarantees nevertheless that over the long run, the fraction of your loss per dollar staked will come arbitrarily close to

$$\frac{E(\text{loss in a doubling round})}{E(\text{amount staked in a doubling round})} = \frac{0.3401}{12.583} = 0.027.$$

This is the same house advantage of 2.7% that we saw earlier! You simply cannot beat the casino over the long run using the doubling strategy. The doubling strategy does rearrange your losses, but over the long run you would get the self-same result if you simply gave away 2.7 cents of every dollar you planned to stake.

3.4.4 The Labouchère system

The Labouchère system is often used in the game of roulette. According to this system, you must decide beforehand how much you want to win, and you make a list of positive numbers whose sum add up to this amount. You bet on red each time. For each bet, you stake an amount equal to the sum of the first and last numbers on your list (if the list consists of just one number, then that number is the amount of the stake). If you win the amount staked, you cross off the amounts you used from your list. If you lose, you add the amount lost to the bottom of your list. You continue in this manner until your list is used up (your target amount has been achieved) or until you have lost all of your money. The Labouchère system is exciting, but it must also be understood that with repeated play, this system will deliver an unavoidable average loss of 2.7 cents per dollar staked. In general, this is not easily calculated, mathematically. For each specific situation, however, it is easy to make a simulation study. Let's say that your goal is to win $250 and that you have a starting capital of $2,500. Your list consists of the numbers 50, 25, 75, 50, 25, 25. The first time then, you will stake $50 + $25 = $75. If you win, your list will be narrowed down to contain the numbers 25, 75, 50, 25, whereas if you lose, the list will be extended to contain the numbers 50, 25, 75, 50, 25, 25, 75. It is worth stating that, at a given moment, the sum of the first and last numbers could be larger than the amount of money you have at that moment. For example, let's say the first and last numbers are 50 and 125, and you only have $150. Naturally, then, you would stake the $150. If you lose, your store of money is depleted and the game is over. If you win, you cross off the last amount on your list and bring the first amount back to 25. In every simulation experiment, you begin with a capital of $2,500 and play the Labouchère system until you have either added $250 to your starting capital or until you have lost all of your money. The outcome of one round is 250 for a win and $-2,500$ for a loss. The total amount staked in a round consists of the individual stakes from the starting capital of $2,500 together with the amounts that are won and subsequently staked anew. If you run a sufficiently large number of experiments and divide the sum of the outcomes by the sum of the amounts staked in the experiments, you will arrive at an estimate for the average loss per dollar staked when the Labouchère system is played repeatedly. In simulation runs of 10,000 and 100,000 experiments, we found the estimates to be 0.0294 and 0.0276, respectively, for the average loss per dollar staked. Indeed, the estimates are very close to the theoretical value of 0.027. For probability P_w, defined as your probability of winning a given round, we found in these two simulation runs the simulated values 0.890 and 0.886. These values are in agreement with the relation

$$1 - \frac{P_w \times 2{,}750}{2{,}500} = 0.027.$$

This relation can be seen by noting that the expected value of your capital at the end of one round is $P_w \times 2{,}750$, and thus $(2{,}500 - P_w \times 2{,}750)/2{,}500$ must give the house percentage of 0.027.

The conclusion is that the Labouchère system will not be of help to you in assuring a win. A nonmathematical but nonetheless convincing proof of the fact that a winning betting system does not exist for the game of roulette is evident from the fact that casinos have never shown any resistance to the use of any such system at the roulette table. In fact, the only sure way to get rich through roulette is to open a casino!

3.5 The 1970 draft lottery

In 1970, during the Vietnam War, the American army used a lottery system based on birth dates to determine who would be called up for service in the military forces. The lottery worked like this: each of the 366 days of the year (including February 29) was printed on a slip of paper. These slips of paper were placed into individual capsules. The capsules were then placed into a large receptacle, which was rotated in order to mix them. Then, the capsules were drawn one by one out of the receptacle. The first date drawn was assigned a draft number of "one," the second date drawn was assigned a draft number of "two," and so on, until each day of the year had been drawn out of the receptacle and assigned a draft number. Draftees were called up for service based on the draft number assigned to their dates of birth, with those receiving low draft numbers being called up first. Table 3.3 gives the numbers assigned to the days of the various months. Directly after the lottery drawing, doubts were raised as to its fairness. In Chapter 2, we discussed the errors made in the randomization procedure used in this lottery. But, for the sake of argument, let's say we are unaware of these errors. Now, based on the results shown in Table 3.3, we must decide whether the lottery can reasonably be said to have been random. How can we do this? We can use a Monte Carlo simulation to test whether the order of the lottery numbers in Table 3.3 can be described as random. First, we aggregate the data in a suitable and insightful way. Table 3.4 provides, for each month, the average value of the numbers representing the days of that month that were chosen. The monthly averages should fluctuate around 183.5 (why?). One glance at Table 3.4 will be enough to raise serious doubts about the fairness of the draft lottery. After May, the monthly averages show an obvious decline. What we now must determine is whether the deviations in Table 3.4 can more reasonably be described as an example of how fate can be fickle or as hard evidence of an unfair lottery. In order to make this determination, let's start out with the hypothesis that the lottery was fair. If we can show that the outcomes in Table 3.4 are extremely improbable under the hypothesis, we can reject our hypothesis and conclude that the lottery was most probably unfair. Many test criteria are possible. One generally applicable test criterion is to consider the sum of the absolute deviations of the outcomes from their expected values. The expected value of the average draft number for a given month is 183.5 for each month. For convenience of notation, denote by $g_1 = 201.2, \ldots, g_{12} = 121.5$

Table 3.3 *Draft numbers assigned by lottery*

Day	Jan.	Feb.	Mar.	Apr.	May	June	July	Aug.	Sep.	Oct.	Nov.	Dec.
1	305	086	108	032	330	249	093	111	225	359	019	129
2	159	144	029	271	298	228	350	045	161	125	034	328
3	251	297	267	083	040	301	115	261	049	244	348	157
4	215	210	275	081	276	020	279	145	232	202	266	165
5	101	214	293	269	364	028	188	054	082	024	310	056
6	224	347	139	253	155	110	327	114	006	087	076	010
7	306	091	122	147	035	085	050	168	008	234	051	012
8	199	181	213	312	321	366	013	048	184	283	097	105
9	194	338	317	219	197	335	277	106	263	342	080	043
10	325	216	323	218	065	206	284	021	071	220	282	041
11	329	150	136	014	037	134	248	324	158	237	046	039
12	221	068	300	346	133	272	015	142	242	072	066	314
13	318	152	259	124	295	069	042	307	175	138	126	163
14	238	004	354	231	178	356	331	198	001	294	127	026
15	017	089	169	273	130	180	322	102	113	171	131	320
16	121	212	166	148	055	274	120	044	207	254	107	096
17	235	189	033	260	112	073	098	154	255	288	143	304
18	140	292	332	090	278	341	190	141	246	005	146	128
19	058	025	200	336	075	104	227	311	177	241	203	240
20	280	302	239	345	183	360	187	344	063	192	185	135
21	186	363	334	062	250	060	027	291	204	243	156	070
22	337	290	265	316	326	247	153	339	160	117	009	053
23	118	057	256	252	319	109	172	116	119	201	182	162
24	059	236	258	002	031	358	023	036	195	196	230	095
25	052	179	343	351	361	137	067	286	149	176	132	084
26	092	365	170	340	357	022	303	245	018	007	309	173
27	355	205	268	074	296	064	289	352	233	264	047	078
28	077	299	223	262	308	222	088	167	257	094	281	123
29	349	285	362	191	226	353	270	061	151	229	099	016
30	164		217	208	103	209	287	333	315	038	174	003
31	211		030		313		193	011		079		100

the observed values for the average draft numbers for the months $1, \ldots, 12$ (see Table 3.4). The sum of the absolute deviations of the outcomes $g_1, \ldots, g_{12}$ from their expected values is

$$\sum_{i=1}^{12} |g_i - 183.5| = 272.4.$$

Is this large? We can answer this by means of a simple model. Determine a random permutation $(n_1, \ldots, n_{366})$ of the days $1, \ldots, 366$. Assign lottery

Table 3.4 *Average draft number per month*

January	201.2
February	203.0
March	225.8
April	203.7
May	208.0
June	195.7
July	181.5
August	173.5
September	157.3
October	182.5
November	148.7
December	121.5

number n_1 to January 1, number n_2 to January 2, etc., ending with lottery number n_{366} for December 31. For this assignment, define the random variable G_i as the average value of the lottery numbers assigned to the days of month i for $i = 1, \ldots, 12$. In order to answer the above question, we need

$$P\left(\sum_{i=1}^{12} |G_i - 183.5| \geq 272.4\right).$$

Deriving a versatile mathematical formula for this probability seems like an endless task. The value for this probability, however, is easily determined with the help of a Monte Carlo simulation. You conduct a large number of independent simulation trials, and in each trial a random permutation of the whole numbers $1, \ldots, 366$ is determined in order to assign lottery numbers to the days of the various months. A procedure for the determination of a random permutation is given in Section 2.9. A simulation trial is considered a "success" when the resulting monthly averages G_i measure up to $\sum_{i=1}^{12} |G_i - 183.5| \geq 272.4$. If you divide the number of successes by the total number of trials, you will come out with an estimate for the probability you are seeking. In a Monte Carlo study with 100,000 simulation runs, we came out with a simulated value of 0.012 for the probability in question.

Still another, yet stronger, indication that the lottery was not fair can be found in a test criterion that bears in mind the established trend of the monthly averages in Table 3.4. You would assign the index number 1 to the month with the highest monthly average, index number 2 to the month with the second highest monthly average, etc. For the 1970 draft lottery, these index numbers are shown in Table 3.5. They result in the permutation $(5,4, \ldots, 12)$ of the numbers $1, 2, \ldots, 12$.

Table 3.5 *Index numbers for the 1970 draft lottery*

Month	1	2	3	4	5	6	7	8	9	10	11	12
Index	5	4	1	3	2	6	8	9	10	7	11	12

Under the hypothesis that the lottery is fair, this permutation would have to be a "random" permutation. How can we test this? First, for a random permutation $\sigma = (\sigma_1, \ldots, \sigma_{12})$ of the numbers $1, \ldots, 12$, we define the distance measure $d(\sigma)$ by

$$d(\sigma) = \sum_{i=1}^{12} |\sigma_i - i|.$$

You can immediately verify that for each permutation σ, it holds that $0 \leq d(\sigma) \leq 72$. For the permutation $\sigma^* = (5,4, \ldots, 12)$ from Table 3.5, it holds that $d(\sigma^*) = 18$. In order to judge whether the value 18 is "small" you must know, for a randomly chosen permutation σ, how likely the distance measure $d(\sigma)$ is less than or equal to 18. Again, you can apply a Monte Carlo simulation in order to find the value for this probability. You generate a large number of random permutations of the numbers $1, \ldots, 12$ and determine the proportion of permutations in which the distance measure $d(\sigma)$ is less than or equal to 18. A Monte Carlo study with 100,000 generated random permutations led us to an estimate of 0.0009 for our sought-after probability. This is strong evidence that the 1970 draft lottery did not proceed fairly.

3.6 Bootstrap method

In the statistical analysis of the 1970 draft lottery from Section 3.5, we used a powerful, generally applicable form of statistical methodology, namely the *bootstrap method.* This new method, developed in 1977 by American statistician Bradley Efron, has modern computer technology to thank for its efficacious calculating power. Conventional statistical methods were, for the most part, developed before we had computers at our disposal. The standard methods, therefore, necessarily relied on simplifying assumptions and relatively simple statistical measures that could be calculated from mathematical formulas. In contrast to these methods, the bootstrap method is letting the data speak for itself by making use of the number-breaking power of modern-day computers, through the use of which calculation-intensive simulations can be made in virtually no time. A typical application can be described by the following situation: in order to test a new skin infection remedy, twenty healthy volunteers are infected with the corresponding ailment. They are then split up into two groups of equal size: a remedy group and a placebo group. The study being a double-blind study, the volunteers are not aware of which group they are assigned to, and the doctors do not have this information either. Each volunteer undergoes daily examinations until the malady is cured. In the remedy group, the values for the number of days required until all patients are cured are 7, 9, 9, 11, 12, 14, 15, 15, 15, and 17. In the placebo group, the values for the number of days until all patients are cured are 9, 11, 11, 11, 12, 15, 17, 18, 18, and 20. In order to test whether the remedy helps or not, we take the difference in the total number of days until cured between the placebo group and the remedy group as test statistic T. For the sample data, the one-sided test statistic T takes on the value $142 - 124 = 18$. In order to make a statistical statement of whether the remedy works or not, we assume that it does not matter whether or not the remedy is used. Under this so-called null-hypothesis, our twenty case studies can be seen as twenty independent drawings from the distribution of the time elapsed until cure is effected. Data from the experiment can be used for the empirical distribution of the time until cure. One of the twenty case studies reports a time value of 7, three of the twenty cases report a value of 9, and so on. Thus, for the time required until cure is effected, the respective probabilities 1/20, 3/20, 4/20, 2/20, 1/20, 4/20, 2/20, 2/20, and 1/20 are assigned to the possible values 7, 9, 11, 12, 14, 15, 17, 18, and 20. According to the bootstrap method, you would now instruct your computer to make a large number of drawings, say 10,000, from this distribution (the array method from Section 2.9 can be used for this purpose). For each of the 10,000 simulation runs, you determine the difference between the sum of the last ten values drawn and the sum of the first ten values drawn. The proportion of the number of simulation runs in which this difference is greater than or equal to 18 gives the bootstrap estimate for the probability that the test statistic T will take on a value greater than or equal to 18 under the null-hypothesis. If this probability is smaller than a previously chosen threshold value, say 0.01, then the null-hypothesis is discarded. Using the original data and performing 10,000 simulation runs, we found a value of 0.135 for the probability $P(T \geq 18)$. This probability is not small enough to reject the null-hypothesis. The conclusion seems to be that the experiment must be redone using larger groups of people before any definitive conclusion can be reached about the remedy's effectiveness.

Another example of the bootstrap method is the prediction of election results based on probability statements made by polled voters. Consider the polling method in which respondents are asked to indicate not which candidate is their favorite, but rather what the various probabilities might be of their voting for each of the candidates in the running. Let's assume that a representative group of 1,000 voters is polled in this way. We then have 1,000 probability distributions over the various political candidates. Next, the computer allows us to draw from these 1,000 probability distributions a large number of times. In this way, we can simulate the probability that a given candidate will receive the most votes or the probability that, in the parliamentary system, a given two parties will receive more than half of the number of votes cast. In Section 12.3 of Chapter 12, we come back to this application.

3.6.1 A statistical test problem

The bootstrap method can also be used to solve Question 9 from Chapter 1. The question was whether someone can credibly claim to have rolled a one 196 times, a two 202 times, a three 199 times, a four 198 times, a five 202 times, and a six 203 times in 1,200 rolls of one fair die. In order to test whether something is credible or not, you must choose a suitable test statistic. A claim of having tossed 100 heads in a row in 100 tosses of a fair coin cannot be said to be incredible based simply on the grounds that the sequence $HH \ldots H$, consisting of 100 heads, has an inconceivably small probability of $(\frac{1}{2})^{100}$ of occurring. Actually, each *specific* sequence of heads and tails of length 100 has a probability of $(\frac{1}{2})^{100}$. No, the claim is incredible on the basis of the test statistic, which counts the total number of heads in 100 tosses of a fair coin. A value of 100 for this test statistic is improbably far away from the expected value of 50. Similarly, in the situation described in Question 9 of Chapter 2, you may observe that each of the outcomes is suspiciously close to its expected value of 200. The sum of the absolute deviations of the outcomes from their expected values of 200 is $4 + 2 + 1 + 2 + 2 + 3 = 14$. This sum is a natural touchstone for the question of whether the outcomes are invented or not. For the case that a single die actually is rolled 1,200 times, define the random variable X as

$$X = \sum_{i=1}^{6} |N_i - 200|,$$

where N_i represents the number of times that i points are rolled. The distribution of this test statistic cannot be calculated by mathematical equations (a related test statistic whose distribution can be approximated by a mathematically tractable distribution will be discussed in Section 12.4). However, the distribution of X can easily be found using Monte Carlo simulation. We can find an estimate for the probability $P(X \leq 14)$ by simulating 1,200 rolls of one die many times. If we divide the number of times that the simulated value of the random variable

X is less than or equal to 14 by the total number of simulation runs, we arrive at an estimate for $P(X \leq 14)$. A simulation study with 100,000 runs leads us to an estimate of 0.0020 for our sought-after probability. This small probability means that the reported outcomes of 1,200 rolls of the die are difficult to explain as a chance variation. In other words, this is a strong indication that the outcomes claimed above are invented. Statistics can never definitely prove that data are fabricated. In statistics, there are no absolute certainties like "water boils at a temperature of 100 degrees Celsius," but statistics does provide answers such as "there is clear evidence against the null-hypothesis."

3.7 Problems

3.1 Is it credible if a local newspaper somewhere in the world reports on a given day that a member of the local bridge club was dealt a hand containing a full suit of 13 clubs?

3.2 Is the probability of a randomly chosen person having his/her birthday fall on a Monday equal to the probability of two randomly chosen people having their birthdays fall on the same day of the week?

3.3 In both the Massachusetts Numbers Game and the New Hampshire Lottery, a four-digit number is drawn each evening from the sequence 0000, 0001, ..., 9999. On Tuesday evening, September 9, 1981, the number 8092 was drawn in both lottery games. Lottery officials declared that the probability of both lotteries drawing the same number on that particular Tuesday evening was inconceivably small and was equal to one in one hundred million. Do you agree with this?

3.4 The national lottery is promoting a special, introductory offer for the upcoming summer season. Advertisements claim that during the four scheduled summer drawings, it will hardly be possible not to win a prize, because four of every ten tickets will win at each drawing. What do you think of this claim?

3.5 What is the probability of a randomly chosen five-digit number lining up in the same order from right to left as it does from left to right?

3.6 The Yankees and the Mets are playing a best-four-of-seven series. The winner takes all of the prize money of $1,000,000. Unexpectedly, the competition must be suspended when the Yankees lead two games to one. How should the prize money be divided between the two teams if the remaining games cannot be played? Assume that the Yankees and the Mets are evenly matched and that the outcomes of the games are independent of each other. (This problem is a variant of the famous "problem of points" that, in 1654, initiated the correspondence between the great French mathematicians Pascal and Fermat.)

3.7 Five friends go out to a pub together. They agree to let a roll of the dice determine who pays for each round. Each friend rolls one die, and the one getting the lowest number of points picks up the tab for that round. If the low number is rolled by more than one friend in any given round, then the tab will be divided among them. At a certain point in the evening, one of the friends decides to go home, however, rather than withdraw from the game he proposes to participate in absentia, and he is assigned a point value of $2\frac{1}{2}$. Afterward, he will be responsible for paying up on the rounds he lost, calculating in an amount for the rounds he won. Is this a fair deal?

3.8 Suppose that a large group of people are undergoing a blood test for a particular illness. The probability that a random person has the illness in question is equal to 0.001. In order to save on the work, it is decided to split the group into smaller groups each consisting of r people. The blood samples of the r people is then mixed and tested all at once. If the test results are favorable, then one test will have been sufficient for that whole group. Otherwise, r extra tests will be necessary in order to test each of the r people individually. What is the expected value of the number of tests that will have to be done for a group of r people? Verify that $r = 32$ is the optimal group size.

3.9 You bet your friend that, of the next fifteen automobiles to appear, at least two will have license plates beginning and ending with the same number. What is your probability of winning?

3.10 What is the probability that the same number will come up at least twice in the next ten spins of a roulette wheel?

3.11 A group of seven people in a hotel lobby are waiting for the elevator to take them up to their rooms. The hotel has twenty-five floors, each floor containing the same number of rooms. Suppose that the rooms of the seven waiting people are randomly distributed around the hotel.

(a) What is the probability of at least two people having rooms on the same floor?

(b) Suppose that you, yourself, are one of the seven people. What is the probability of at least one of the other six people having a room on the same floor as you?

3.12 The birthday problem and those cited in Problems 3.9–3.11 can be described as a special case of the following model. Randomly, you drop n balls in c compartments such that each ball is dropped independently of the others. It is assumed that $c > n$. What is the probability p_n that at least two balls will drop into the same compartment?

(a) Verify that the probability p_n is given by

$$p_n = 1 - \frac{c \times (c-1) \times \cdots \times (c-n+1)}{c^n}.$$

(b) Prove the approximation formula

$$p_n \approx 1 - e^{-\frac{1}{2}n(n-1)/c}$$

for c sufficiently large in comparison with n (use the fact that $e^{-x} \approx 1 - x$ for x close to 0).

(c) Verify that, with a fixed c, the value n must be chosen as

$$n \approx 1.18\sqrt{c}$$

in order to get a "50-50" chance of at least two balls dropping into the same compartment.

3.13 Suppose that someone has played bridge thirty times a week on average over a period of fifty years. Apply the result from Problem 3.12(b) to calculate the probability that this person has played exactly the same hand at least twice during the span of the fifty years.

3.14 In the Massachusetts Numbers Game, a four-digit number is drawn from the numbers $0000, 0001, \ldots, 9999$ every evening (except Sundays). Let's assume that the same lottery takes place in ten other states each evening.

(a) What is the probability that the same number will be drawn in two or more states next Tuesday evening?

(b) What is the probability that on some evening in the coming 300 drawings, the same number will be drawn in two or more states?

3.15 Of the unclaimed prize monies from the previous year, a lottery has purchased 500 automobiles to raffle off as bonus prizes among its 2.4 million subscribing members. Bonus winners are chosen by a computer programmed to choose 500 random numbers from among the 2.4 million registration numbers belonging to the subscribers. The computer is not programmed to avoid choosing the same number more than one time. What is the probability that someone will win two or more automobiles?

3.16 You conduct the following experiment ten times. In each round, you ask a random passer-by when his/her birthday is and you continue asking passers-by until you encounter two people who have the same birth date. What is the probability of at least one round being completed after interviewing twenty-three or fewer passers-by?

3.17 A company has 110 employees in service. Solve the following two problems.

(a) Using simulation, determine the probability of there being twelve or more separate occasions when two or more employees have the same birthday.

(b) Using simulation, determine the probability that, in each of the twelve months, two or more employees have the same birthday.

3.18 A commercial radio station is advertising a particular call-in game that will be played in conjunction with the introduction of a new product. The game is to be played every day for a period of thirty days. The game is only open to listeners between the ages of fifteen and thirty. Each caller will be the possible winner of one million dollars. The game runs as follows. At the beginning of each day, the radio station randomly selects one date (day/month/year) from within a fifteen-year span, that span consisting of the period from fifteen to thirty years ago. Listeners whose birthday fall on the current day will be invited to call in to the station. At the end of the day, one listener will be chosen at random from among all of the listeners that called in that day. If that person's birth date matches the predetermined date picked by the radio station exactly, he/she will win one million dollars. What is the probability of someone winning the prize money during the thirty-day run of the game?

3.19 Twenty-two soccer players have been selected to try out for a world-class team that will play a benefit match at Wembley. In order to pull the team together, twenty soccer coaches are asked to nominate two players from the pool of twenty-two. Assume that each of the twenty-two players has an equal probability of being chosen by each of the twenty coaches and assume that the coaches make their choices independently of one another.

(a) What is the probability that a given player will not be chosen by any of the coaches?

(b) What is the expected value of the number of players that will not be nominated?

(c) Using simulation, determine the probability of the most often-nominated player being nominated at least seven times.

3.20 The following variation can be played in the game of craps. After the first roll of the dice, the player wins immediately if 2 or 3 points are rolled, loses immediately if 7 or 11 points are rolled, and gets his/her stake refunded if 12 points are rolled. In every other case, the player rolls again until either a sum of seven, or the sum of the first roll appears. A roll of seven translates as a win, and the sum of the first roll translates as a loss. Calculate the house percentage for this craps variant.

3.21 In a television game show, the contestant can win a small prize, a medium prize, and a large prize. The large prize is a sports car. Each of the three prizes is "locked up" in a separate box. There are five keys randomly arranged in front of the contestant. One opens the lock to the small prize, another to the medium prize, another to the large prize. Another key is a dud that does not open any of the locks. The final key is the "master key" that opens all three locks. The contestant has a chance to choose up to two keys. For that purpose, the contestant is asked two quiz questions. For each correct answer, he/she can select one key. The probability of correctly answering any given quiz question is 0.5. The contestant tries the keys he/she has gained (if any) on all three doors. What is the probability that the contestant wins the sports car?

3.22 In a particular game, you begin by tossing a die. If the toss results in i points, then you go on to toss i dice together. If the sum of the points resulting from the toss of the i dice is greater than (less than) twelve, you win (lose) one dollar, and if the sum of those points is equal to twelve, you neither win nor lose anything. Use either simulation or a theoretical approach to determine the expected value of your net win in one round of this game.

3.23 In the popular English game of Hazard, a player must first determine which of the five numbers from $5, \ldots, 9$ will be the "main" point. The player does this by rolling two dice until such time as the point sum equals one of these five numbers. The player then rolls again. He/she wins if the point sum of this roll corresponds with the "main" point as follows: main 5 corresponds with a point sum of 5, main 6 corresponds with a point sum of 6 or 7, main 7 corresponds with sum 7 or 11, main 8 corresponds with sum 8 or 12, and main 9 corresponds with sum 9. The player loses if, having taken on a main point of 5 or 9, he/she then rolls a sum of 11 or 12, or by rolling a sum of 11 against a main of 6 or 8, or by rolling a sum of 12 against a main of 7. In every other situation, the sum thrown becomes the player's "chance" point. From here on the player rolls two dice until either the "chance" point (player wins) or the "main" point (player loses) reappears. Verify that the probability of the player winning is equal to 0.5228, where the main and the chance points contribute 0.1910 and 0.3318, respectively, to the probability of winning. What is the house percentage if the house pays the player $1\frac{1}{2}$ and 2 dollars per dollar staked for a main point win and a chance point win, respectively?

3.24 Go back and take another look at Problem 2.31 from Chapter 2. Let's assume that candidate A scores G points in the first spin of the wheel. Using conditional probabilities, show that the probability of A losing to B by stopping after the first spin is equal to $1 - \frac{G}{100} + \frac{G}{100}(1 - \frac{G}{100})$ and that the probability of A losing to B by spinning the wheel a second time is equal to

$$\frac{G}{100} + \frac{1}{20}\sum_{j=1}^{n(G)}\left[1 - \left(\frac{G+5j}{100}\right)^2\right] \qquad \text{with } n(G) = 20 - \frac{G}{5}.$$

3.25 A gang of thieves has gathered at their secret hideaway. Just outside, a beat-cop lurking about realizes that he has happened upon the notorious hideaway and takes it upon himself to arrest the gang leader. He knows that the villains, for reasons of security, will exit the premises one by one in a random order, and that as soon as he were to arrest one of them, the others would be alerted and would flee. For this reason, the agent plans only to make an arrest if he can be reasonably sure of arresting the top man himself. Fortunately, the cop knows that the gang leader is the tallest member of the gang, and he also knows that the gang consists of ten members. How can he maximize his probability of arresting the gang leader?

(a) Suppose a strategy whereby the cop always passes over the first $s-1$ gang members that exit the hideaway, and then arrests the first gang member that is taller than the members who have previously exited the premises. Argue that this strategy will allow the cop to arrest the gang leader with a probability of

$$p(s,n) = \sum_{k=s}^{n}\left(\frac{s-1}{k-1} \times \frac{1}{n}\right),$$

where $n (= 10)$ is the number of gang members.

(b) For fixed n, analyze the difference function $p(s+1,n) - p(s,n)$ and demonstrate that the probability $p(s,n)$ is maximal for the unique value of s, which satisfies

$$\frac{1}{s} + \frac{1}{s+1} + \cdots + \frac{1}{n-1} < 1 \leq \frac{1}{s-1} + \frac{1}{s} + \cdots + \frac{1}{n-1}.$$

Using the approximate expression for $\sum_{i=1}^{n}\frac{1}{i}$ stated in Example 3.2, verify that the optimal value of s satisfies $\ln(\frac{n}{s}) \approx 1$ when n is sufficiently large. Next, prove that the optimal value of s and the corresponding probability $p(s,n)$ are given by $s^* \approx \frac{n}{e}$ and $p(s^*,n) \approx \frac{1}{e}$ for large n, where $e = 2.7183\ldots$. The maximal probability of arresting the gang leader is, then, by good approximation, equal to 36.8% regardless of the magnitude of the gang (for $n = 10$ gang members the precise value of the probability is equal to 0.3987 and is achieved for $s^* = 4$).

3.26 The game "Casino War" is played with a deck of cards compiled of six ordinary decks of 52 playing cards. Each of the cards is worth the face value shown (color is irrelevant). The player and the dealer each receive one card. If the player's card has a higher value than the dealer's, he wins double the amount he staked. If the dealer's card is of a higher value, then the player loses the amount staked. If the cards are of an equal value, then there is a clash and the player doubles his original bet. The dealer then deals one card to the player, one card to himself. If the value of the player's card is higher than the dealer's, he wins twice his original stake, otherwise he loses his original stake and the amount of the added raise. Using either simulation or a theoretical approach, determine the house percentage on this game.

3.27 Red Dog is a casino game played with a deck of 52 cards. Suit plays no role in determining the value of each card. An ace is worth 14, king 13, queen 12, jack 11, and numbered cards are worth the number indicated on the card. After staking a bet, a player is dealt two cards. If these two cards have a "spread" of one or more, a third card is dealt. The spread is defined as the number of points between the values of the two cards dealt (e.g., if a player is dealt a 5 and a 9, he has a spread of three). When a player has a spread of at least one, he may choose to double his initial stake before the third card is dealt. At this point, the third card is dealt. If the value of the third card lies between the two cards dealt earlier, the player gets a payoff of s times his final stake plus the final stake itself, where $s = 5$ for a spread of 1, $s = 4$ for a spread of 2, $s = 2$ for a spread of 3, and $s = 1$ for a spread of 4 or more. In cases where the value of the two cards dealt is sequential (e.g., 7 and 8), no third card is dealt and the player gets his initial stake back. If the values of the two cards dealt are equal, the player immediately gets a third card. If this third card has the same value as the other two, the player gets a payoff of 11 times his initial stake plus the stake itself. The player applies the following simple strategy. The initial stake is only doubled if the spread equals 7 or more. Can you explain why it is not rational to double the stake if the spread is less than 7? Using computer simulation, determine the house percentage for Red Dog.

3.28 You are playing rounds of a certain game against an opponent until one of you has won all of the other one's betting money. At the start of each round, each of you stakes one dollar. The probability of winning any given round is equal to p, and the winner of a round gets the other player's dollar. Your starting capital is a dollars, and your opponent's starting capital is equal to b dollars. What is the probability of your winning all of the money? The renowned *gambler's formula* is

$$P(\text{you win all the money}) = \frac{1 - [(1-p)/p]^a}{1 - [(1-p)/p]^{a+b}},$$

with $p \neq \frac{1}{2}$ (otherwise your probability of winning is equal to $a/(a+b)$). In order to prove this formula, argue first the recursion relation

$$P_i = pP_{i+1} + (1-p)P_{i-1} \quad \text{for} \quad i = 1, \ldots, a+b-1,$$

in which P_k is defined as the probability of your eventually winning all of the money, when your capital is k dollars and your opponent's capital is $a+b-k$ dollars ($P_0 = 0$ and $P_{a+b} = 1$). Next, verify through substitution that the above formula is correct.

3.29 Suppose you go to the local casino with \$50 in your pocket, and it is your goal to multiply your capital to \$250. You are playing (European) roulette, and you stake a fixed amount on red for each spin of the wheel. What is the probability of your reaching your goal when you stake fixed amounts of \$5, \$10, \$25 and \$50, respectively, on each spin of the wheel? *Hint*: apply the gambler's formula from Problem 3.27.

3.30 A fair die is rolled repeatedly. Let p_n be the probability that the sum of scores will ever be n. Use the law of conditional probabilities to find a recursion equation for the p_n. Verify numerically that p_n tends to $\frac{1}{3.5} = 0.2857$ as n gets large. Can you explain this result?

3.31 A fair coin is tossed k times. Let a_k denote the probability of having no two heads in a row in the sequence of tosses. Use the law of conditional probabilities to obtain a recurrence relation for a_k. Calculate a_k for $k = 5$, 10, 25, and 50.

3.32 A drunkard is wandering back and forth on a road. At each step, he moves two units distance to the north with a probability of $\frac{1}{2}$, or one unit to the south with a probability $\frac{1}{2}$. Let a_k denote the probability of the drunkard ever returning to his point of origin if the drunkard is k units distance removed from his point of origin. Use the law of conditional probabilities to argue that $a_k = \frac{1}{2}a_{k+2} + \frac{1}{2}a_{k-1}$ for $k \geq 1$. Next show that $a_k = q^k$ for all k with $q = \frac{1}{2}(\sqrt{5}-1)$. Could you give a probabilistic explanation of why a_k must be of the form q^k for some $0 < q < 1$? Use the result for the drunkard's walk to prove that the probability of the number of heads ever exceeding twice the number of tails is $\frac{1}{2}(\sqrt{5}-1)$ if a fair coin is tossed over and over.

3.33 In Section 3.5, the permutation (5, 4, 1, 3, 2, 6, 8, 9, 10, 7, 11, 12) of the numbers $1, \ldots, 12$ is analyzed. This permutation has an increase factor of 55. The increase factor of a permutation $(k_1, \ldots, k_n)$ of the numbers $1, \ldots, n$ is defined as the number of pairs (k_i, k_j) with $i < j$ for which $k_j > k_i$. Use simulation to find the probability that a random permutation of the numbers $1, \ldots, 12$ has an increase factor of 55 or more. What is the expected value of the increase factor of a random permutation of the integers $1, \ldots, n$?

3.34 Using simulation, compare the distribution of your end capital for two different betting systems in the game of European roulette. You begin with 100 chips and you bet consistently on red. The one betting system is a flat system in which you stake one chip each time for a sum of 100 times. The other system is a Labouchère system in which you work with the list $(1, 2, 3, 4, 5)$ and your number of bets is limited to 100, where you start with list $(1, 2, 3, 4, 5)$ again if you complete it successfully before betting 100 times.

3.35 In the last 250 drawings of Lotto 6/45, the numbers $1, \ldots, 45$ were drawn

$$46,31,27,32,35,44,34,33,37,42,35,26,41,38,40,$$

$$38,23,27,31,37,28,25,37,33,36,32,32,36,33,36,$$

$$22,31,29,28,32,40,31,30,28,31,37,40,38,34,24$$

times, respectively. Using simulation, determine whether these results are suspicious, statistically speaking.

3.36 Jeu de Treize was a popular card game in seventeenth century France. This game was played as follows. One person is chosen as dealer and the others are players. Each player puts up a stake. The dealer takes a full deck of 52 cards and shuffles them thoroughly. Then the dealer turns over the cards one at a time, calling out "one" as he turns over the first card, "two" as he turns over the second, "three" as he turns over the third, and so on up to the thirteenth. A match occurs if the number the dealer is calling corresponds to the card he turns over, where "one" corresponds to an ace of any suit, "two" to a two of any suit, "three" to a three of any suit, ... , "thirteen" to a king of any suit. If the dealer goes through a sequence of 13 cards without a match, the dealer pays the players an amount equal to their stakes, and the deal passes to the player sitting to his right. If there is a match, the dealer collects the player's stakes and the players put up new stakes for the next round. Then, the dealer continues through the deck and begins over as before, calling out "one," and then "two," and so on. If the dealer runs out of cards, he reshuffles the full deck and continues the count where he left off. Use computer simulation to find the probability that the dealer wins k or more consecutive rounds, for $k = 0, 1, \ldots, 8$. Also, verify by computer simulation that the expected number of rounds won by the dealer is equal to 1.803.

4
Rare events and lotteries

How does one calculate the probability of throwing heads more than fifteen times in 25 tosses of a fair coin? What is the probability of winning a lottery prize? Is it exceptional for a city that averages eight serious fires per year to experience twelve serious fires in one particular year? These kinds of questions can be answered by the probability distributions that we will be looking at in this chapter. These are the binomial distribution, the Poisson distribution, and the hypergeometric distribution. A basic knowledge of these distributions is essential in the study of probability theory. This chapter gives insight into the different types of problems to which these probability distributions can be applied. The binomial model refers to a series of independent trials of an experiment that has *two* possible outcomes. Such an elementary experiment is also known as a *Bernoulli experiment*, after the famous Swiss mathematician Jakob Bernoulli (1654–1705). In most cases, the two possible outcomes of a Bernoulli experiment will be specified as "success" or "failure." Many probability problems boil down to determining the probability distribution of the total number of successes in a series of independent trials of a Bernoulli experiment. The Poisson distribution is another important distribution and is used, in particular, to model the occurrence of *rare* events. When you know the expected value of a Poisson distribution, you know enough to calculate all of the probabilities of that distribution. You will see that this characteristic of the Poisson distribution is exceptionally useful in practice. The hypergeometric distribution goes hand in hand with a model known as the "urn model." In this model, a number of red and white balls are selected out of an urn without any being replaced. The hypergeometric probability distribution enables you to calculate your chances of winning in lotteries.

In calculating the binomial, Poisson, and hypergeometric probabilities, tables are no longer needed if you have access to a computer. There are many software modules available for this purpose.

4.1 The binomial distribution

The binomial probability distribution is the most important of all the discrete probability distributions. The following simple probability model underlies the binomial distribution: a certain chance experiment has two possible outcomes ("success" and "failure"), the outcome "success" having a given probability of p and the outcome "failure" a given probability of $1 - p$. An experiment of this type is called a Bernoulli experiment. Consider now the compound experiment that consists of n independent trials of the Bernoulli experiment. Define the random variable X by

X = the total number of successes in n independent trials of the Bernoulli experiment.

The distribution of X is then calculated thus:

$$P(X = k) = \binom{n}{k} p^k (1-p)^{n-k} \quad \text{for} \quad k = 0, 1, \ldots, n.$$

This discrete distribution is called the *binomial distribution* and is derived as follows. Let's say that a success will be recorded as a "one" and a failure as a "zero." The sample space of the compound experiment is made up of all the possible sequences of zeros and ones to a length of n. The n trials of the Bernoulli experiment are physically independent and thus the probability assigned to an element of the sample space is the product of the probabilities of the individual outcomes of the trials. A specific sequence with k ones and $n-k$ zeros gets assigned a probability of $p^k(1-p)^{n-k}$. The total number of ways by which k positions can be chosen for a 1 and $n-k$ positions can be chosen for a 0 is $\binom{n}{k}$ (see the Appendix). Using the addition rule, the formula for $P(X = k)$ follows.

The expected value of the binomial variable X is given by

$$E(X) = np.$$

The proof is simple. Write $X = Y_1 + \cdots + Y_n$, where Y_i is equal to 1 if the ith trial is a success and 0 otherwise. Noting that $E(Y_i) = 0 \times (1-p) + 1 \times p = p$ and using the fact that the expected value of a sum of random variables is the sum of the expected values, the desired result follows.

The binomial probability model has many applications, in illustration of which we offer four examples.

Example 4.1 Daily Airlines flies from Amsterdam to London every day. The price of a ticket for this extremely popular flight route is \$75. The aircraft has a passenger capacity of 150. The airline management has made it a policy to sell 160 tickets for this flight in order to protect themselves against no-show passengers. Experience has shown that the probability of a passenger being a no-show is equal to 0.1. The booked passengers act independently of each other. Given this overbooking strategy, what is the probability that some passengers will have to be bumped from the flight?

This problem can be treated as 160 independent trials of a Bernoulli experiment with a success rate of $\frac{9}{10}$, where a passenger who shows up for the flight is counted as a success. Use the random variable X to denote number of passengers that show up for a given flight. The random variable X is binomially distributed with the parameters $n = 160$ and $p = \frac{9}{10}$. The probability in question is given by $P(X > 150)$. If you feed the parameter values $n = 160$ and $p = \frac{9}{10}$ into a software module for a binomial distribution, you get the numerical value $P(X > 150) = 0.0359$. Thus, the probability that some passengers will be bumped from any given flight is 3.6%.

Example 4.2 In a desperate attempt to breathe new life into the commercial television network "Gamble 7" and to acquire wider cable access, network management has decided to broadcast a lottery called "Choose Your Favorite Spot." Here is how the lottery works: individual participants purchase lottery tickets that show a map of the Netherlands split into four regions, each region listing 25 cities. They choose and place a cross next to the name of one city in each of the four regions. In the weekly television broadcast of the lottery show, one city is randomly chosen for each region. If Gamble 7 has cable access in the cities whose names were drawn, it will make a donation to the cultural coffers of the local government of those cities. In order to determine the prize amount for individual participants in the lottery, Gamble 7 wants to know the probability of one participant correctly guessing the names of four, three, or two of the cities drawn. What are these probabilities?

What we have here is four trials of a Bernoulli experiment (four times a selection of a city), where the probability of success is $\frac{1}{25}$. This means that the binomial probability model, with $n = 4$ and $p = \frac{1}{25}$, is applicable. In other terms:

$$P(\text{you have } k \text{ cities correct}) = \binom{4}{k}\left(\frac{1}{25}\right)^k\left(\frac{24}{25}\right)^{4-k}, \qquad k = 0, \ldots, 4.$$

This leads to the numerical values

$$P(\text{you have 4 cities correct}) = 2.56 \times 10^{-6}$$

$$P(\text{you have 3 cities correct}) = 2.46 \times 10^{-4}$$

$$P(\text{you have 2 cities correct}) = 8.85 \times 10^{-3}.$$

Example 4.3 Gordie the Gambler, a familiar figure in the cafés of central Amsterdam, offers café customers a game of chance called Chuck-a-Luck. To play this game, a customer chooses one number from the numbers $1, \ldots, 6$. A die is then rolled three times. If the customer's number does not come up at all in the three rolls, the customer pays Gordie 100 dollars. If the chosen number comes up one, two, or three times, Gordie pays the customer \$100, \$200, or \$300, respectively. How remunerative is this game for Gordie?

This game seems at first glance to be more favorable for the customer. Many people think that the chosen number will come up with a probability of $\frac{1}{2}$. This is actually not the case, even if the expected value of the number of times the chosen number comes up is equal to $\frac{1}{2}$. The number of times the customer's number comes up is seen as the number of successes in $n = 3$ independent trials of a Bernoulli experiment with a probability of success of $p = \frac{1}{6}$. This gives

$$P(\text{the chosen number comes up } k \text{ times}) = \binom{3}{k}\left(\frac{1}{6}\right)^k \left(\frac{5}{6}\right)^{3-k}$$

for $k = 0, 1, 2, 3$. From this, it follows that

$$\begin{aligned}&\text{average win for Gordie per wager}\\ &= 100 \times \tfrac{125}{216} - 100 \times 3 \times \tfrac{25}{216} - 200 \times 3 \times \tfrac{5}{216} - 300 \times \tfrac{1}{216}\\ &= 100 \times \tfrac{17}{216} = 7.87 \text{ dollars.}\end{aligned}$$

Not a bad profit return for a small businessman!

Example 4.4 You must play a competition (tennis, for example) against a competitor who is stronger than you are. The competition will consist of a previously agreed on number of games, where the probability p of your winning any given game is smaller than 0.5. It is settled that an *even* number of games will be played during the competition. The champion of the competition is the one to win more than half of the agreed upon number of games. Because you are at a disadvantage, you will be permitted to say how many games the competition will consist of with the restriction that you must make it an even number of games. How do you choose the number of games to play such that you will have the highest probability of becoming champion when $p = 0.45$? Assume that the end results of the various games will be independent from one another.

Just off the top of their heads, nearly everyone says that the best choice is to confine the competition to $n = 2$ games. In this way, you would limit your disadvantage to a minimum of exposure. Surprisingly enough, this is not the right decision when an *even* number of games must be played (if an *uneven* number of games had been played, $n = 1$ would indeed have been the best option). This surprising result can be shown by using the binomial probability model. In a competition consisting of n games (n being even), the probability that you will win more than $\frac{1}{2}n$ games is equal to the probability of achieving more than $\frac{1}{2}n$ successes in n independent trials of a Bernoulli experiment with a success probability of $p = 0.45$. Using a software module for binomial probabilities, you will come up with the results shown in Table 4.1 for different values of n. According to the results shown in this table, you will do best to choose for $n = 10$ games when $p = 0.45$. The probability of winning

Table 4.1 *The probability of becoming champion*

n(= number of games)	P(more than $\frac{1}{2}n$ successes)
2	0.2025
4	0.2415
6	0.2553
8	0.2604
10	0.2616
12	0.2607
14	0.2586

the championship is then 26.16%. Your competitor will very likely think it unwise of you to commit yourself to ten games; possibly, you can make use of this faulty reasoning to stipulate a favorable return for yourself on each dollar staked on the winner. It seems difficult to give an intuitive explanation of the fact that choosing $n = 2$ games is only optimal if the probability p of your winning any given game is close enough to zero. Using the mathematical formula for the probability of more than $\frac{1}{2}n$ successes, it can be determined algebraically that the best choice for n is the even number closest to the number $\frac{1}{1-2p}$. This means that in an even number of games it is only optimal to play $n = 2$ games when $\frac{1}{1-2p} \leq 3$, or rather when the probability of winning is $p \leq \frac{1}{3}$.

4.2 The Poisson distribution

In 1837, the famous French mathematician Siméon-Denis Poisson (1781–1840) published his *Recherches sur la Probabilité des Jugements en Matière Criminelle et en Matière Civile*. Indirectly, this work introduced a probability distribution that would later come to be known as the Poisson distribution, and this would develop into one of the most important distributions in probability theory. In this section, the Poisson distribution will be revealed in all its glory. The first issue at hand will be to show how this distribution is realized, namely as a limiting distribution of the binomial distribution. In case of a very large number of independent trials of a Bernoulli experiment with a very small probability of success, the binomial distribution gives way to the Poisson distribution. This insight is essential in order to apply the Poisson distribution in practical situations. In the course of this account, we will offer illustrative applications of the Poisson distribution. Finally, we will delve into the Poisson process. This random process is closely allied with the Poisson distribution and describes the occurrence of events at random points in time.

4.2.1 The origin of the Poisson distribution

A random variable X is *Poisson distributed* with parameter λ if

$$P(X=k)=e^{-\lambda}\frac{\lambda^k}{k!} \quad \text{for} \quad k=0,1,\ldots,$$

where $e=2.7182\ldots$ is the base of the natural logarithm. The Poisson distribution is characterized by just a single parameter λ, where λ is a positive real number. The expected value of the Poisson distribution is equal to this parameter λ. This follows from

$$\begin{aligned}E(X)&=0\times P(X=0)+1\times P(X=1)+2\times P(X=2)+\cdots\\&=\lambda e^{-\lambda}+2\frac{\lambda^2}{2!}e^{-\lambda}+3\frac{\lambda^3}{3!}e^{-\lambda}+\cdots\\&=\lambda e^{-\lambda}\left(1+\frac{\lambda}{1!}+\frac{\lambda^2}{2!}+\cdots\right)=\lambda e^{-\lambda}e^{\lambda}=\lambda,\end{aligned}$$

where we make use of the well-known power series $e^x=1+\frac{x}{1!}+\frac{x^2}{2!}+\cdots$ for every real number x; see the Appendix.

Many practical phenomena can be described according to the Poisson distribution. Evidence of this lies in the following important result:

> **in a** *very large* **number of independent repetitions of a Bernoulli experiment having a** *very small* **probability of success, the total number of successes is approximately Poisson distributed with the expected value $\lambda = np$, where $n =$ the number of trials and $p =$ the probability of success.**

To give a precise mathematical formulation of this result, let Z represent a binomially distributed random variable with the parameters n and p. In other words, Z represents the number of successes in n independent repetitions of a Bernoulli experiment with a success probability of p. Assume now that n becomes *very large* and p becomes *very small* so that np remains equal to the constant λ. The following is then true:

$$\lim_{n\to\infty,\,p\to 0} P(Z=k)=e^{-\lambda}\frac{\lambda^k}{k!} \quad \text{for} \quad k=0,1,\ldots.$$

Proving this is not difficult. Since $p = \frac{\lambda}{n}$,

$$P(Z = k) = \binom{n}{k}\left(\frac{\lambda}{n}\right)^k\left(1 - \frac{\lambda}{n}\right)^{n-k}$$

$$= \frac{n!}{k!(n-k)!}\frac{\lambda^k}{n^k}\frac{(1-\lambda/n)^n}{(1-\lambda/n)^k}$$

$$= \frac{\lambda^k}{k!}(1 - \frac{\lambda}{n})^n\left[\frac{n!}{n^k(n-k)!}\right](1 - \frac{\lambda}{n})^{-k}.$$

Now, let's look at the different terms separately. Assign a fixed value to k of $0 \leq k \leq n$. The term $\frac{n!}{n^k(n-k)!}$ is equal to

$$\frac{n(n-1)\cdots(n-k+1)}{n^k} = \left(1 - \frac{1}{n}\right)\cdots\left(1 - \frac{k-1}{n}\right).$$

With a *fixed* k, this term approaches 1 as $n \to \infty$, as does the term $(1 - \lambda/n)^{-k}$. The function e^x has the property that $\lim_{n\to\infty}(1 + b/n)^n = e^b$ for every real number b (see the Appendix). This results in $\lim_{n\to\infty}(1 - \lambda/n)^n = e^{-\lambda}$, which proves the equation for the limit of $P(Z = k)$.

To give you an idea of how quickly the binomial distribution approaches the Poisson distribution, refer to Table 4.2, where the probabilities of $P(X = k)$ are given for $k = 0,1,2,3,4$, and 5 using both a Poisson-distributed random variable X with expected value $\lambda = 1$, and for a binomial distributed random variable X with expected value $np = 1$, where n is equal to 25, 100, 500, and 1,000, respectively.

The Poisson approximation is characterized by the pleasant fact that one does not need to know the precise number of trials and the precise value of the probability of success; it is enough to know what the product of these two values is. This product is the expected value of the total number of successes.

Table 4.2 *Binomial probabilities and Poisson probabilities*

k	$n = 25$	$n = 100$	$n = 500$	$n = 1{,}000$	Pois(1)
0	0.3604	0.3660	0.3675	0.3677	0.3679
1	0.3754	0.3697	0.3682	0.3681	0.3679
2	0.1877	0.1849	0.1841	0.1840	0.1839
3	0.0600	0.0610	0.0613	0.0613	0.0613
4	0.0137	0.0149	0.0153	0.0153	0.0153
5	0.0024	0.0029	0.0030	0.0030	0.0031

The Poisson distribution is uniquely determined by its expected value. This fact is extremely useful for practical purposes.

The importance of the Poisson distribution cannot be emphasized enough. As is often remarked, the French mathematician Poisson did not recognize the huge practical importance of the distribution that would later be named after him. In his book, he dedicates just one page to this distribution. It was L. von Bortkiewicz, in 1898, who first discerned and explained the importance of the Poisson distribution in his book *Das Gesetz der Kleinen Zahlen* (*The Law of Small Numbers*). One unforgettable example from this book applies the Poisson model to the number of Prussian cavalry deaths attributed to fatal horse kicks (in each of the years between 1875 and 1894). Here, indeed, one encounters a very large number of trials (the Prussian cavalrymen), each with a very small probability of "success" (fatal horse kick). The Poisson distribution is applicable to many other situations from daily life, such as the number of serious traffic accidents that occur yearly in a certain area, the weekly number of winners in a football pool, the number of serious earthquakes occurring in one year, the number of damage claims filed yearly with an insurance company, and the yearly number of mail carriers that are bitten, etc.

4.2.2 Applications of the Poisson model

In this section, we will discuss a number of applications of the Poisson model. The examples are taken from everyday life.

Example 4.5 The Pegasus Insurance Company has introduced a policy that covers certain forms of personal injury with a standard payment of $100,000. The yearly premium for the policy is $25. On average, 100 claims per year lead to payment. There are more than one million policyholders. What is the probability that more than 15 million dollars will have to be paid out in the space of a year?

In fact, every policyholder conducts a personal experiment in probability after purchasing this policy, which can be considered to be "successful" if the policyholder files a rightful claim during the ensuing year. This example is characterized by an extremely large number of independent probability experiments each having an extremely small probability of success. This means that a Poisson distribution with an expected value of $\lambda = 100$ can be supposed for the random variable X, which is defined as the total number of claims that will be approved for payment during the year of coverage. The probability of having to pay out more than 15 million dollars within that year is equal to $P(X > 150)$. Entering $\lambda = 100$ into a software module for the Poisson distribution gives

$P(X > 150) = 1.23 \times 10^{-6}$ – not a probability the insurance executives need worry about.

Example 4.6 During the last few years in Gotham City, a provincial city with more than 100,000 inhabitants, there have been eight serious fires per year, on average. Last year, by contrast, twelve serious fires blazed, leading to great consternation among the populace of the ordinarily tranquil city. The newspaper serving Greater Gotham, *the Gotham Echo*, went wild, carrying inflammatory headlines declaring "50% more fires" and demanding the resignation of the local fire chief. Is all this uproar warranted?

In a city as large as Gotham City, it is reasonable to assume that the number of fires occurring within one year has a Poisson distribution (why?). In order to determine whether twelve fires occurring in the past year is exceptional, one must know the probability of a Poisson distributed random variable X with expected value $\lambda = 8$ taking on a value greater than 11. Entering $\lambda = 8$ in a software module for the Poisson distribution gives a result of

$$P(X > 11) = 0.112.$$

The question that follows is whether this probability of 11.2% is, in fact, so small that the occurrence of twelve or more fires must be qualified as exceptional. The answer to this question is subjective: some would say yes, some no. It is common practice, in statistics, to limit oneself to probabilities of less than 5% when speaking of exceptional outcomes. A statistician would, in this case, give the benefit of the doubt to the local fire brigade.

Example 4.7[1] The following item was reported in the February 14, 1986 edition of *The New York Times*: "A New Jersey woman wins the New Jersey State Lottery twice within a span of four months." She won the jackpot for the first time on October 23, 1985 in the Lotto 6/39. Then she won the jackpot in the new Lotto 6/42 on February 13, 1986. Lottery officials declare that the probability of winning the jackpot twice in one lifetime is approximately one in 17.1 trillion. What do you think of this statement?

The claim made in this statement is easily challenged. The officials' calculation proves correct only in the extremely farfetched case scenario of a given person entering a six-number sequence for Lotto 6/39 and a six-number sequence for Lotto 6/42 just one time in his/her life. In this case, the probability

[1]This example is based on the article "Jumping to coincidences: defying odds in the realm of the preposterous," by J. A. Hanley, in *Journal of the American Statistical Association*, 1992, 46, 197–202.

of getting all six numbers right, both times, is equal to

$$\frac{1}{\binom{39}{6}} \times \frac{1}{\binom{42}{6}} = \frac{1}{1.71 \times 10^{13}}.$$

But this result is far from miraculous when you begin with an extremely large number of people who have been playing the lottery for a long period of time, each of whom submit more than one entry for each weekly draw. For example, if every week 50 million people randomly submit 5 six-number sequences to one of the (many) Lottos 6/42, then the probability of one of them winning the jackpot twice in the coming four years is approximately equal to 63%. The calculation of this probability is based on the Poisson distribution and goes as follows. The probability of your winning the jackpot in any given week by submitting 5 six-number sequences is

$$\frac{5}{\binom{42}{6}} = 9.531 \times 10^{-7}.$$

The number of times that a given player will win a jackpot in the next 200 drawings of a Lotto 6/42, then, is Poisson distributed with expected value

$$\lambda_0 = 200 \times \frac{5}{\binom{42}{6}} = 1.983 \times 10^{-4}.$$

For the next 200 drawings, this means that

$$P(\text{any given player wins the jackpot two or more times})$$
$$= 1 - e^{-\lambda_0} - e^{-\lambda_0}\lambda_0 = 1.965 \times 10^{-8}.$$

Subsequently, we can conclude that the number of people under the 50 million mark, who win the jackpot two or more times in the coming four years, is Poisson distributed with expected value

$$\lambda = 50{,}000{,}000 \times (1.965 \times 10^{-8}) = 0.9825.$$

The probability in question, that at some point in the coming four years at least one of the 50 million players will win the jackpot two or more times, can be given as $1 - e^{-\lambda} = 0.626$. A few simplifying assumptions are used to make this calculation, such as the players choose their six-number sequences randomly. This does not influence the conclusion that it may be expected once in a while, within a relatively short period of time, that *someone* will win the jackpot two times.

A legal problem

The Poisson model also provides an explanation for the legal problem posed in Question 6 of Chapter 1. The conclusion made by the prosecutor is inaccurate. The prosecutor argues that "the probability of finding anyone in the city at all who conforms to the description of the perpetrator is negligible. The suspect does conform to this description, thus it is nearly one hundred percent certain that he is the perpetrator." This is a textbook example of the faulty use of probabilities. The probability that the suspect is innocent of the crime is altogether different from the probability that an arbitrary person bears the physical attributes in question. What we are actually looking for is the probability that, among all persons with the physical attributes in question, the arrested person is the perpetrator. On average, one out of every million adult men has the physical attributes in question, and 150,000 adult men live in the city. Hence, the number of men in the city, excluding the suspect, having the attributes is Poisson distributed with expected value $\lambda = 149{,}999 \times 10^{-6} = 0.15$. The conditional probability of the suspect being the perpetrator is equal to $\frac{1}{k+1}$ when, in addition to the suspect, there are k persons in the city with the physical attributes in question. Thus, by the law of conditional probabilities, the unconditional probability that the suspect is the perpetrator is given by

$$\sum_{k=0}^{\infty} \frac{1}{k+1} e^{-0.15} \frac{(0.15)^k}{k!} = \frac{1}{0.15} \sum_{k=0}^{\infty} e^{-0.15} \frac{(0.15)^{k+1}}{(k+1)!}$$

$$= \frac{e^{-0.15}}{0.15} (e^{0.15} - 1) = 0.9286.$$

In other words, there is a rather convincing probability of 7.1% that the suspect is *not* the perpetrator!

4.2.3 Poisson model for weakly dependent trials

The Poisson distribution is derived for the situation of many independent trials each having a small probability of success. In case the independence assumption is not satisfied, but there is a "weak" dependence between the trial outcomes, the Poisson model may still be useful as an approximation. In surprisingly many probability problems, the Poisson approximation method enables us to obtain quick estimates for probabilities that are otherwise difficult to calculate. This approach requires that the problem is reformulated in the framework of a series of (weakly dependent) trials. The idea of the method is first illustrated by the birthday problem.

The birthday problem revisited

The birthday problem deals with the question of determining the probability of at least two people in a randomly formed group of m people having their birthdays on the same day. This probability can be approximated with the help of the Poisson model. To place the birthday problem in the context of a series of trials, some creativity is called for. The idea is to consider all of the possible combinations of two people and to trace whether in any of those combinations both people have birthdays on the same day. Only when such a combination exists can it be said that two or more people out of the whole group have birthdays on the same day. What you are doing, in fact, is conducting $n = \binom{m}{2}$ experiments. Every experiment has the same probability of success $p = \frac{1}{365}$ in showing the probability that two given people will have birthdays on the same day. Assume that the random variable X indicates the number of experiments where both people have birthdays on the same day. The probability that, in a group of m people, two or more people will have birthdays on the same day is then equal to $P(X \geq 1)$. Although the outcomes of the experiments are dependent on one another, this dependence is considered to be weak because of the vast number (365) of possible birth dates. It is therefore reasonable to approximate the distribution of X using a Poisson distribution with expected value $\lambda = np$. In particular, $P(X \geq 1) \approx 1 - e^{-\lambda}$. In other words, the probability that, within a randomly formed group of m people, two or more people will have birthdays on the same day is approximately equal to

$$1 - e^{-\frac{1}{2}m(m-1)/365}.$$

This results in an approximate value of $1 - e^{-0.69315} = 0.5000$ for the probability that, in a group of 23 people, two or more people will have their birthdays on the same day. This is an excellent approximation for the exact value 0.5073 of this probability. The approximation approach with $\binom{23}{2} = 253$ experiments and a success probability of $\frac{1}{365}$ explains why a relatively small group of 23 people is sufficient to give approximately a 50% probability of encountering two people with birthdays on the same day. The exact solution for the birthday problem does not provide this insight. The birthday problem is not the only problem in which the Poisson approximation method is a useful tool for a quick assessment of the magnitude of certain probabilities.

The exact solution to the birthday problem is easily derived, and the Poisson approximation is not necessarily required. This is different for the "almost" birthday problem: what is the probability that, within a randomly formed group of m people, two or more people will have birthdays within one day of each other? The derivation of an exact formula for this probability is far from simple, but a Poisson approximation is particularly simple to give. You must reconsider all the possible combinations of two people, that is, you must run $n = \binom{m}{2}$ experiments. The probability of success in a given experiment is now equal to $p = \frac{3}{365}$ (the probability that two given people will have birthdays within one day of each other). The number of successful experiments is approximately Poisson distributed with an expected value of $\lambda = np$. In particular, the probability that two or more people will have birthdays within one day of each other is approximately equal to

$$1 - e^{-\frac{3}{2}m(m-1)/365}.$$

For $m = 14$, the approximate value is $1 - e^{-0.74795} = 0.5267$ (the exact value of the probability is 0.5375). The Poisson approximation method can be used to find solutions to many variants of the birthday problem.

A scratch-and-win lottery

A lottery organization distributes one million tickets every week. At one end of the ticket, there is a visible printed number consisting of six digits, say 070469. At the other end of the ticket, another six-digit number is printed, but this number is hidden by a layer of scratch-away silver paint. The ticket holder scratches the paint away to reveal the underlying number. If the number is the same as the number at the other end of the ticket, it is a winning ticket. The two six-digit numbers on each of the one million tickets printed each week are randomly generated in such a way that no two tickets are printed with the same visible numbers or the same hidden numbers. Assuming that all tickets are sold each week, the following questions are of interest to the lottery organizers. What is the probability distribution of the number of winners in any given week? In particular, what is the average number of winners per week?

The surprising answer is that the probability distribution of the number of winners in any given week is practically indistinguishable from a Poisson distribution with an expected value of 1. Even more astonishingly, the Poisson distribution with an expected value of 1 applies to any scratch lottery, regardless of whether the lottery issues one million six-digit tickets or 100 two-digit tickets. This is an astounding result that few will believe at first glance! However, the phenomenon can easily be explained by the Poisson-approximation approach. To do so, let's assume a scratch lottery that issues n different tickets with the printed numbers $1, \ldots, n$ each week. Use the random variable X to denote the number of winners in any given week. The random variable X can be seen as the number of successes in n trials. In each trial, the printed number and the hidden number are compared. The success probability for each trial is $\frac{1}{n}$. If n is large enough, the dependence between the trials is weak enough to approximate the probability distribution of X by a Poisson distribution with an expected value of $\lambda = n \times \frac{1}{n} = 1$. In particular, the probability of no winner in any given week is approximately $\frac{1}{e} = 0.368$. It turns out that the Poisson distribution is indeed an excellent approximation to the exact probability distribution of X. The exact probability distribution will be given in Example 7.7 of Chapter 7. A numerical comparison of the exact distribution with the Poisson distribution reveals that $n = 10$ is sufficiently large in order for the Poisson probabilities to match the exact probabilities in at least 8 decimals.

The scratch-and-win lottery problem is one of the many manifestations of the so-called *hat-check problem*. To explain this problem, imagine that, at a country wedding in France, all male guests throw their berets in a corner. After the reception, each guest takes a beret without bothering to check if it is his. The probability that at least one guest goes home with his own beret is approximately $1 - \frac{1}{e} = 0.632$. The origin of matching problems like the scratch-and-win lottery problem and the hat-check problem can be found in the book *Essay d'Analyse sur les Jeux de Hasard*, written in 1708 by Pierre Rémond de Montmort (1678–1719). In his book, Montmort solved a variant of the original card game *Jeu de Treize*, which is described in Exercise 3.36. Montmort simplified this game by assuming that the deck of cards has only 13 cards of one suit. The dealer shuffles the cards and turns them up one at a time, calling out "Ace, two, three, ..., king." A match occurs if the card that is turned over matches the rank called out by the dealer as he turns it over. The dealer wins if a match occurs. The probability of a match occurring is approximately $1 - \frac{1}{e} = 0.632$. A related problem was discussed in Marilyn vos Savant's column in *Parade* magazine of August 21, 1994. An ordinary deck of 52 cards is thoroughly shuffled. The dealer turns over the cards one at a time, counting as he goes "ace, two, three, ..., king, ace, two, ...," and so on, so that the dealer ends up calling out the thirteen ranks four times each. A match occurs if the card that comes up matches the rank called out by the dealer as he turns it over. Using the Poisson-approximation method, it is easy to calculate an estimate of the probability of the occurrence of a match. There are $n = 52$ trials, and the success probability for each trial is $p = \frac{4}{52}$. The probability distribution of the number of matches is then approximated by a Poisson distribution with an expected value of $\lambda = 52 \times \frac{4}{52} = 4$. In particular, the probability of the dealer winning is approximated by $1 - e^{-4} = 0.9817$. This is again an excellent approximation. The exact value of the probability of the dealer winning is 0.9838, as can be calculated using the inclusion-exclusion rule in Chapter 7.

A lottery problem

What is the probability that, in 30 lottery drawings of six numbers from the numbers $1, \ldots, 45$, not each of these 45 numbers will be drawn at least once? This is the question that appears in Problem 4 of Chapter 1. To calculate this probability, a simple Poisson approximation can be given. We will look to the lottery drawings as 45 trials. The ith trial is limited to the number i and is considered successful when the number i does not come up in any of the 30 drawings. For each trial, the probability of the pertinent number not being drawn in any of the 30 drawings is equal to $p = (\frac{39}{45})^{30} = 0.0136635$. This calculation uses the fact that the probability of a specific number i not coming up in a given drawing is equal to $\frac{44}{45} \times \frac{43}{44} \times \cdots \times \frac{39}{40} = \frac{39}{45}$. Although a slight dependence does exist between the trials, it seems reasonable to estimate the distribution of the amount of numbers that will not come up in 30 drawings by using a Poisson distribution with an expected value of $\lambda = 45 \times 0.0136635 = 0.61486$. This gives a surprising result: the probability that not each of the 45 numbers will come up in 30 drawings is approximately equal to

$$1 - e^{-0.61486} = 0.4593.$$

The exact value of the probability is 0.4722, as can be calculated using the inclusion-exclusion rule in Chapter 7. The methodology used for the lottery problem can be applied to a larger class of problems that are related to the coupon collector's problem set forth in Section 3.2.

4.2.4 The Poisson process[2]

The Poisson process is inseparably linked to the Poisson distribution. This process is used to count events that occur randomly in time. Examples include: the arrival of clients at a bank, the occurrence of serious earthquakes, telephone calls to a call center, the occurrence of power outages, urgent calls to an emergency center, etc. When does the process of counting events qualify as a Poisson process? To specify this, it is convenient to consider the Poisson process in terms of customers arriving at a facility. As such, it is necessary to begin with the assumption of a population *unlimited in size* of potential customers, in which the customers act independently of one another. The process of customer arrivals at a service facility is called a *Poisson process* if the process possesses the following properties:

[2]This section is earmarked for the more advanced student and may be set aside for subsequent readings by the novice.

a. the customers arrive one by one
b. the numbers of arrivals during any two nonoverlapping time intervals are independent of one another
c. the number of arrivals during any given time interval has a Poisson distribution of which the expected value is proportional to the duration of the interval.

Regarding property b, it is noted that independence of two events A and B is defined by $P(AB) = P(A)P(B)$ (see also Chapter 8). Letting

$$\alpha = \text{the expected value of the number of arrivals during a given time interval of unit length,}$$

then property c demands that, for each $t > 0$, it is true that

$$P(k \text{ arrivals during a given time interval of duration } t) = e^{-\alpha t}\frac{(\alpha t)^k}{k!} \quad \text{for} \quad k = 0, 1, \ldots .$$

Also, by property b, the joint probability of j arrivals during a given time interval of length t and k arrivals during another given time interval of length u is equal to $e^{-\alpha t}\frac{(\alpha t)^j}{j!} \times e^{-\alpha u}\frac{(\alpha u)^k}{k!}$, provided that the two intervals are nonoverlapping.

The number α is called the *arrival intensity* of the Poisson process. The assumptions of the Poisson process are natural assumptions that hold in many practical situations.

A construction of the Poisson process

An alternative approach to the Poisson process is outlined as follows. Split the time axis up into intervals of length Δt with Δt very small. Assume also that during a given interval of length Δt, the probability that precisely one customer will arrive is equal to $\alpha\Delta t$, and the probability that no customer will arrive is equal to $1 - \alpha\Delta t$, independently of what has happened before the interval in question. In this way, if a given interval of length t is split up into n smaller intervals each having length Δt, then the number of arrivals during the interval of length t has a binomial distribution with parameters $n = \frac{t}{\Delta t}$ and $p = \alpha\Delta t$. Now, let $\Delta t \to 0$, or equivalently $n \to \infty$. Because the Poisson distribution is a limiting case of the binomial distribution, it follows that the number of arrivals during an interval of length t has a Poisson distribution with an expected value of $np = \alpha t$. This construction of the Poisson process is especially useful and may be extended to include the situation in which customer arrival intensity is dependent on time.

The construction of a Poisson process on the line can be generalized to a Poisson process in the plane or other higher dimensional spaces. The Poisson model defines a random way to distribute points in a higher dimensional space. Examples are defects on a sheet of material and stars in the sky.

Relationship between the Poisson process and the exponential distribution

In a Poisson arrival process, the number of arrivals during a given time interval is a discrete random variable, but the time between two successive arrivals can take on any positive value and is thus a so-called *continuous random variable*. This can be seen in the following:

$$P(\text{time between two successive arrivals is greater than } y)$$

$$= P(\text{during an interval of duration } y \text{ there are no arrivals})$$

$$= e^{-\alpha y} \quad \text{for} \quad \text{each } y > 0.$$

Thus, in a Poisson arrival process with an arrival intensity α, the time T between two successive arrivals has the probability distribution function

$$P(T \leq y) = 1 - e^{-\alpha y} \quad \text{for} \quad y \geq 0.$$

This continuous distribution is known as the *exponential distribution* (see also Chapter 10). The expected value of the interarrival time T is $\frac{1}{\alpha}$. Given property b of the Poisson process, it will not come as a surprise to anyone that the intervals between the arrivals of successive clients are independent from each other. A more surprising property of the Poisson process is as follows: for every fixed point in time, the waiting period from that point until the first arrival after that point has the *same* exponential distribution as the interarrival times, regardless of how long it has been since the last client arrived before that point in time. This extremely important *memoryless property* of the Poisson process can be shown with the help of property b of the Poisson process, which says that the number of arrivals in non-overlapping intervals are independent from one another. The memoryless property is characteristic for the Poisson process.

Example 4.8 Out in front of Central Station, multiple-passenger taxicabs wait until they have either acquired four passengers or a period of ten minutes has passed since the first passenger stepped into the cab. Passengers arrive according to a Poisson process with an average of one passenger every three minutes.

a. You are the first passenger to get into a cab. What is the probability that you will have to wait ten minutes before the cab gets underway?
b. You were the first passenger to get into a cab, and you have been waiting there for five minutes. In the meantime, two other passengers have entered

the cab. What is the probability that you will have to wait another five minutes before the cab gets underway?

The answer to Question a rests on the observation that you will only have to wait ten minutes if, during the next ten minutes, fewer than three other passengers arrive. This gives us:

$$P(\text{you must wait ten minutes})$$

$$= P(0, 1, \text{ or } 2 \text{ passengers arrive within the next ten minutes})$$

$$= e^{-10/3} + e^{-10/3}\frac{(10/3)^1}{1!} + e^{-10/3}\frac{(10/3)^2}{2!} = 0.3528.$$

Solving Question b rests on the memoryless property of the Poisson process. The waiting period before the arrival of the next passenger is exponentially distributed with an expected value of three minutes. You will have to wait another five minutes if this waiting period takes longer than five minutes. Thus, the probability of having to wait another five minutes is then $e^{-5/3} = 0.1889$.

Clustering of arrival times

Customer arrival times reveal a tendency to cluster. This is clearly shown in Figure 4.1. This figure gives simulated arrival times in the time interval (0, 45) for a Poisson process with arrival intensity $\alpha = 1$. A mathematical explanation of the clustering phenomenon can be given. As shown before, the interarrival time T has probability distribution function $P(T \leq y) = 1 - e^{-\alpha y}$ for $y \geq 0$. The derivative of the function $F(y) = 1 - e^{-\alpha y}$ is given by $f(y) = \alpha e^{-\alpha y}$. By definition, $f(y) = \lim_{\Delta y \to 0} [F(y + \Delta y) - F(y)] / \Delta y$. This implies that

$$P(y < T \leq y + \Delta y) \approx f(y)\Delta y \quad \text{for} \quad \Delta y \text{ small}$$

(see also Chapter 10). The function $f(y) = \alpha e^{-\alpha y}$ is largest at $y = 0$ and decreases from $y = 0$ onward. In other words, the point y at which $P(y < T \leq y + \Delta)$ is largest for fixed but small $\Delta y > 0$ is the point $y = 0$. Thus, short interarrival times occur relatively frequently, and this suggests that a random series

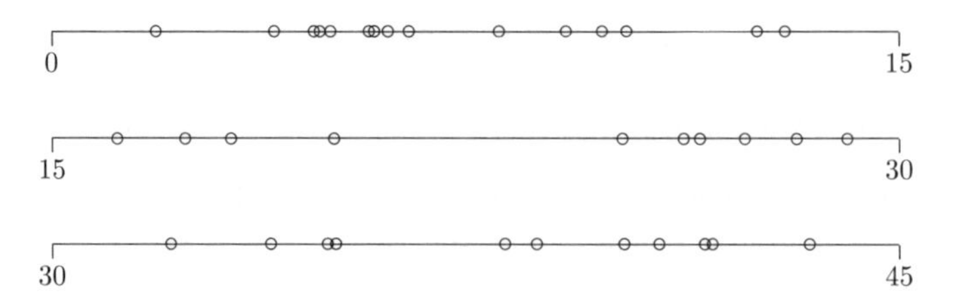

Figure 4.1 Arrival times of a Poisson process.

of arrivals will show a considerable tendency to cluster. The phenomenon of clustered arrival times casts an interesting light on a series of murders in Florida that caused a great deal of turmoil. In the period between October 1992 and October 1993, nine tourists of international origins were murdered in Florida. The murders were attributed to the fact that foreign tourists could easily be recognized as such because they drove rental cars. This could well have been one explanation for the explosion of murders, but it is also quite possible that one can only speak of a "normal" probability event when it is observable over a greater period of time. Assume that for each day there is a 1% probability of a foreign tourist being murdered somewhere in Florida. The random process showing the occurrence of foreign tourist murders in Florida over time can reasonably be modeled as a Poisson process with an intensity of 3.65 murders per year. Now, what is the probability that somewhere within a time frame of say, ten years, there will be one 12-month period containing nine or more foreign tourist murders? There is no easy formula for computing this probability, but a solution can easily be found by means of computer simulation (it will be explained later how to simulate arrival times in a Poisson process). The probability is approximately 36%. Over a period of twenty years, the probability of such a series of murders increases to approximately 60%. This contrasts with the probability of nine or more murders in a *given* 12-month period of 0.0127, a much smaller probability than the ones obtained for a moving time frame. In the latter situation, the clustering phenomenon compounds the "law of coincidences": if you give something enough of a chance to happen, it eventually will.

Example 4.9 In a given city, traffic accidents occur according to a Poisson process with an average of $\lambda = 10$ accidents per week. In a certain week, seven accidents have occurred. What is the probability that exactly one accident has occurred on each day of that week? Can you explain beforehand why this probability must be small?

Let the random variable $N(t)$ denote the number of accidents occurring in the time interval $(0,t)$, where a day is taken as time unit. Letting the epoch $t = u - 1$ correspond to the beginning of day u for $u = 1, 2, \ldots, 7$, the probability we are seeking is given by

$$P(N(u) - N(u-1) = 1 \quad \text{for} \quad u = 1, \ldots, 7 \mid N(7) = 7)$$

with the convention $N(0) = 0$. By the properties b and c of the Poisson process, the random variables $N(1)$, $N(2) - N(1), \ldots, N(7) - N(6)$ are independent and have a Poisson distribution with expected value $\lambda/7$. Also, by property c, the random variable $N(7)$ is Poisson distributed with expected value λ. Thus, the desired probability is equal to

$$\frac{P(N(u)-N(u-1)=1 \text{ for } u=1,\ldots,7)}{P(N(7)=7)}$$
$$=\frac{P(N(1)=1)\times P(N(2)-N(1)=1)\times\cdots\times P(N(7)-N(6)=1)}{P(N(7)=7)}$$
$$=\frac{e^{-\lambda/7}(\lambda/7)\times e^{-\lambda/7}(\lambda/7)\times\cdots\times e^{-\lambda/7}(\lambda/7)}{e^{-\lambda}\lambda^7/7!}=\frac{7!}{7^7}.$$

Hence, the desired probability is equal to 0.0162. Indeed, a small probability. The tendency of Poisson arrivals to cluster explains why the probability is so small. Incidentally, the probability $7!/7^7$ is the same as the probability of getting exactly one random number in each of the seven intervals $(0,\frac{1}{7}),(\frac{1}{7},\frac{2}{7}),\ldots,(\frac{6}{7},1)$ when drawing seven independent random numbers from $(0,1)$. It can be proved that there is a close relationship between the Poisson arrival process and the uniform distribution: under the condition that exactly r arrivals have occurred in the fixed time interval $(0,t)$, then the r arrival epochs will be statistically indistinguishable from r random points that are independently chosen in the interval $(0,t)$.

This characteristic property of the Poisson process on the line extends to the Poisson process in the plane or other higher dimensional spaces: under the condition that exactly r entities (e.g., stars) are contained in a given bounded region, then the positions of the r entities will be distributed as r random points that are independently chosen in the region. This is a useful result for simulating a Poisson process in the plane or other higher dimensional spaces. A simpler procedure to simulate a Poisson process on the line will be discussed in the next section.

Simulating a Poisson process

There are several ways to simulate arrival times of a Poisson process. The easiest method is based on the result that the Poisson process with arrival intensity α can be equivalently defined by assuming single arrivals with interarrival times that are independent and have an exponential distribution with expected value $\frac{1}{\alpha}$. In Chapter 10, the reader will be asked to show that the random variable $X=-\frac{1}{\alpha}\ln(U)$ is exponentially distributed with expected value $\frac{1}{\alpha}$ if U is uniformly distributed on (0,1). This leads to the following algorithm for generating an interarrival time:

1. Generate a random number u between 0 and 1.
2. Take $x=-\frac{1}{\alpha}\ln(u)$ as the inter-arrival time.

This simple procedure gives the reader the power to verify the probabilities cited in the example of the Florida murders by means of a simulation study.

Merging and splitting Poisson processes

In applications of the Poisson process, it is frequently necessary to link two Poisson processes together, or to thin out one Poisson process. For example, consider a call center that functions as the telephone information facility for two completely different business organizations. Calls come in for the first company A according to a Poisson process with arrival intensity λ_A, and, independently of that, calls come in for the other company B according to a Poisson process with arrival intensity λ_B. The merging of these two arrival processes can be shown to give us a Poisson process with arrival intensity $\lambda_A + \lambda_B$. It can also be shown that any future telephone call will be for company A with a probability of $\frac{\lambda_A}{\lambda_A+\lambda_B}$ and will be for company B with a probability of $\frac{\lambda_B}{\lambda_A+\lambda_B}$.

In order to show how a Poisson process can be split up, we will refer to the example of a Poisson process with intensity λ that describes the occurrence of earthquakes in a certain region. Assume that the magnitudes of the earthquakes are independent from one another. Any earthquake is classified as being a high-magnitude earthquake with probability p and as being a low-magnitude earthquake with probability $1 - p$. Then, the process describing the occurrence of high-magnitude earthquakes is a Poisson process with intensity λp, and the occurrence of low-magnitude earthquakes is described by a Poisson process with intensity $\lambda(1 - p)$. It is surprising to find that these two Poisson processes are independent from one another!

4.3 The hypergeometric distribution

The *urn* model is at the root of the hypergeometric distribution. In this model, you have an urn that is filled with R red balls and W white balls. You must randomly select n balls out of the urn without replacing any. What is the probability that, out of the n selected balls, r balls will be red? When the random variable X represents the number of red balls among the selected balls, this probability is given as follows:

$$P(X = r) = \frac{\binom{R}{r}\binom{W}{n-r}}{\binom{R+W}{n}} \quad \text{for} \quad r = 0, 1, \ldots, n.$$

This is called the *hypergeometric distribution* with parameters R, W, and n. This comes with the understanding that $P(X = r) = 0$ for impossible combinations, or rather for values of r when $r > R$ or $n - r > W$. In skimming over the above formula, imagine that the R red balls are numbered $1, \ldots, R$ and the W white balls are numbered $R+1, \ldots, R+W$. There are, in total, $\binom{R+W}{n}$ different ways to select n balls from the $R+W$ balls in the urn, whereas there are $\binom{R}{r} \times \binom{W}{n-r}$ different ways to select r balls from the R red balls and $n-r$ balls from the W white balls. Each of these outcomes is equally probable. When you divide the number of favorable outcomes by the total number of possible outcomes, you get the above formula.

The hypergeometric distribution has the expected value

$$E(X) = n \frac{R}{R+W}.$$

The proof is simple. Write $X = Y_1 + \cdots + Y_n$, where Y_i is equal to 1 if the ith drawn ball is red and 0 otherwise. For reasons of symmetry, each of the random variables Y_i has the same distribution as Y_1. Noting that $E(Y_1) = 1 \times \frac{R}{R+W}$ and using the fact that the expected value of a sum is the sum of the expected values, the desired result follows.

The hypergeometric distribution is often used when calculating the probability of winning prize money in a lottery.[3] The examples that follow show that when gambling with money, one is better off in a casino than taking part in a lottery. Lotteries often sell themselves by using psychological tricks to make one's chances of winning appear higher than they are in reality. Lottery organizers are perennial peddlers of hope! Providing hope to the masses is their ironclad sales objective. The purchasers of this laudable commodity, however, ordinarily see no more than 50% of their outlay return in the form of prize money.

Example 4.10 In the game "Lucky 10," twenty numbers are drawn from the numbers $1, \ldots, 80$. One plays this game by ticking one's choice of 10, 9, 8, 7, 6, 5, 4, 3, 2, or 1 number(s) on the game form. Table 4.3 indicates what the payoff rate is, depending on how many numbers are ticked and how many of those are correct. Also we give the chance of winning for each of the various combinations in Table 4.3. How are these chances of winning calculated and what are the expected payments per dollar staked when one ticks 10, 9, 8, 7, 6, 5, 4, 3, 2, or 1 number(s)?

[3] The modern lottery with prize money attached has its origins in the Netherlands: the oldest known lotteries of this kind have been traced as far back as 1444–1445, to the state lotteries of Brugge and Utrecht. Indeed, the local sovereign in Brugge, Philips de Goede (Philips the Good) moved quickly to require lottery organizers to obtain a license requiring them to hand one third of the lottery profits over to him. Because such a high percentage of ticket sale monies went to the "house" (i.e., Philips de Goede), taking part in the lottery soon became tantamount to making a voluntary tax contribution.

Table 4.3 *Winning combinations in Lucky 10*

Player's choice	Match	Payoff	Chance of winning
10 numbers	10/10	\$100,000	1.12×10^{-7}
	9/10	\$4,000	6.12×10^{-6}
	8/10	\$200	1.35×10^{-4}
	7/10	\$30	1.61×10^{-3}
	6/10	\$8	1.15×10^{-2}
	5/10	\$2	5.14×10^{-2}
	4/10	\$1	1.47×10^{-1}
	0/10	\$1	4.58×10^{-2}
9 numbers	9/9	\$25,000	7.24×10^{-7}
	8/9	\$2,000	3.26×10^{-5}
	7/9	\$200	5.92×10^{-4}
	6/9	\$15	5.72×10^{-3}
	5/9	\$3	3.26×10^{-2}
	4/9	\$1	1.14×10^{-1}
8 numbers	8/8	\$15,000	4.35×10^{-6}
	7/8	\$250	1.60×10^{-4}
	6/8	\$20	2.37×10^{-3}
	5/8	\$10	1.83×10^{-2}
	4/8	\$2	8.15×10^{-2}
7 numbers	7/7	\$2,000	2.44×10^{-5}
	6/7	\$80	7.32×10^{-4}
	5/7	\$12	8.64×10^{-3}
	4/7	\$2	5.22×10^{-2}
	3/7	\$1	1.75×10^{-1}
6 numbers	6/6	\$1,000	1.29×10^{-4}
	5/6	\$25	3.10×10^{-3}
	4/6	\$6	2.85×10^{-2}
	3/6	\$1	1.30×10^{-1}
5 numbers	5/5	\$200	6.45×10^{-4}
	4/5	\$8	1.21×10^{-2}
	3/5	\$3	8.39×10^{-2}
4 numbers	4/4	\$20	3.06×10^{-3}
	3/4	\$5	4.32×10^{-2}
	2/4	\$1	2.13×10^{-1}
3 numbers	3/3	\$16	1.39×10^{-2}
	2/3	\$2	1.39×10^{-1}
2 numbers	2/2	\$2	6.01×10^{-2}
	1/2	\$1	3.80×10^{-1}
1 number	1/1	\$2	2.50×10^{-1}

In the case of ten numbers being ticked on the entry form, the following calculations apply (the same procedure is applicable in all of the other cases). Imagine that the twenty numbers drawn from the numbers $1, \ldots, 80$ are identified as $R = 20$ red balls in an urn and that the remaining sixty, non-chosen numbers are identified as $W = 60$ white balls in the same urn. You have ticked ten numbers on your game form. The probability that you have chosen r numbers from the red group is simply the probability that r red balls will come up in the random drawing of $n = 10$ balls from the urn when no balls are replaced. This gives

$$P(r \text{ numbers correct out of 10 ticked numbers}) = \frac{\binom{20}{r}\binom{60}{10-r}}{\binom{80}{10}}.$$

Let us abbreviate this probability as $p_{r,10}$. With the data provided in Table 4.3, you will get an expected payoff of:

$$\begin{aligned}
&E(\text{payoff per dollar staked on ten ticked numbers}) \\
&\quad = 100{,}000 \times p_{10,10} + 4{,}000 \times p_{9,10} + 200 \times p_{8,10} + 30 \times p_{7,10} \\
&\qquad + 8 \times p_{6,10} + 2 \times p_{5,10} + 1 \times p_{4,10} + 1 \times p_{0,10}.
\end{aligned}$$

When you enter the parameter values $R = 20,\ W = 60$, and $n = 10$ into a software module for the hypergeometric distribution, you get the numerical value of $p_{r,10}$ for each r. When these numerical values are filled in, you find

$$\begin{aligned}
&E(\text{payoff per dollar staked on ten ticked numbers}) \\
&\quad = 0.499 \text{ dollars}.
\end{aligned}$$

In other words, the house percentage in the case of ten ticked numbers is 50.1%. The other house percentages in Table 4.4 are calculated in the same way.

In Table 4.4, the expected payoff per dollar staked on total of ticked numbers is indicated. This is eye-opening information that you will not find on the Lucky 10 game form. The percentage of monies withheld on average by the lotto organizers says a lot. These house percentages linger in the neighborhood of 50% (and consider, on top of that, that many a lottery prize goes unclaimed!). That is quite a difference from the house percentage of 2.7% at a casino roulette wheel! But then, of course, when you play Lucky 10 you are not only lining the pockets of the lotto organizers, but you are also providing support for some worthy charities.

Example 4.11 The "New Amsterdam Lottery" offers the game "Take Five." In this game, players must tick five different numbers from the numbers $1, \ldots, 39$.

Table 4.4 *The average payoff on Lucky 10*

Total numbers ticked	Average payoff per dollar staked	House percentage
10	0.499	50.1%
9	0.499	50.1%
8	0.499	50.1%
7	0.490	51.0%
6	0.507	49.3%
5	0.478	52.2%
4	0.490	51.0%
3	0.500	50.0%
2	0.500	50.0%
1	0.500	50.0%

The lottery draws five distinct numbers from the numbers $1, \ldots, 39$. For every one dollar staked, the payoff is \$100,000 for five correct numbers, \$500 for four correct numbers, and \$25 for three correct numbers. For two correct numbers, the player wins a free game. What is the house percentage for this lottery?

The hypergeometric model with $R = 5$, $W = 34$, and $n = 5$ is applicable in this case. This gives

$$P(\text{you have precisely } k \text{ numbers correct}) = \frac{\binom{5}{k}\binom{34}{5-k}}{\binom{39}{5}}.$$

The numerical value of the probability is 1.74×10^{-6}, 2.95×10^{-4}, 0.00974, and 0.10393, respectively, for $k = 5, 4, 3$, and 2. The expected payoff per dollar staked is denoted with E. This results in

$$E = 1.74 \times 10^{-6} \times 100{,}000 + 2.95 \times 10^{-4} \times 500 + 0.00974 \times 25 + 0.10393 \times E,$$

from which it follows that

$$E = 0.456 \text{ dollars}.$$

This means that the house percentage of the lottery is 54.4%. Nothing new under the sun: house percentages of lotteries the world over tend to be on the hefty side.

Table 4.5 *The winning combinations in the Powerball Lottery*

You match	Payoff	Chance of winning
5 white + Powerball	jackpot	8.30×10^{-9}
5 white	\$ 100,000	3.40×10^{-7}
4 white + Powerball	\$ 5,000	1.99×10^{-6}
4 white	\$ 100	0.0000816
3 white + Powerball	\$ 100	0.0000936
3 white	\$ 7	0.0038372
2 white + Powerball	\$ 7	0.0014350
1 white + Powerball	\$ 4	0.0080721
0 white + Powerball	\$ 3	0.0142068

Example 4.12[4] In the Powerball Lottery, five white balls are drawn from a drum containing 53 white balls numbered from $1, \ldots, 53$, and one red ball (Powerball) is drawn from 42 red balls numbered from $1, \ldots, 42$. This lottery is played in large sections of North America. On the game form, players must tick five "white" numbers from the numbers $1, \ldots, 53$ and one red number from the numbers $1, \ldots, 42$. The winning combinations with the corresponding payoffs and win probabilities are given in Table 4.5. The prizes are based on fixed monetary amounts except for the jackpot, which varies in its amounts and sometimes has to be divided among a number of winners. The amount of cash in the jackpot increases continuously until such time as it is won.

The calculation of the chances of winning rests on the hypergeometric distribution and the product formula for probabilities. The probability of choosing k white balls and the red Powerball correctly on one ticket is given as

$$\frac{\binom{5}{k}\binom{48}{5-k}}{\binom{53}{5}} \times \frac{1}{42},$$

while the probability of choosing just k white balls correctly is equal to

$$\frac{\binom{5}{k}\binom{48}{5-k}}{\binom{53}{5}} \times \frac{41}{42}.$$

The probability of winning the jackpot on one ticket is inconceivably small: 1 in 121 million. It is difficult to represent, in real terms, just how small this probability is. It can best be attempted as follows: if you enter twelve Powerball

[3]This example and the ensuing discussion are based on the teaching aid "Using lotteries teaching a chance course," available at *www.dartmouth.edu/~chance.*

tickets every week, then you will need approximately 134,000 years in order to have about a 50% chance of winning the jackpot at some time in your life (you can verify this for yourself by using the Poisson distribution!).

The Powerball game costs the player one dollar per ticket. The expected payoff for one ticket depends on the size of the jackpot and the total number of entries. The winning combinations, except the jackpot, make the following contribution to the expected value of the payoff for one ticket:

$$\begin{aligned}&100{,}000 \times 0.0000003402 + 5{,}000 \times 0.000001991 + 100 \times 0.00008164\\&\quad + 100 \times 0.00009359 + 7 \times 0.0038372 + 7 \times 0.001435\\&\quad + 4 \times 0.0080721 + 3 \times 0.0142068 = 0.1733 \text{ dollars.}\end{aligned}$$

In order to determine how much the jackpot contributes to the expected payoff, let's assume the following: there is a jackpot of 100 million dollars and 150 million tickets have been randomly filled out and entered. In calculating the jackpot's contribution to the expected payoff for any given ticket, you need the probability distribution of the number of winners of the jackpot among the remaining $n = 149{,}999{,}999$ tickets. The probability that any given ticket is a winning ticket is $p = 8.2969 \times 10^{-9}$. Thus, the probability distribution of the number of winning tickets is a Poisson distribution with an expected value of $\lambda = np = 1.2445$. This means that the contribution of the jackpot to the expected payoff of any given ticket is equal to

$$p \times \left(\sum_{k=0}^{\infty} \frac{1}{k+1} e^{-\lambda} \frac{\lambda^k}{k!} \right) \times 10^8 = 0.4746 \text{ dollars.}$$

The value of the expected payoff for any one dollar ticket, then, is equal to $0.1733 + 0.4746 = 0.6479$ dollars when the jackpot contains \$100 million and 150 million tickets are randomly filled out and entered.

Choosing lottery numbers

There is no reasonable way to improve your chances of winning at Lotto except to fill in more tickets. That said, it is to one's advantage not to choose "popular" numbers, i.e., numbers that a great many people might choose, when filling one's ticket in. If you are playing Lotto 6/45, for example, and you choose 1-2-3-4-5-6 or 7-14-21-28-35-42 as your six numbers, then you can be sure that you will have to share the jackpot with a huge number of others should it come up as the winning sequence. People use birth dates, lucky numbers, arithmetical sequences, etc., in order to choose lottery numbers. This is nicely illustrated in an empirical study done in 1996 for the Powerball Lottery. At that time, players of the Powerball Lottery chose six numbers from the numbers

Table 4.6 *Relative frequencies of numbers chosen*

37	0.010	34	0.014	18	0.022	21	0.025	10	0.029
38	0.011	40	0.015	30	0.023	15	0.025	4	0.029
43	0.012	32	0.015	19	0.023	25	0.026	8	0.030
45	0.012	35	0.016	27	0.023	1	0.026	12	0.030
39	0.012	33	0.018	24	0.024	22	0.026	11	0.031
44	0.012	20	0.019	14	0.024	13	0.026	3	0.033
41	0.013	29	0.020	26	0.024	23	0.027	5	0.033
36	0.013	28	0.020	16	0.024	6	0.028	9	0.033
42	0.014	31	0.020	17	0.024	2	0.029	7	0.036

$1, \ldots, 45$ (before 1997 the Powerball Lottery consisted of the selection of five white balls out of a drum containing 45 white balls, and one red ball out of a drum containing 45 red balls). In total, a good 100,000 hand-written ticket numbers were analyzed. The relative frequencies of numbers chosen are given, in increasing order, in Table 4.6. No statistical tests are necessary in order to recognize that these people did not choose their numbers randomly. Table 4.6 indicates that people often use birth dates in choosing lottery numbers: the numbers 1 through 12, which may refer to both days of the month and months of the year, are frequently chosen. In lotteries where the majority of numbers in a series must be filled in by hand, it appears that the number of winners is largest when most of the six numbers drawn fall in the lower range.[5] When it comes to choosing nonpopular numbers in betting games, racetrack betting offers the reverse situation to the lottery: at the end of the day, when the last races are being run, one does well to bet on the favorites. The reason for this is that gamblers facing a loss for the day and hoping to recover that loss before it is too late, will most often place bets on nonfavorites with a high payoff.

The fact that people do not choose their number sequences randomly decreases the probability that they will be the only winner, should they be lucky enough to win. On January 14, 1995, the UK National Lottery had a record number of 133 winners sharing a jackpot of £16,293,830. In this lottery, six numbers must be ticked from the numbers $1, \ldots, 49$. Before the drawing in question, players had filled in 69.8 million tickets, the vast majority by hand. Had all of the tickets been filled in randomly, the probability of 133 or more winners would be somewhere on the order of 10^{-136} (verify this using the Poisson

[5]In the case of the majority of tickets being required to be filled in by hand, intelligent number choices can be found in N. Henze and H. Riedwyl's *How to Win More*, A. K. Peters, Massachusetts, 1998. These choices do not increase one's chances of winning, but they do increase the expected payoff for someone who is lucky enough to win.

distribution with an expected value of $69{,}800{,}000/\binom{49}{6} = 4.99$). This inconceivably small probability indicates again that people do not choose their lottery numbers randomly. The winning sequence of the draw on Saturday January 14, 1995 was 7-17-23-32-38-42. The popularity of this sequence may be explained from the fact that the numbers 17, 32, and 42 were winning numbers in the draw two weeks before the draw on January 14, 1995.

Ticking the number sequences 1-2-3-4-5-6 or 7-14-21-28-35-42 is about the most foolish thing you can do in a lottery. In the improbable case that those six numbers actually come up winners, you can be quite sure that a massive number of players will be sharing the jackpot with you. This is what happened on June 18, 1983, in the Illinois Lotto Game, when 78 players won the jackpot with the number sequence 7-13-14-21-28-35. Today, most lotto games offer players "Quick Pick" or "Easy Pick" opportunities to choose their numbers randomly with the aid of a computer.[6] As the percentage of plays using random play grows, the number of winners becomes more predictable. The game then becomes less volatile and exorbitantly high jackpots are seen less frequently. As more handwritten tickets are entered into a lottery, the probability of a rollover of the jackpot will get larger. This is plausible if one considers the extreme case of all players choosing the same six numbers. In such an extreme case, the jackpot will never be won. Lottery officials are, to a certain point, not unhappy to see rollovers of the jackpot, as it will naturally be accompanied by increased ticket sales!

As a final point, we will apply the hypergeometric distribution to Question 7 from Chapter 1.

A coincidence problem

The coincidence problem presented in Question 7 of Chapter 1 can be solved according to the hypergeometric model. A bit of imagination will show that this problem reflects the hypergeometric model with a drum containing $R = 500$ red marbles and $W = 999{,}498$ white marbles, from which $n = 500$ marbles will be chosen. The probability we are looking for to solve our coincidence problem is equal to the probability of at least one red marble being drawn. This probability is equal to 0.2214. Stated in other terms, there is approximately a 22% probability that the two people in question have a common acquaintance. This answer, together with the answer to Question 1 of Chapter 1, remind us that events are often less "coincidental" than we may tend to think.

[6] Approximately 70% of the tickets entered in the Powerball Lottery are nowadays Easy Picks. Players of the German Lotto, by contrast, chose no more than 4% of their number sequences by computer.

4.4 Problems

4.1 During World War II, London was heavily bombed by V-2 guided ballistic rockets. These rockets, luckily, were not particularly accurate at hitting targets. The number of direct hits in the southern section of London has been analyzed by splitting the area up into 576 sectors measuring one quarter of a square kilometer each. The average number of direct hits per sector was 0.9323. The relative frequency of the number of sectors with k direct hits is determined for $k = 0, 1, \ldots$. In your opinion, which distribution is applicable in determining the frequency distribution of the number of direct hits? Is it a Poisson distribution or a geometric distribution?

4.2 What are the probabilities of getting at least one six in one throw of six dice, at least two sixes in one throw of twelve dice, and at least three sixes in one throw of eighteen dice? [7] Do you think these probabilities are the same?

4.3 In an attempt to increase his market share, the maker of Aha Cola has formulated an advertising campaign to be released during the upcoming European soccer championship. The image of an orange ball has been imprinted on the underside of approximately one out of every one thousand cola can pop-tops. Anyone turning in such a pop-top on or before a certain date will receive a free ticket for the soccer tournament finale. Fifteen hundred cans of cola have been purchased for a school party, and all fifteen hundred cans will be consumed on the evening in question. What is the probability that the school party will deliver one or more free tickets?

4.4 A game of chance played historically by Canadian Indians involved throwing eight flat beans into the air and seeing how they fell. The beans were symmetrical and were painted white on one side and black on the other. The bean thrower would win one point if an odd number of beans came up white, two points if either zero or eight white beans came up, and would lose one point for any other configurations. Does the bean-thrower have the advantage in this game?

4.5 One hundred and twenty-five mutual funds have agreed to take part in an elimination competition being sponsored by Four Leaf Clover investment magazine. The competition will last for two years and will consist of seven rounds. At the beginning of each quarter, each fund remaining in the competition will put together a holdings portfolio. Funds will go through to the next round if, at the end of the quarter, they have performed

[7] In a letter dated November 22, 1693, the gambler Samuel Pepys posed this question to Isaac Newton. It was not a trivial question for Newton.

above the market average. Funds finishing at or below market average will be eliminated from the competition. We can assume that the funds' successive quarterly performances are independent from one another and that there is a probability of $\frac{1}{2}$ that a fund will perform above average during any given quarter. Calculate the probability that at least one fund will come through all seven rounds successfully. Calculate the probability that three or more funds will come through all seven rounds.

4.6 In 1989, American investment publication *Money Magazine* assessed the performance of 277 important mutual funds over the previous ten years. For each of those ten years they looked at which mutual funds performed better than the S&P (Standard & Poor's) index. Research showed that five of the 277 funds performed better than the S&P index for eight or more years. Verify that the expected value of the number of funds performing better than the S&P index for eight years or more is equal to 15.2, when the investment portfolios of each fund have been compiled by a blindfolded monkey throwing darts at the Wall Street Journal. Assume that each annual portfolio has a 50% probability of performing better than the S&P index.

4.7 What is the fewest number of dice one can roll such that, when they are rolled simultaneously, there will be at least a 50% probability of rolling two or more sixes?

4.8 A military early-warning installation is constructed in a desert. The installation consists of five main detectors and a number of reserve detectors. If fewer than five detectors are working, the installation ceases to function. Every two months, an inspection of the installation is mounted and at that time all detectors are replaced by new ones. There is a probability of 0.05 that any given detector will cease to function during the period between inspections. The detectors function independently of one another. How many reserve detectors are needed to ensure a probability of less than 0.1% that the system will cease to function between inspections?

4.9 In the World Series Baseball final, two teams play a series consisting of a possible seven games until such time that one of the two teams has won four games. In one such final, two unevenly matched teams are pitted against each other and the probability that the weaker team will win any given game is equal to 0.45. Taking into consideration that the results of the various games are independent from each other, calculate the probability of the weaker team winning the final.

4.10 Operating from within a tax-haven, some quick-witted businessmen have started an Internet Web site called Stockgamble. Through this Web site, interested parties can play the stock markets in a number of countries. Each of the participating stock markets lists 24 stocks available in their country.

The game is played on a daily basis, and for each market the six stocks that have performed the best are noted at the end of each day. Participants each choose a market and click on six of the 24 stocks available. The minimum stake is \$5 and the maximum stake is \$1,000. The payoff is 100 times the stake if all six of the top performing stocks have been clicked on. What would the expected pay-off be per dollar staked if this "game of skill" was purely a game of chance?

4.11 In a game called "26," a player chooses one number from the numbers $1, \ldots, 6$ as point number. After this, the player rolls a collection of ten dice thirteen times in succession. If the player's point number comes up 26 times or more, the player receives five times the amount staked on the game. Is this game to the player's advantage?

4.12 Daily Airlines flies every day from Amsterdam to London. The price for a ticket on this popular route is \$75. The aircraft has a capacity of 150 passengers. Demand for tickets is greater than capacity, and tickets are sold out well in advance of flight departures. The airline company sells 160 tickets for each flight to protect itself against no-show passengers. The probability of a passenger being a no-show is $q = 0.1$. No-show passengers are refunded half the price of their tickets. Passengers that do show up and are not able to board the flight due to the overbooking are refunded the full amount of their tickets plus an extra \$425 compensation. What is the average daily return for the airline? What is the probability that more passengers will turn up for a flight than the aircraft has the seating capacity for?

4.13 Decco is played with an ordinary deck of 52 playing cards. It costs \$1 to play this game. Having purchased a ticket on which the 52 playing cards of an ordinary deck are represented, each player ticks his choice of one card from each of the four suits (the ten of hearts, jack of clubs, two of spades, and ace of diamonds, for example). On the corresponding television show, broadcast live, one card is chosen randomly from each of the four suits. If the four cards chosen by a player on his/her ticket are the same as the four chosen on the show, the player wins \$5,000. A player having three of the four cards correct wins \$50. Two correct cards result in a win of \$5. One correct card wins the player a free playing ticket for the next draw. What is the house percentage of this exciting game?

4.14 Consider an experiment with three possible outcomes 1,2, and 3, which occur with respective probabilities of p_1, p_2, and $p_3 = 1 - p_1 - p_2$. For a given value of n, n independent trials of the experiment are performed. The random variable X_i gives the number of times that the outcome i

occurs for $i = 1,2,3$. Verify that

$$P(X_1 = k_1, X_2 = k_2) = \binom{n}{k_1}\binom{n-k_1}{k_2} p_1^{k_1} p_2^{k_2} p_3^{n-k_1-k_2}$$

for all $k_1, k_2 \geq 0$ with $k_1 + k_2 \leq n$. This distribution is called the *multinomial distribution* (with $r = 3$ possible outcomes).

4.15 For the upcoming drawing of the Telenet Lottery, five extra prizes have been added to the pot. Each prize consists of an all-expenses paid vacation trip. The five winners of the extra prize may choose from among three possible destinations $i = 1$, 2, and 3. The winners choose independently of each other. The probability that a given winner chooses destination i is equal to p_i with $p_1 = 0.5$, $p_2 = 0.3$, and $p_3 = 0.2$. What is the probability that both destination 2 and destination 3 will be chosen? What is the probability that not all three destinations will be chosen?

4.16 A particular scratch-lottery ticket has 16 painted boxes on it, each box having one of the numbers 1, 2, 5, 10, 100, or 1,000 hidden under the paint. When the paint is scratched off and it appears that a same number shows up in seven or more boxes, the player wins an amount equal to that number multiplied by the purchasing price of the ticket. It is understood that, in cases where more than one number appears seven or more times, the higher number will serve as the winner. The preprinted number assortments are established randomly in accordance with the premise that on average 25% of the numbers will be a 1, 20% of the numbers will be a 2, another 20% will be a 5, 15% will be a 10, 10% will be a 100, and 10% will be a 1,000. Use the multinomial distribution to determine the house percentage in this game.

4.17 A particular game is played with five poker dice. Each die displays an ace, king, queen, jack, ten, and nine. Players may bet on two of the six images displayed. When the dice are thrown and the bet-on images turn up, the player receives three times the amount wagered. In all other cases, the amount of the wager is forfeited. Is this game advantageous for the player?

4.18 **(a)** A die is rolled until a six appears for the third time. What is the probability distribution of the number of rolls required?

(b) Independent trials of a Bernouilli experiment with success probability of p are done until a success occurs for the rth time. What is the probability distribution of the number of trials required? *Remark*: this distribution is known as the *negative binomial distribution*.

4.19 The keeper of a certain king's treasure receives the task of filling each of 100 urns with 100 gold coins. While fulfilling this task, he substitutes one lead coin for one gold coin in each urn. The king suspects deceit on the part of the sentry and has two methods at his disposal of auditing the contents of the urns. The first method consists of randomly choosing one coin from each of the 100 urns. The second method consists of randomly choosing four coins from each one of 25 of the 100 urns. Which method provides the largest probability of uncovering the deceit?

4.20 On bridge night, the cards are dealt round seven times. Only two times do you receive an ace. From the beginning, you had your doubts as to whether the cards were being shuffled thoroughly. Are these doubts confirmed?

4.21 In the famous problem of Chevalier de Méré, players bet first on the probability that a six will turn up at least one time in four rolls of a fair die; subsequently, players bet on the probability that a double six will turn up in 24 rolls of a fair pair of dice. In a generalized version of the de Méré problem, the dice are rolled a total of $4 \times 6^{r-1}$ times; each individual roll consists of r fair dice being rolled simultaneously. A king's roll results in all of the r dice rolled turning up sixes. Demonstrate how, in the generalized de Méré problem, the probability of at least one king's roll converges to $1 - e^{-2/3} = 0.4866$ if $r \to \infty$.

4.22 In the kingdom of Lightstone, the game of Lotto 6/42 is played. In Lotto 6/42, six numbers out of the numbers $1, \ldots, 42$ are drawn. At the time of an oil sheik's visit to Lightstone, the jackpot for the next drawing is listed at 27.5 million dollars. The oil sheik decides to take a gamble and orders his retinue to fill in 15 million tickets in his name. These 15 million tickets do not have to be filled in by hand; rather a Lotto computer fills them in by randomly generating 15 million sequences of six distinct numbers (note that this manner of "random picks" allows for the possibility of the same sequence being generated more than once). Suppose that the local people have purchased ten million tickets for the same jackpot, and assume that the sequences for these tickets are also the result of "random picks." Each ticket costs $1.

(a) What is the probability that the oil sheik will win the jackpot and what is the probability that the oil sheik will be the only winner?

(b) What is the probability that the oil sheik will win back his initial outlay?

4.23 The Brederode Finance Corporation has begun the following advertising campaign in Holland. Each new loan application submitted is accompanied by a chance to win a prize of $25,000. Every month, 100 zip codes will be drawn in a lottery. In Holland, each house address has a unique zip code and there are about 2,500,000 zip codes. Each serious applicant whose zip code is drawn will receive a \$25,000 prize. Considering that Brederode Finance Corporation receives 200 serious loan applications each month, calculate the distribution of the monthly amount that they will have to give away.

4.24 You are at an assembly where 500 other persons are also present. The organizers of the assembly are raffling off a prize to be shared by all of those present whose birthday falls on that particular day. What is the probability that you will win the prize?

4.25 An organization running the Lotto 6/45 analyzes 100,000 tickets that were filled-in by hand. On each ticket of the Lotto 6/45, six different numbers from the numbers $1, \ldots, 45$ are filled in. A particular pick of six numbers occurred eight times in the one-hundred thousand tickets. What is the probability of the same sequence of six numbers turning up eight or more times in the 100,000 tickets? Assuming that the tickets are randomly filled in, calculate a Poisson approximation for this variant of the birthday probability.

4.26 In the Massachusetts Numbers Game, one number is drawn each day from the 10,000 four-digit number sequence $0000, 0001, \ldots, 9999$. Calculate a Poisson approximation for the probability that the same number will be chosen two or more times in the upcoming 625 drawings. Before making the calculations in this variant of the birthday problem, can you say why this probability cannot be negligibly small?

4.27 What is a Poisson approximation for the probability that in a randomly selected group of 25 persons, three or more will have birthdays on a same day. What is a Poisson approximation for the probability that three or more persons from the group will have birthdays falling within one day of each other? Can you think of an alternative Poisson approximation?

4.28 Ten married couples are invited to a bridge party. Bridge partners are chosen at random, without regard to gender. What is the probability of at least one couple being paired as bridge partners? Calculate a Poisson approximation for this probability.

4.29 A group of 25 students is going on a study trip of 14 days. Calculate a Poisson approximation for the probability that during this trip two or more students from the group will have birthdays on the same day.

4.30 Three people each write down the numbers $1, \ldots, 10$ in a random order. Calculate a Poisson approximation for the probability that the three people all have one number in the same position.

4.31 What is the probability of two consecutive numbers appearing in any given lotto drawing of six numbers from the numbers $1, \ldots, 45$? Calculate a Poisson approximation for this probability. Also calculate a Poisson approximation for the probability of three consecutive numbers appearing in any given drawing of the Lotto 6/45.

4.32 Calculate a Poisson approximation for the probability that in a randomly selected group of 2,287 persons all of the 365 possible birthdays will be represented.

4.33 Sixteen teams remain in a soccer tournament. A drawing of lots will determine which eight matches will be played. Before the drawing takes place, it is possible to place bets with bookmakers over the outcome of the drawing. You are asked to predict all eight matches, paying no regard to the order of the two teams in each match. Calculate a Poisson distribution for the number of correctly predicted matches.

4.34 Calculate a Poisson approximation for the probability that in a thoroughly shuffled deck of 52 playing cards, it will occur at least one time that two cards of the same face value will succeed one another in the deck (two aces, for example). In addition, make the same calculation for the probability of three cards of the same face value succeeding one another in the deck.

4.35 A company has 75 employees in service. The administrator of the company notices, to his astonishment, that there are seven days on which two or more employees have birthdays. Verify, by using a Poisson approximation, whether this is so astonishing after all.

4.36 Argue that the following two problems are manifestations of the 'hat-check' problem:

(a) In a particular branch of a company, the fifteen employees have agreed that, for the upcoming Christmas party, each employee will bring one present without putting any name on it. The presents will be distributed blindly among them during the party. What is the probability of not one person ending up with his/her own present?

(b) A certain person is taking part in a blind taste test of ten different wines. The person has been made aware of the names of the ten wine producers beforehand, but does not know what order the wines will be served in. He may only name a wine producer one time. After the tasting session is over, it turns out that he has correctly identified five of the ten wines. Do you think he is a connoisseur?

4.37 A businessman parks his car illegally for one hour, twice a day, along the banks of an Amsterdam canal. During the course of an ordinary day, parking attendants monitor the streets according to a Poisson process with an average of α rounds per hour. What is the probability that the businessman will be ticketed and fined on any given day?

4.38 In the first five months of the year 2000, the tram hit and killed seven pedestrians in Amsterdam, each case caused by the pedestrian's own carelessness. In preceding years, such accidents occurred at a rate of 3.7 times per year. Simulate a Poisson process to estimate the probability that within a period of ten years, a block of five months will occur during which seven or more fatal tram accidents happen (you can simplify the problem by assuming that all months have the same number of days). Would you say that the disproportionately large number of fatal tram accidents in the year 2000 is the result of bad luck or would you categorize it in other terms?

4.39 Calls arrive at a computer controlled exchange according to a Poisson process at a rate of two calls per second. Use computer simulation to find the probability that during the busy hour there will be some period of 30 seconds in which 90 or more calls arrive.

4.40 During the course of a summer day, tourist buses come and go in the picturesque town of Edam according to a Poisson process with an average arrival rate of five buses per hour. Each bus stays approximately two hours in Edam, which is famous for its cheese. What is the distribution of the number of buses to be found in Edam at 4 P.M.? What would the answer be if each bus stayed one hour with a probability of $\frac{1}{4}$ and two hours with a probability of $\frac{3}{4}$?

4.41 Paying customers (i.e., those who park legally) arrive at a large parking lot according to a Poisson process with an average of 45 automobiles per hour. Independently of this, nonpaying customers (i.e., those who park illegally) arrive at the parking lot according to a Poisson process with an average of five automobiles per hour. The length of parking time has the same distribution for legal as for illegal parking customers. At a given moment in time, there are 75 automobiles parked in the parking lot. What is the probability that fifteen or more of these 75 automobiles are parked illegally?

4.42 Suppose that emergency response units are distributed throughout a large area according to a two-dimensional Poisson process. That is, the number of response units in any given bounded region has a Poisson distribution whose expected value is proportional to the area of the region, and the numbers of response units in disjoint regions are independent. An incident occurs at some arbitrary point. Argue that the probability of having at least one response unit within a distance r is $1 - e^{-\alpha\pi r^2}$ for some constant $\alpha > 0$ (this probability distribution is called the Rayleigh distribution). More generally, letting $R_0 = 0$ and denoting by R_i the distance from the point of incident to the ith nearest response unit, argue that $R_i^2 - R_{i-1}^2$ for $i = 1, 2, \ldots$ are independent random variables each having an exponential distribution with expected value $\frac{1}{\pi\alpha}$.

4.43 In a blind drawing, you pick ten distinct numbers from the numbers $1, \ldots, 100$. What is the probability that five of the numbers you pick will be less than or equal to 50 and the other five will be greater than 50?

4.44 Ten identical pairs of shoes are jumbled together in one large box. Without looking, someone picks four shoes out of the box. What is the probability that, among the four shoes chosen, there will be both a left and a right shoe?

4.45 A psychologist claims that he can determine from a person's handwriting whether the person is left-handed or not. You do not believe the psychologist and therefore present him with 50 handwriting samples, of which 25 were written by left-handed people and 25 were written by right-handed people. You ask the psychologist to say which 25 were written by left-handed people. Will you change your opinion of him if the psychologist correctly identifies 18 of the 25 left-handers?

Ticked	Numbers correct	You win	Chance of winning
10 numbers	10/10	$2,000,000	4.66×10^{-7}
	9/10	$50,000	2.12×10^{-5}
	8/10	$2,000	3.89×10^{-4}
	7/10	$500	3.83×10^{-3}
	6/10	$50	2.25×10^{-2}
9 numbers	9/9	$1,000,000	2.58×10^{-6}
	8/9	$10,000	9.69×10^{-5}
	7/9	$1,000	1.46×10^{-3}
	6/9	$50	1.17×10^{-2}
8 numbers	8/8	$250,000	1.33×10^{-5}
	7/8	$5,000	4.11×10^{-4}
	6/8	$100	5.03×10^{-3}
7 numbers	7/7	$50,000	6.47×10^{-5}
	6/7	$500	1.62×10^{-3}
	5/7	$100	1.58×10^{-2}
6 numbers	6/6	$10,000	2.96×10^{-4}
	5/6	$500	5.91×10^{-3}
5 numbers	5/5	$2,500	1.28×10^{-3}
	4/5	$100	2.00×10^{-2}
4 numbers	4/4	$500	5.28×10^{-3}
	3/4	$50	6.22×10^{-2}

4.46 Keno, a game of Chinese origin, is wildly popular in many countries. The game has many variants. In France, it is played with seventy numbers. The numbers 1 through 70 are printed on the ticket, and players may tick

4, 5, 6, 7, 8, 9, or 10 of these numbers. Each ticket costs \$10. Drawings for the game occur daily, and during these drawings 20 numbers from 1 through 70 are randomly selected. In the table, all of the various winning combinations are relative pay-offs are listed. The probabilities of the various combinations occurring are also listed. Verify the correctness of these probabilities with the help of the hypergeometric model. What is the expected pay-off per \$10 staked if you tick four numbers? What is the answer if you tick ten numbers?

4.47 Take another look at the lottery problem in Section 3.5. If we divide the lottery numbers 1, . . . , 366 into three equal groups, then we can see from Table 3.3 that 17 or more days in December fall into the first group of low numbers 1, . . . , 122. In a fair drawing, what would be the probability of 17 or more days in December falling into the first group?

4.48 Computer chips are made according to two different production processes. The probability is the same for both processes that a finished chip will be rejected. A test sample of the first production process includes 100 chips and a test sample from the second process includes 50 chips. In total, 30 chips are identified as being defective. What is the distribution of the number of defective chips in the test sample from the first production process?

4.49 In the German "Lotto am Samstag," six regular numbers and one reserve number are drawn from the numbers 1, . . . , 49. On the lottery ticket, players must tick six different numbers out of the numbers 1, . . . , 49. There is also an area of the lottery ticket reserved for something known as the Super Number. This is a number chosen from the sequence 0, 1, . . . , 9. So, in addition to drawing the regular and reserve numbers, a Super Number between 0, 1, . . . , 9 is also drawn in Lotto am Samstag. The Super Number only comes into play in combination with six correctly chosen regular numbers. The eight winning combinations are: six regular numbers correct + Super Number (jackpot), six regular numbers correct, five regular numbers correct and the reserve number, four regular numbers correct, three regular numbers correct + the reserve number, and three regular numbers correct.

(a) Calculate the probability of winning on one ticket for each of these combinations.

(b) You purchase twelve tickets every week for the German Lotto am Samstag. How many years will you need in order to have at least a 50% chance of ever winning the jackpot in your lifetime?

4.50 In Lottoland, there is a daily lottery in which six (standard) numbers plus one bonus number are drawn from the numbers 1, . . . , 45. In addition to this, one color is randomly drawn out of six colors. On the lottery ticket, six numbers and one color must be chosen. The players use the computer for a random selection of the six numbers and the color. Each ticket costs $1.50. The number of tickets sold is about the same each week. The prizes are allotted as shown in the table. The jackpot begins with 4 million dollars; this is augmented each week by another half million dollars if the jackpot is not won. The lottery does not publish information regarding ticket sales and intake, but does publish a weekly listing in the newspaper of the number of winners for each of the six top prizes. The top six prizes from the table had 2, 10, 14, 64, 676, and 3,784 winners over the last 50 drawings.

6 + color	Jackpot*
6	$1 million*
5 + bonus number + color	$250,000*
5 + bonus number	$150,000*
5 + color	$2,500
5	$1,000
4 + bonus number + color	$375
4 + bonus number	$250
4 + color	$37.50
4	$25
3 + bonus number + color	$15
3 + bonus number	$10
3 + color	$7.50
3	$5

* Prize is divided by multiple winners.

(a) Estimate the amount of the weekly intake.
(b) Estimate the average number of weeks between jackpots being won and estimate the average size of the jackpot when it is won.
(c) Estimate the percentage of the intake that gets paid out as prize money.

5

Probability and statistics

Chapter 2 was devoted to the law of large numbers. This law tells you that you may estimate the probability of a given event A in a chance experiment by simulating many independent repetitions of the experiment. Then, the probability $P(A)$ is estimated by the proportion of trials in which the event A occurred. This estimate has an error. No matter how many repetitions of the experiment are performed, the law of large numbers will not tell you exactly how close the estimate is to the true value of the probability $P(A)$. How to quantify the error? For that purpose, you can use standard tools from statistics. Note that simulation is analogous to a sampling experiment in statistics. An important concept in dealing with sample data is the central limit theorem. This theorem states that the histogram of the data will approach a bell-shaped curve when the number of observations becomes very large. The central limit theorem is the basis for constructing confidence intervals for simulation results. The confidence interval provides a probability statement about the magnitude of the error of the sample average. A confidence interval is useful not only in the context of simulation experiments, but in situations that also crop up in our daily lives. Consider the example of estimating the unknown percentage of the voting population that will vote for a particular political party. Such an estimate can be made by doing a random sampling of the voting population at large. Finding a confidence interval for the estimate is then essential: this is what allows you to judge how confident one might be about the prediction of the opinion poll.

The concepts of normal curve and standard deviation are at the center of the central limit theorem. The normal curve is a bell-shaped curve that appears in numerous applications of probability theory. It is a sort of universal curve for displaying probability mass. The normal curve is symmetric around the expected value of the underlying probability distribution. The peakedness of the curve is measured in terms of the standard deviation of the probability distribution. The standard deviation is a measure for the spread of a random variable around its expected value. It says something about how likely certain deviations from the expected value are. When independent random variables each having the same distribution are averaged together, the standard deviation is reduced according to the square root law. This law is at the heart of the central limit theorem.

The concept of standard deviation is of great importance in itself. In finance, standard deviation is a key concept and is used to measure the volatility (risk) of investment returns and stock returns. It is common wisdom in finance that diversification of a portfolio of stocks generally reduces the total risk exposure of the investment. In the situation of similar but independent stocks, the volatility of the portfolio is reduced according to the square root of the number of stocks in the portfolio. The square root law also provides useful insight in inventory control. Aggregation of independent demands at similar retail outlets by replacing the outlets with a single large outlet reduces the total required safety stock. The safety stock needed to protect against random fluctuations in the demand is then reduced according to the square root of the number of retail outlets aggregated.

In the upcoming sections, we take a look at the normal curve and standard deviation before the central limit theorem and its application to confidence intervals.

5.1 The normal curve

In many practical situations, histograms of measurements approximately follow a bell-shaped curve. A histogram is a bar chart that divides the range of values covered by the measurements into intervals of the same width, and shows the proportion of the measurements in each interval. For example, let's say you have the height measurements of a very large number of Dutch men between 20 and 30 years of age. To make a histogram, you break up the range of values covered by the measurements into a number of disjoint adjacent intervals each having the same width, say width Δ. Let f_j be the proportion of the measurements that are in the interval $[j\Delta, (j+1)\Delta]$. Then define the function $f_\Delta(x)$ to be equal to f_j for $j\Delta \leq x < (j+1)\Delta$. The step function $f_\Delta(x)$ determines the graph of the histogram (see Figure 5.1). Making the width Δ of the intervals smaller and smaller, the graph of the histogram will begin to look more and more like the bell-shaped curve shown in Figure 5.2.

The bell-shaped curve in Figure 5.2 can be described by a function $f(x)$ of the form

$$f(x) = \frac{1}{\sigma\sqrt{2\pi}} e^{-\frac{1}{2}(x-\mu)^2/\sigma^2}.$$

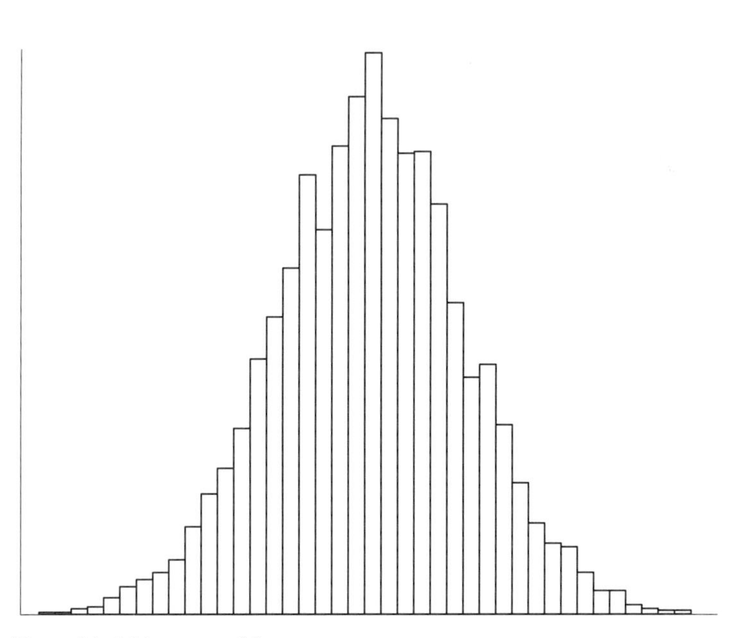

Figure 5.1 A histogram of data.

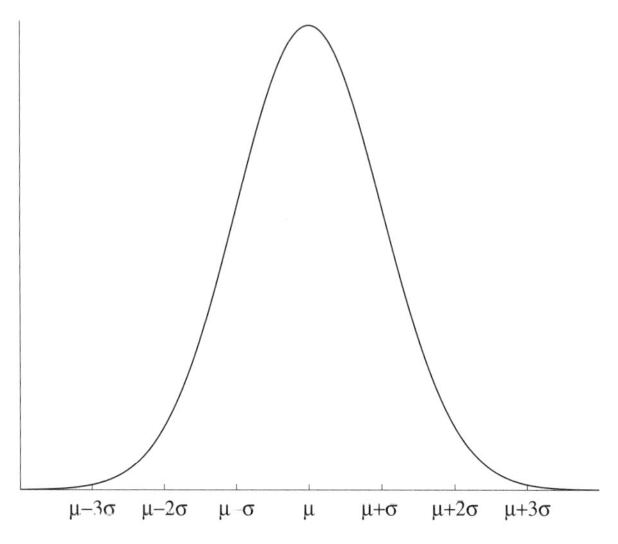

Figure 5.2 The normal curve.

This function is defined on the real line and has two parameters μ and σ, where μ (the location parameter) is a real number and σ (the shape parameter) is a positive real number. The characteristic bell-shaped curve in Figure 5.2 is called the *normal curve*. It is also known as the Gaussian curve (of errors), after the famous mathematician/astronomer Carl Friedrich Gauss (1777–1855), who showed in a paper from 1809 that this bell curve is applicable with regard to the accidental errors that occur in the taking of astronomical measurements. It is usual to attribute the discovery of the normal curve to Gauss. However, the normal curve was discovered by the mathematician Abraham de Moivre (1667–1754) around 1730 when solving problems connected with games of chance. The pamphlet "Approximato ad Summani Terminorum Binomi $(a+b)^n$ in Seriem Expansis" containing this discovery was first made public in 1738 in the second edition of De Moivre's masterwork *Doctrine of Chance*. Also a publication of Pierre Simon Laplace (1749–1829) from 1778 contains the normal curve function and emphasizes its importance. De Moivre anticipated Laplace and the latter anticipated Gauss. One could say that the normal curve is a natural law of sorts, and it is worth noting that each of the three famous mathematical constants $\sqrt{2}$, $\pi = 3.141\ldots$ and $e = 2.718\ldots$ play roles in its makeup. Many natural phenomena, such as the height of men, harvest yields, errors in physical measurements, luminosity of stars, returns on stocks, etc., can be described by a normal curve. The Belgian astronomer and statistician Adolphe Quetelet (1796–1894) was the first to recognize the universality of the normal curve and he fitted it to a large collection of data taken from all corners of science, including economics and the social sciences. Many in the eighteenth and nineteenth centuries considered the normal curve a God-given law. The universality of the bell-shaped Gaussian curve explains the popular use of the name normal curve for it. Later on in the text we shall present a mathematical explanation of the frequent occurrence of the normal curve with the help of the central limit theorem. But first we will give a few notable facts about the normal curve. It has a peak at the point $x = \mu$ (the height of the peak is $(\sigma\sqrt{2\pi})^{-1}$) and is symmetric around this point. Second, the total area under the curve is 1. Of the total area under the curve, approximately 68% is concentrated between points $\mu - \sigma$ and $\mu + \sigma$, and approximately 95% is concentrated between $\mu - 2\sigma$ and $\mu + 2\sigma$. Nearly the entire area is concentrated between points $\mu - 3\sigma$ and $\mu + 3\sigma$. For example, if the height of a certain person belonging to a particular group is normally distributed with parameters μ and σ, then it would be exceptional for another person from that same group to measure in at a height outside of the interval $(\mu - 3\sigma, \mu + 3\sigma)$.

5.1.1 Probability density function

Before giving further properties of the normal curve, it is helpful, informally, to discuss the concept of a probability density function. The function $f(x)$ describing the normal curve is an example of a probability density function. Any nonnegative function for which the total area under the graph of the function equals 1 is called a *probability density function*. In Chapter 10, the concept of probability density function will be discussed in detail. A density function can be seen as a limiting case of a probability histogram when the width of the base intervals of the histogram tends to zero. Any probability density function underlies a so-called *continuous random variable*. The random variable describing the height of a randomly chosen person is an example of a continuous random variable if it is assumed that the height can be measured in infinite precision. Another example of a continuous random variable is the annual rainfall in a certain area. Taking these examples in mind, you may accept the fact that a continuous random variable X cannot be defined by assigning probabilities to individual values. For any number a, the probability that X takes on the value a is 0. Instead, a continuous random variable is described by assigning probabilities to intervals via a probability density function. In Chapter 10, it will be proved that the probability $P(a \leq X \leq b)$, being the probability that the continuous random variable X takes on a value between a and b, satisfies

$$P(a \leq X \leq b) = \text{the area under the graph of the density function } f(x) \text{ between points } a \text{ and } b$$

for any real numbers a and b when $f(x)$ is the probability density function of X. Readers who are familiar with integral calculus will recognize the area under the graph of $f(x)$ between a and b as the integral of $f(x)$ from a to b. Mathematically,

$$P(a \leq X \leq b) = \int_a^b f(x)\,dx.$$

Any introductory course in integral calculus shows that the area under the graph of $f(x)$ between a and b can be approximated through the sum of the areas of small rectangles by dividing the interval $[a, b]$ into narrow subintervals of equal width. In particular, taking $a = v$ and $b = v + \Delta$ for Δ small, the area under the graph of $f(x)$ between v and $v + \Delta$ is approximately equal to $f(v)\Delta$ when Δ is small enough. In other words, $f(v)\Delta$ is approximately equal to the probability that the random variable X takes on a value in a small interval around v of width Δ. Because $f(x)\Delta \approx P(x \leq X \leq x + \Delta)$ for Δ small, it is reasonable to define the *expected value* of a continuous random variable X by

$$E(X) = \int_{-\infty}^{\infty} x f(x)\,dx.$$

This definition parallels the definition $E(X) = \sum_x x P(X = x)$ for a discrete random variable X.

5.1.2 Normal density function

A continuous random variable X is said to have a *normal distribution* with parameters μ and σ if

$$P(a \leq X \leq b) = \frac{1}{\sigma\sqrt{2\pi}} \int_a^b e^{-\frac{1}{2}(x-\mu)^2/\sigma^2}\,dx$$

for any real numbers a and b with $a \leq b$. The corresponding normal density function is given by

$$f(x) = \frac{1}{\sigma\sqrt{2\pi}} e^{-\frac{1}{2}(x-\mu)^2/\sigma^2} \quad \text{for } -\infty < x < \infty.$$

The notation X is $N(\mu, \sigma^2)$ is often used as a shorthand for X is normally distributed with parameters μ and σ. Theoretically, a normally distributed random variable has the whole real line as its range of possible values. However, a normal distribution can also be used for a nonnegative random variable provided that the normal distribution assigns a negligible probability to the negative axis.

In Chapter 14, it will be shown for an $N(\mu, \sigma^2)$ random variable X that

$$E(X) = \mu \qquad \text{and} \qquad E\left[(X - \mu)^2\right] = \sigma^2.$$

Thus, the parameter μ gives the expected value of X, and the parameter σ gives an indication of the spread of the random variable X around its expected value. In the next section, it will be seen that σ is the standard deviation of X.

An important result is:

if a random variable X is normally distributed with parameters μ and σ, then for each two constants $a \neq 0$ and b the random variable $U = aX + b$ is normally distributed with parameters $a\mu + b$ and $|a|\sigma$.

This result, being a special case of a more general result in Section 10.3, states that any linear combination of a normally distributed random variable X is again normally distributed. In particular, the random variable

$$Z = \frac{X - \mu}{\sigma}$$

is normally distributed with parameters 0 and 1. A normally distributed random variable Z with parameters 0 and 1 is said to have a *standard normal* distribution. The shorthand notation Z is $N(0, 1)$ is often used. The special notation

$$\Phi(z) = \frac{1}{\sqrt{2\pi}} \int_{-\infty}^{z} e^{-\frac{1}{2}x^2}\, dx$$

is used for the cumulative probability distribution function $P(Z \leq z)$ of Z. The derivative of $\Phi(z)$ is the standard normal density function which is given by

$$\phi(z) = \frac{1}{\sqrt{2\pi}} e^{-\frac{1}{2}z^2} \quad \text{for} \quad -\infty < z < \infty.$$

The quantity $\Phi(z)$ gives the area under the graph of the standard normal density function left from the point $x = z$. No closed form of the cumulative distribution function $\Phi(z)$ exists. In terms of calculations, the integral for $\Phi(z)$ looks terrifying, but mathematicians have shown that the integral can be approximated with extreme precision by the quotient of two suitably chosen polynomials. This means that in practice the calculation of $\Phi(x)$ for a given value of x presents no difficulties at all and can be accomplished very quickly.

All calculations for an $N(\mu, \sigma^2)$ distributed random variable X can be reduced to calculations for the $N(0, 1)$ distributed random variable Z by using the linear transformation $Z = (X - \mu)/\sigma$. Writing $P(X \leq a) = P((X - \mu)/\sigma \leq (a - \mu)/\sigma)$ and noting that $\Phi(z) = P(Z \leq z)$, it follows that

$$P(X \leq a) = \Phi\left(\frac{a - \mu}{\sigma}\right).$$

An extremely useful result is the following:

> **the probability that a normally distributed random variable will take on a value that lies $z > 0$ or more standard deviations above the expected value is equal to $1 - \Phi(z)$, as is the probability of a value that lies z or more standard deviations below the expected value.**

This important result is the basis for a rule of thumb that is much used in statistics when testing hypotheses (see Section 5.6). The proof of the result is easy. Letting Z denote the standard normal random variable, it holds that

$$P(X \geq \mu + z\sigma) = P\left(\frac{X - \mu}{\sigma} \geq z\right) = P(Z \geq z) = 1 - P(Z < z)$$
$$= 1 - \Phi(z).$$

The reader should note that $P(Z < z) = P(Z \leq z)$, because Z is a continuous random variable and so $P(Z = z) = 0$ for any value of z. Since the graph of the normal density function of X is symmetric around $x = \mu$, the area under this graph left from the point $\mu - z\sigma$ is equal to the area under the graph right from the point $\mu + z\sigma$. In other words, $P(X \leq \mu - z\sigma) = P(X \geq \mu + z\sigma)$. This completes the proof of the above result.

5.1.3 Percentiles

In applications of the normal distribution, percentiles are often used. For a fixed number p with $0 < p < 1$, the $100p\%$ *percentile* of a normally distributed random variable X is defined as the number x_p for which

$$P(X \leq x_p) = p.$$

In other words, the area under the graph of the normal density function of X left from the percentile point x_p is equal to p. The percentiles of the $N(\mu, \sigma^2)$ distribution can be expressed in terms of the percentiles of the $N(0, 1)$ distribution. The $100p\%$ percentile of the standard normal distribution is denoted as z_p and is thus the solution of the equation

$$\Phi(z_p) = p.$$

It is enough to tabulate the percentiles of the standard normal distribution. If the random variable X has an $N(\mu, \sigma^2)$ distribution, then it follows from

$$P(X \leq x_p) = P\left(\frac{X - \mu}{\sigma} \leq \frac{x_p - \mu}{\sigma}\right) = \Phi\left(\frac{x_p - \mu}{\sigma}\right)$$

that its $100p\%$ percentile x_p satisfies $(x_p - \mu)/\sigma = z_p$. Hence,

$$x_p = \mu + \sigma z_p.$$

A much used percentile of the standard normal distribution is the 95% percentile

$$z_{0.95} = 1.6449.$$

Let's illustrate the use of percentiles by means of the following example: of the people calling in for travel information, how long do 95% of them spend on the line with an agent when the average length of a telephone call is normally distributed with an expected value of four minutes and a standard deviation of half a minute? The 95% percentile of the call-conclusion time is $4 + 0.5 \times 1.6449 = 4.82$ minutes. In other words, 95% of the calls are concluded within 4.82 minutes.

In inventory control, the normal distribution is often used to model the demand distribution. Occasionally, one finds oneself asking experts in the field for educated guesses with regard to the expected value and standard deviation of the normal demand distribution. But even such experts often have difficulty with the concept of standard deviation. They can, however, provide an estimate (educated guess) for the average demand, and they can usually even estimate the threshold level of demand that will only be exceeded with a 5% chance, say. Let's say you receive an estimated value of 75 for this threshold, against an estimated value of 50 for the average level of demand. From this, you can immediately derive what the expected value μ and the standard deviation σ of the normal demand distribution are. Obviously, the expected value μ is 50. The standard deviation σ follows from the relationship $x_p = \mu + \sigma z_p$ with $x_p = 75$ and $z_p = 1.6449$. This gives $\sigma = 15.2$. The same idea of estimating μ and σ through an indirect approach may be useful in financial analysis. Let the random variable X represent the price of a stock next year. Suppose that an investor expresses his/her belief in the future stock price by assessing that there is a 25% probability of a stock price being below \$80 and a 25% probability of a stock price being above \$120. Estimates for the expected value μ and standard deviation σ of the stock price next year are then obtained from the equations $(80 - \mu)/\sigma = z_{0.25}$ and $(120 - \mu)/\sigma = z_{0.75}$, where $z_{0.25} = -0.67449$ and $z_{0.75} = 0.67449$. This leads to $\mu = 100$ and $\sigma = 5.45$.

5.2 The concept of standard deviation

The expected value of a random variable X is an important feature of this variable. Say, for instance, that the random variable X represents the winnings in a certain game. The law of large numbers teaches us that the average win per game will be equal to $E(X)$ when a very large number of independent repetitions are completed. However, the expected value reveals little about the value of X

in any one particular game. To illustrate, say that the random variable X takes on the two values 0 and 5,000 with corresponding probabilities 0.9 and 0.1. The expected value of the random variable is then 500, but this value tells us nothing about the value of X in any one game. The following example shows the danger of relying merely on average values in situations involving uncertainty.

5.2.1 Pitfalls for averages[1]

A retired gentleman would like to place \$100,000 in an investment fund in order to ensure funding for a variety of purposes over the coming 20 years. How much will he be able to draw out of the account at the end of each year such that the initial investment capital, which must remain in the fund for 20 years, will not be disturbed? In order to research this issue, our man contacts Legio Risk, a well-known investment fund corporation. The advisor with whom he speaks tells him that the average rate of return has been 14% for the past 20 years (the one-year rate of return on a risky asset is defined as the beginning price of the asset minus the end price divided by the beginning price). The advisor shows him that with a *fixed* yearly return of 14%, he could withdraw \$15,098 at the end of each of the coming 20 years given an initial investment sum of \$100,000 (one can arrive at this sum by solving x from the equation $(1+r)^{20}A - \sum_{k=0}^{19}(1+r)^k x = 0$ yielding $x = [r(1+r)^{20}A]/[(1+r)^{20} - 1]$ with $A = \$100{,}000$ and $r = 0.14$). This is music to the ears of our retired friend, and he decides to invest \$100,000 in the fund. His wife does not share his enthusiasm for the project, and cites Roman philosopher and statesman Pliny the Elder to support her case: the only certainty is that nothing is certain. Her husband ignores her concerns and says that there will be nothing to worry about so long as the average value of the rate of return remains at 14%. Can our retiree count on a yearly payoff of \$15,098 at the end of each of the coming 20 years if the rate of return fluctuates, from year to year, around 14% such that the average rate of return really is 14%? The answer is a resounding no! In this case, there is a relatively high chance of the capital being used up before the 20-year term is over (on the other hand, there is also a chance that after 20 years a hefty portion of the initial investment will still be left). In situations of uncertainty you cannot depend on average values. Statisticians like to tell the story of the man who begins walking across a particular lake, having ascertained beforehand that it is, on average, 30 centimeters deep. Suddenly, he encounters an area where the lake is approximately 3 meters deep, and, a nonswimmer, he

[1] This example is borrowed from Sam Savage in his article "The flaw of averages," October 8, 2000 in the *San Jose Mercury News*.

falls in and drowns. In Figure 5.3, we illustrate the consequences of uncertainty, using a simple probability model for the development of the rate of return. If last year the rate of return was $r\%$, then over the course of the coming year the rate of return will be $r\%$, $(1+f)r\%$, or $(1-f)r\%$ with respective probabilities p, $\frac{1}{2}(1-p)$, and $\frac{1}{2}(1-p)$. In this model, the expected value of the rate of return is the same for every year and is equal to the initial value of the rate of return. When we assume an initial investment capital of \$100,000 and the knowledge of a 14% rate of return for the year preceding the initial investment, we arrive at the data presented in Figure 5.3. This figure displays the distribution of the invested capital after 15 years, when at the end of each of those 15 years a total of \$15,098 is withdrawn from the fund (when there is less than \$15,098 in the fund, the entire amount remaining is withdrawn). The nonshaded distribution corresponds to $p = 0.8$ and $f = 0.1$, and the shaded distribution corresponds to $p = 0.5$ and $f = 0.2$. These distributions are calculated by simulation (4 million runs). The nonshaded distribution is less spread out than the shaded distribution. Can you explain this? In addition, the simulation reveals surprising pattern similarities between the random walk describing the course of the invested capital at the end of each year and the random walk describing

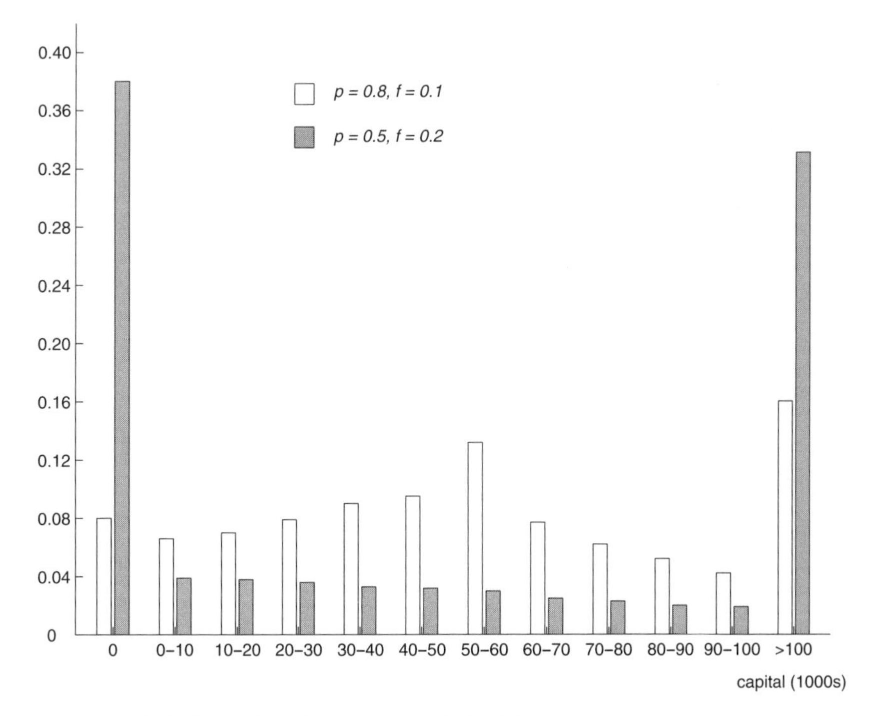

Figure 5.3 Distribution of invested capital after 15 years.

the difference between the number of times heads comes up and the number of times tails comes up in the experiment of the recurring coin toss (see Problem 5.11) and the arc-sine law in Section 2.1. The fair coin is a familiar figure in the world of finance!

5.2.2 Variance and standard deviation

Let X be any random variable with expected value

$$\mu = E(X).$$

A measure of the spread of the random variable X around the expected value μ is the variance. The *variance* of X is denoted and defined by

$$\sigma^2(X) = E\left[(X-\mu)^2\right].$$

Another common notation for the variance of X is var(X). Why not use $E(|X-\mu|)$ as the measuring gauge for the spread? The answer is simply that it is much easier to work with $E\left[(X-\mu)^2\right]$ than with $E(|X-\mu|)$. The variance $\sigma^2(X) = E\left[(X-\mu)^2\right]$ can also be seen in the famed Chebyshev's inequality:

$$P(|X-\mu| \geq a) \leq \frac{\sigma^2(X)}{a^2}$$

for every constant $a > 0$. This inequality is generally applicable regardless of what form the distribution of X takes. It can even be sharpened to

$$P\big(X > \mu + a\big) \leq \frac{\sigma^2(X)}{\sigma^2(X)+a^2} \quad \text{and} \quad P\big(X < \mu - a\big) \leq \frac{\sigma^2(X)}{\sigma^2(X)+a^2}$$

for every constant $a > 0$. This one-sided version of Chebyshev's inequality is a practical and useful result. In practical situations, it commonly occurs that only the expected value $E(X)$ and the variance $\sigma^2(X)$ of the distribution of X are known. In such situations, you can still establish an upper limit for a probability of the form $P\{X > \mu + a\}$ or $P\{X < \mu - a\}$. For example, imagine that X is the return on a certain investment and that you only know that the return has an expected value of 100 and a variance of 150. In this case, the probability of the return X taking on a value less than 80 will always be capped off at $150/(150+20^2) = 0.273$, regardless of what the distribution of X is.

The variance $\sigma^2(X)$ does not have the same dimension as the values of the random variable X. For example, if the values of X are expressed in dollars, then the dimension of $\sigma^2(X)$ will be equal to (dollars)^2. A measure for the spread that has the same dimension as the random variable X is the *standard deviation*. It is defined as

$$\sigma(X) = \sqrt{E\left[(X-\mu)^2\right]}.$$

Referring back to the distribution in Figure 5.3, the nonshaded distribution corresponding to the case of $p = 0.8$ and $f = 0.1$ has an expected value of approximately \$58,000 and a standard deviation of approximately \$47,000, whereas the shaded distribution corresponding to the case of $p = 0.5$ and $f = 0.2$ has an expected value of approximately \$142,000 and a standard deviation of approximately \$366,000. The results for the case of $(p = 0.5,\ f = 0.2)$ are quite surprising and go against intuitive thinking! The explanation lies in the sharp movement of the yearly rate of return. This comes out in a standard deviation of the capital after 15 years that is relatively large with regard to the expected value (so we get, for example, a strong 6% probability of an invested capital of more than \$500,000 after 15 years lining up right next to a probability of 38% that the capital will be depleted after 15 years).

In the field of investment, smaller standard deviations are considered to be highly preferable when the expected value remains stable. Nevertheless, it is not always wise to base decisions on expected value and standard deviation alone. Distributions having the same expected value and the same standard deviation may display strong differences in the tails of the distributions. We illustrate this in the following example: investment A has a 0.8 probability of a \$2,000 profit and a 0.2 probability of a \$3,000 loss. Investment B has a 0.2 probability of a \$5,000 profit and a 0.8 probability of a zero profit. The net profit is denoted by the random variable X for investment A and by the random variable Y for investment B. Then

$$E(X) = 2{,}000 \times 0.8 - 3{,}000 \times 0.2 = \$1{,}000$$

and

$$\sigma(X) = \sqrt{(2{,}000 - 1{,}000)^2 \times 0.8 + (-3{,}000 - 1{,}000)^2 \times 0.2} = \$2{,}000.$$

Similarly, $E(Y) = \$1{,}000$ and $\sigma(Y) = \$2{,}000$ (verify!). Hence both investments have the same expected value and the same standard deviation for the net profit. In this situation, it is important to know the entire distribution in order to choose wisely between the two investments.

We will now present a number of properties of the standard deviation.

Property 1. *For every two constants a and b,*

$$\sigma^2(aX + b) = a^2\sigma^2(X).$$

This property comes as the result of applying the definition of variance and using the fact that the expected value of a sum is the sum of the expected values, we leave its derivation to the reader. To illustrate Property 1, let's say that an investor has a portfolio half of which is made up of liquidities and half of equities. The liquidities show a fixed 4% return. The equities show an uncertain return with an expected value of 10% and a standard deviation of 25%. The return on the portfolio, then, has an expected value of $\frac{1}{2} \times 4\% + \frac{1}{2} \times 10\% = 7\%$ and a standard deviation of $\sqrt{\frac{1}{4} \times 625\%} = 12.5\%$.

In contrast to expected value, it is not always the case with variance that the variance of the sum of two random variables is equal to the sum of the variances of the two individual random variables. In order to give a formula for the variance of the sum of two random variables, we need the concept of covariance. The covariance of two random variables X and Y is denoted and defined by

$$\text{cov}(X,Y) = E\big[(X - E(X))(Y - E(Y))\big].$$

The value of $\text{cov}(X, Y)$ gives an indication of how closely connected the random variables X and Y are. If random variable Y tends to take on values smaller (larger) than $E(Y)$ whenever X takes on values larger (smaller) than $E(X)$, then $\text{cov}(X, Y)$ will usually be negative. Conversely, if the random variables X and Y tend to take on values on the same side of $E(X)$ and $E(Y)$, then $\text{cov}(X, Y)$ will usually be positive. Two random variables X and Y are said to be *positively (negatively) correlated* if the covariance has a positive (negative) value.[2] We can now state Property 2 whose proof can be found in Chapter 11.

Property 2. *For every two random variables X and Y,*

$$\sigma^2(X + Y) = \sigma^2(X) + \sigma^2(Y) + 2\,\text{cov}(X,Y).$$

5.2.3 Independent random variables

Random variables X and Y are said to be *uncorrelated* if $\text{cov}(X, Y) = 0$. In Chapter 11, it will be shown that a sufficient (but not necessary) condition for uncorrelatedness of two random variables X and Y is that X and Y are independent random variables. The concept of independent random variables

[2]The correlation coefficient of two random variables X and Y is defined by $\rho(X, Y) = \frac{\text{cov}(X,Y)}{\sigma(X)\sigma(Y)}$. This is a dimensionless quantity with $-1 \leq \rho(X, Y) \leq 1$. The correlation coefficient measures how strongly X and Y are correlated. The farther $\rho(X, Y)$ is from 0, the stronger the correlation between X and Y.

is very important. Intuitively, two random variables X and Y are independent if learning that Y has taken on the value y gives no additional information about the value that X will take on and, conversely, learning that X has taken on the value x gives no additional information about the value that Y will take on. In the experiment of throwing two dice, the two random variables giving the number of points shown by the first die and the second die are independent, but the two random variables giving the largest and the smallest number shown are dependent. Formally, independence of random variables is defined in terms of independence of events. Two random variables X and Y are said to be *independent* if the event of X taking on a value less than or equal to a and the event of Y taking on a value less than or equal to b are independent for all possible values of a and b. Remember that events A and B are independent if and only if $P(AB) = P(A)P(B)$. Since $\text{cov}(X, Y) = 0$ if X and Y are independent, Property 2 implies that:

$$\sigma^2(X + Y) = \sigma^2(X) + \sigma^2(Y) \quad \text{for} \quad \text{independent } X \text{ and } Y.$$

5.2.4 Illustration: investment risks

Property 2 quantifies an important fact that investment experience supports: spreading investments over a variety of funds (diversification) diminishes risk. To illustrate, imagine that the random variable X is the return on every invested dollar in a local fund, and random variable Y is the return on every invested dollar in a foreign fund. Assume that random variables X and Y are independent and both have a normal distribution with expected value 0.15 and standard deviation 0.12. If you invest all of your money in either the local or the foreign fund, the probability of a negative return on your investment is equal to the probability that a normally distributed random variable takes on a value that is $\frac{0.15}{0.12} = 1.25$ standard deviations below the expected value. This probability is equal to 0.106. Now imagine that your money is equally distributed over the two funds. Then, the expected return remains at 15%, but the probability of a negative return falls from 10.6% to 3.9%. To explain this, we need the fact that the sum of two independent *normally* distributed random variables is *normally* distributed (see Chapter 14). This means that $\frac{1}{2}(X + Y)$ is normally distributed with expected value 0.15 and standard deviation

$$\sqrt{\frac{1}{4}(0.12)^2 + \frac{1}{4}(0.12)^2} = \frac{0.12}{\sqrt{2}} = 0.0849.$$

The probability that a normally distributed random variable takes on a value that is $\frac{0.15}{0.0849} = 1.768$ standard deviations below the expected value is equal to

0.039. By distributing your money equally over the two funds, you reduce your downward risk, but you also reduce the probability of doubling your expected return (this probability also falls from 10.6% to 3.9%). In comparison with the distributions of X and Y, the probability mass of $\frac{1}{2}(X+Y)$ is concentrated more around the expected value and less at the far ends of the distribution. The centralization of the distribution as random variables are averaged together is a manifestation of the central limit theorem.

The example is based on the assumption that returns X and Y are independent from each other. In the world of investment, however, risks are more commonly reduced by combining *negatively correlated* funds (two funds are negatively correlated when one tends to go up as the other falls). This becomes clear when one considers the following hypothetical situation. Suppose that two stock market outcomes ω_1 en ω_2 are possible, and that each outcome will occur with a probability of $\frac{1}{2}$. Assume that domestic and foreign fund returns X and Y are determined by $X(\omega_1) = Y(\omega_2) = 0.25$ and $X(\omega_2) = Y(\omega_1) = -0.10$. Each of the two funds then has an expected return of 7.5%, with equal probability for actual returns of 25% and -10%. The random variable $Z = \frac{1}{2}(X+Y)$ satisfies $Z(\omega_1) = Z(\omega_2) = 0.075$. In other words, Z is equal to 0.075 with certainty. This means that an investment that is equally divided between the domestic and foreign funds has a guaranteed return of 7.5%.

We conclude this section with another example showing that you cannot always rely on averages only.

5.2.5 Waiting-time paradox[3]

You are in Manhattan for the first time. Having no prior knowledge of the bus schedules, you happen upon a bus stop located on Fifth Avenue. According to the timetable posted, buses are scheduled to run at ten-minute intervals. So, having reckoned on a waiting period of five minutes, you are dismayed to find that after waiting for more than twenty, there is still no bus in sight. The following day you encounter a similar problem at another busy spot in the city. How is this possible? Is it just bad luck? No, you have merely encountered the bus waiting paradox: when arrival/departure times at the various stops cannot be strictly governed (due to traffic problems, for example), then a person arriving randomly at a bus stop may wind up waiting longer than the average time scheduled between the arrival of two consecutive buses! It is only when buses run *precisely* at ten-minute intervals that the average wait will equal the

[3]This section is highly specialized and may be skipped over.

expected five-minute period. We can elucidate the waiting-time paradox further by looking at a purely fictional example. Suppose that buses run at 30-minute intervals with a probability of 20%, and at one-second intervals with a probability of 80%. The average running time, then, should be six minutes, but the average waiting period for the person arriving randomly at the bus stop is approximately 15 minutes! The paradox can be explained by the fact that one has a higher probability of arriving at the bus stop during a long waiting interval than during a short one. A simple mathematical formula handsomely shows the effect of variability in running times on the average wait for a person turning up randomly at a bus stop. This formula involves the concept of coefficient of variation of a random variable. The coefficient of variation is the ratio of the standard deviation and the expected value. If the random variable T represents the amount of time elapsing between two consecutive busses, then the coefficient of variation of T is denoted and defined by

$$c_T = \frac{\sigma(T)}{E(T)}.$$

The coefficient of variation is dimensionless and is often a better measure for variability than the standard deviation (a large value of the standard deviation does not necessarily imply much variability when the expected value is large as well). Supposing that buses run at independent intervals that are distributed as the random variable T, it can be proved that

$$\frac{1}{2}\left(1 + c_T^2\right) E(T)$$

gives the average time a person must wait for a bus if the person arrives at the bus stop at a random point in time. If the buses run *precisely* on schedule ($c_T = 0$), then the average wait period is equal to $\frac{1}{2}E(T)$ as may be expected. Otherwise, the average wait period is always larger than $\frac{1}{2}E(T)$. The average wait period is even larger than $E(T)$ if the interstop running time T has $c_T > 1$!

Variability is also the reason why a small increase in demand for an already busy cash register at the supermarket leads to a disproportionately large increase in the queue for that cashier. This is nicely explained by the Pollaczek-Khintchine formula from queueing theory:

$$L_q = \frac{1}{2}\left(1 + c_S^2\right) \frac{[\lambda E(S)]^2}{1 - \lambda E(S)}.$$

This formula refers to the situation in which customers arrive at a service facility according to a Poisson process with intensity λ (the Poisson process was discussed in Section 4.2.4). The service times of the customers are independent of each other and have an expected value of $E(S)$ and a coefficient of variation of c_S. There is a single server who can handle only one customer at a time. Assuming that the average number of arrivals during a service time is less than 1, it can be shown that the long-run average number of customers waiting in queue is given by the Pollaczek-Khintchine formula for L_q. This formula clearly shows the danger of increasing the load on a highly loaded system. Normalizing the average service time as $E(S) = 1$ and assuming a highly loaded system with $\lambda = 0.9$, then a 5% increase in the arrival rate λ leads to a 100.5% increase in the average queue size. In stochastic service systems, one should never try to balance the input with the service capacity of the system! This is an important lesson from the Pollaczek-Khintchine formula.

5.3 The square-root law

This section deals with a sequence $X_1, X_2, \ldots, X_n$ of independent random variables each having the same probability distribution with standard deviation σ. Letting X be a random variable defined on the sample space of a random experiment, it is helpful to think of $X_1, X_2, \ldots, X_n$ as the representants of X in n independent repetitions of the experiment. A repeated application of the formula $\sigma^2(X+Y) = \sigma^2(X) + \sigma^2(Y)$ for two independent random variables X and Y gives

$$\sigma^2(X_1 + \cdots + X_n) = \sigma^2(X_1) + \cdots + \sigma^2(X_n) = n\sigma^2.$$

Consequently,

Property 3. *For each $n \geq 1$,*

$$\sigma\,(X_1 + X_2 + \cdots + X_n) = \sigma\sqrt{n}.$$

Properties 1 and 3 make it immediately apparent that:

Property 4. *For every $n \geq 1$,*

$$\sigma\left(\frac{X_1 + X_2 + \cdots + X_n}{n}\right) = \frac{\sigma}{\sqrt{n}}.$$

This property is called *the square-root law*. In finance, diversification of a portfolio of stocks generally achieves a reduction in the overall risk exposure with no reduction in expected return. Suppose that you split an investment budget equally between n similar but independent funds instead of concentrating it all in only one. Then, Property 4 states that that the standard deviation of the rate of return falls by a factor $1/\sqrt{n}$ in comparison with the situation that the full budget is invested in a single fund. Insurance works according to the same mechanism.

The sample mean of the data $X_1, X_2, \ldots, X_n$ is denoted and defined by

$$\overline{X}(n) = \frac{X_1 + X_2 + \cdots + X_n}{n}.$$

We know already, based on the law of large numbers, that the sample mean becomes more and more concentrated around the expected value $\mu = E(X)$ as n increases. The square root law specifies further that:

> **the standard deviation of the sample mean $\overline{X}(n)$ is proportional to $\frac{1}{\sqrt{n}}$ when n is the sample size.**

In other words, in order to reduce the standard deviation of the sample mean *by half*, a sample size *four times as large* is required. The central limit theorem to be discussed in the next section specifies precisely how the probability mass of the sample mean $\overline{X}(n)$ is distributed around the expected value $\mu = E(X)$ when the sample size n is large.

In Figure 5.4, we give an experimental demonstration of the square-root law. A standard normal distribution is taken for the underlying random variable X. For each of the respective sample sizes $n = 1,4,16$, and 64, there are 100 outcomes of the sample average $\overline{X}(n)$ simulated. Figure 5.4 shows that the bandwidths within which the simulated outcomes lie are indeed reduced by an approximate factor of 2 when the sample sizes are increased by a factor of 4.

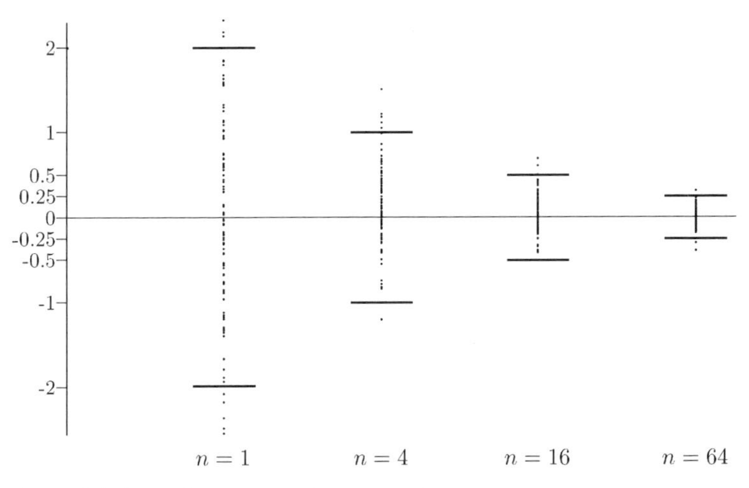

Figure 5.4 Simulation outcomes for the sample mean.

5.4 The central limit theorem

The central limit theorem is without a doubt the most important finding in the fields of probability theory and statistics. This theorem postulates that the sum (or average) of a sufficiently large number of independent random variables approximately follows a normal distribution. Suppose that $X_1, X_2, \ldots, X_n$ represents a sequence of independent random variables, each having the same distribution as the random variable X. Think of X as a random variable defined on the sample space of a random experiment and think of $X_1, X_2, \ldots, X_n$ as the representants of X in n independent repetitions of the experiment. The notation

$$\mu = E(X) \qquad \text{and} \qquad \sigma = \sigma(X)$$

is used for the expected value and the standard deviation of the random variable X. Mathematically, the central limit theorem states:

Central Limit Theorem. *For any real numbers a and b with $a < b$,*

$$\lim_{n\to\infty} P\left(a \leq \frac{X_1 + X_2 + \cdots + X_n - n\mu}{\sigma\sqrt{n}} \leq b\right) = \Phi(b) - \Phi(a),$$

where the standard normal distribution function $\Phi(x)$ is given by

$$\Phi(x) = \frac{1}{\sqrt{2\pi}} \int_{-\infty}^{x} e^{-\frac{1}{2}y^2}\, dy.$$

Thus, the standardized variable $(X_1 + X_2 + \cdots + X_n - n\mu)/(\sigma\sqrt{n})$ has an approximately standard normal distribution. A mathematical proof of the central limit theorem will be outlined in Chapter 14. In Section 5.1, it was pointed out that $V = \alpha W + \beta$ has a normal distribution when W is normally distributed and α, β are constants with $\alpha \neq 0$. In ordinary words, we would be able to reformulate the central limit theorem as follows:

> **if $X_1, X_2, \ldots, X_n$ are independent random variables each having the same distribution with expected value μ and standard deviation σ, then the sum $X_1 + X_2 + \cdots + X_n$ approximately has a normal distribution with expected value $n\mu$ and standard deviation $\sigma\sqrt{n}$ when n is sufficiently large.**

In terms of averaging random variables together, the central limit theorem tells us that:

> **if $X_1, X_2, \ldots, X_n$ are independent random variables each having the same distribution with expected value μ and standard deviation σ, then the sample mean $\overline{X}(n) = \frac{1}{n}(X_1 + X_2 + \cdots + X_n)$ has an approximately normal distribution with expected value μ and standard deviation $\frac{\sigma}{\sqrt{n}}$ when n is sufficiently large.**

This remarkable finding holds true no matter what form the distribution of the random variables X_k takes. How large n must be before the normal approximation is applicable depends, however, on the form of the underlying distribution of the X_k. We return to this point in Section 5.5, where we show that it makes a big difference whether or not the probability mass of the underlying distribution is symmetrically accrued around the expected value.

In the central limit theorem, it is essential that the random variables X_k are independent, but it is not necessary for them to have the same distribution. When the random variables X_k exhibit different distributions, the central limit theorem still holds true in general terms when we replace $n\mu$ en $\sigma\sqrt{n}$ with $\sum_{k=1}^{n}\mu_k$ and $(\sum_{k=1}^{n}\sigma_k^2)^{\frac{1}{2}}$, where $\mu_k = E(X_k)$ and $\sigma_k = \sigma(X_k)$. This generalized version of the central limit theorem elucidates the reason that, in practice, so many random phenomena, such as the rate of return on a stock, the cholesterol level of an adult male, the duration of a pregnancy, etc., are approximately normally distributed. Each of these random quantities can be seen as the result of a large number of small independent random effects.

The central limit theorem has an interesting history. The first version of this theorem was postulated by the French-born English mathematician Abraham de Moivre, who, in a remarkable article published in 1733, used the normal distribution to approximate the distribution of the number of heads resulting from many tosses of a fair coin. This finding was far ahead of its time, and was nearly forgotten until the famous French mathematician Pierre-Simon Laplace rescued it from obscurity in his monumental work *Théorie Analytique des Probabilités*, which was published in 1812. Laplace expanded De Moivre's finding by approximating the binomial distribution with the normal distribution. But as with De Moivre, Laplace's finding received little attention in his own time. It was not until the nineteenth century was at an end that the importance of the central limit theorem was discerned, when, in 1901, Russian mathematician Aleksandr Lyapunov defined it in general terms and proved precisely how it worked mathematically. Nowadays, the central limit theorem is considered to be the unofficial sovereign of probability theory.

5.4.1 Deviations

Do you believe a friend who claims to have tossed heads 5,250 times in 10,000 tosses of a fair coin? The central limit theorem provides an answer to this question.[4] For independent random variables $X_1, \ldots, X_n$, the central limit theorem points out how probable deviations of the sum $X_1 + X_2 + \cdots + X_n$ are from

[4]It appears that, for many students, there is a world of difference between a technical understanding of the central limit theorem and the ability to use it in solving a problem at hand.

its expected value. The random variable $X_1 + \cdots + X_n$ is approximately normally distributed with expected value $n\mu$ and standard deviation $\sigma\sqrt{n}$. Also, as pointed out in Section 5.1, the probability of a normally distributed random variable taking on a value z standard deviations above or below the expected value is equal to $1 - \Phi(z) + 1 - \Phi(z) = 2\{1 - \Phi(z)\}$. Thus, for any constant $c > 0$,

$$P(|\,(X_1 + X_2 + \cdots + X_n) - n\mu\,| > c\sigma\sqrt{n}\,) \approx 2\{1 - \Phi(c)\}$$

when n is sufficiently large. In particular,

$$P(|\,(X_1 + \cdots + X_n) - n\mu\,| > c\sigma\sqrt{n}\,) \approx \begin{cases} 0.317 & \text{for} \quad c = 1 \\ 0.046 & \text{for} \quad c = 2 \\ 2.7 \times 10^{-3} & \text{for} \quad c = 3 \\ 6.3 \times 10^{-5} & \text{for} \quad c = 4 \\ 5.7 \times 10^{-7} & \text{for} \quad c = 5. \end{cases}$$

Thus, the outcome of the sum $X_1 + \cdots + X_n$ will seldom be three or more standard deviations removed from the expected value $n\mu$. Coming back to the issue of whether or not the claim of having tossed 5,250 heads in 10,000 fair coin tosses is plausible, the answer is no. This cannot be explained as a chance variation. In order to find this assertion, one should calculate the probability of 5,250 or more heads appearing in 10,000 tosses of a fair coin. The number of times the coin lands heads can be written as the sum $X_1 + X_2 + \cdots + X_{10{,}000}$, where

$$X_i = \begin{cases} 1 & \text{if the } i\text{th toss turns heads} \\ 0 & \text{otherwise.} \end{cases}$$

Using the fact that $E(X_i) = 0 \times \frac{1}{2} + 1 \times \frac{1}{2} = \frac{1}{2}$, the standard deviation of the Bernoulli variable X_i is

$$\sigma = \sqrt{\left(0 - \frac{1}{2}\right)^2 \times \frac{1}{2} + \left(1 - \frac{1}{2}\right)^2 \times \frac{1}{2}} = \frac{1}{2}.$$

Tossing 5,250 or more heads, then, lies

$$\frac{(5{,}250 - 5{,}000)}{\frac{1}{2}\sqrt{10{,}000}} = 5$$

or more standard deviations above the expected value of 5,000 heads. The chance of this happening is approximately 1 in 3.5 million. Generalizing, for large n, one could say with 99.7% certainty that in n fair coin tosses the coin will land heads between $n - 1.5\sqrt{n}$ and $n + 1.5\sqrt{n}$ times (verify!).

5.5 Graphical illustration of the central limit theorem

The mathematical proof of the central limit theorem is far from simple and is also quite technical. Moreover, the proof gives no insight into the issue of how large n must actually be in order to get an approximate normal distribution for the sum $X_1 + X_2 + \cdots + X_n$. Insight into the working of the central limit theorem can best be acquired through empirical means. Simulation can be used to visualize the effect of adding random variables. For any fixed value of n, one runs many simulation trials for the sum $X_1 + X_2 + \cdots + X_n$ and creates a histogram by plotting the outcomes of the simulation runs. Then, for increasing values of n, it will be seen that the histogram approaches the famous bell-shaped curve. The disadvantage of this empirical approach is that the law of large numbers interferes with the central limit theorem. For fixed value of n, one needs many simulation trials before the simulated distribution of $X_1 + X_2 + \cdots + X_n$ is sufficiently close to its actual distribution. This complication can be avoided by taking a different approach. In the case where the random variables $X_1, X_2, \ldots$ have a discrete distribution, it is fairly simple to calculate the probability mass function of the sum $X_1 + X_2 + \cdots + X_n$ exactly for any value of n. This can be done by using the convolution formula for the sum of discrete random variables. The convolution formula will be discussed in Chapter 9. In this way, you can determine empirically how large n must be in order to ensure that the probability histogram of the sum $X_1 + \cdots + X_n$ will take on the bell shape of the normal curve. You will see that the answer to the question of how large n must be strongly hinges on how "symmetrical" the probability mass of the random variable X_i is distributed around its expected value. The more *skewed* the probability mass is, the *larger* n must be in order for the sum of $X_1 + X_2 + \cdots + X_n$ to be approximately normally distributed. This can be nicely illustrated by using the random experiment of rolling a (biased) die. Let's assume that one roll of the die turns up j points with a given probability p_j for $j = 1, \ldots, 6$. Playing with the probabilities p_j, one can construct both a symmetrical die and an asymmetrical die. Define the random variable X_k by

$$X_k = \text{number of points obtained by the } k\text{th roll of the die.}$$

The random variables $X_1, X_2, \ldots$ are independent and are distributed according to the probability mass function $(p_1, \ldots, p_6)$. The sum $X_1 + \cdots + X_n$ gives the total number of points that have been obtained in n rolls of the die. Figures 5.5 and 5.6 show the probability histogram of the sum $X_1 + \cdots + X_n$ for $n = 5$, 10, 15, and 20 rolls of the die. This is shown for both the unbiased die with the symmetrical distribution $p_1 = \cdots = p_6 = \frac{1}{6}$ and a biased die with

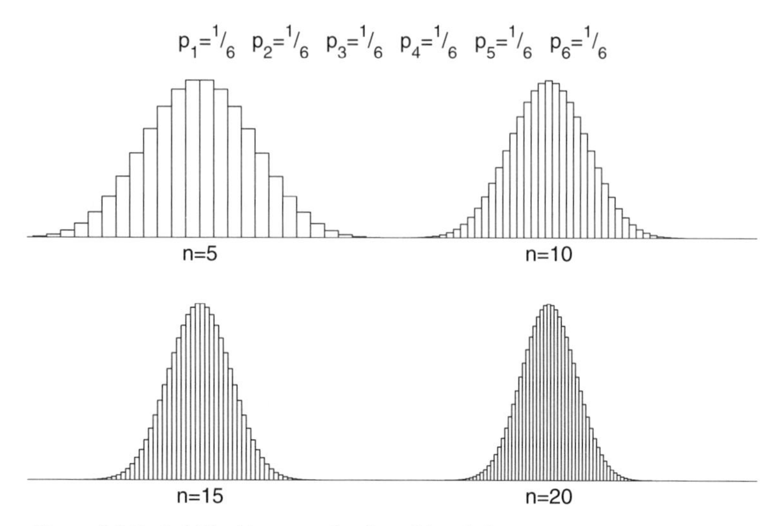

Figure 5.5 Probability histogram for the unbiased die.

the asymmetrical distribution $p_1 = 0.2$, $p_2 = 0.1$, $p_3 = p_4 = 0$, $p_5 = 0.3$ and $p_6 = 0.4$. The two figures speak for themselves. It is quite apparent that for both distributions the diagram of $X_1 + \cdots + X_n$ ultimately takes on a normal bell-shaped curve, but that it occurs much earlier in the case of a symmetrical distribution than it does in the case of an asymmetrical distribution.

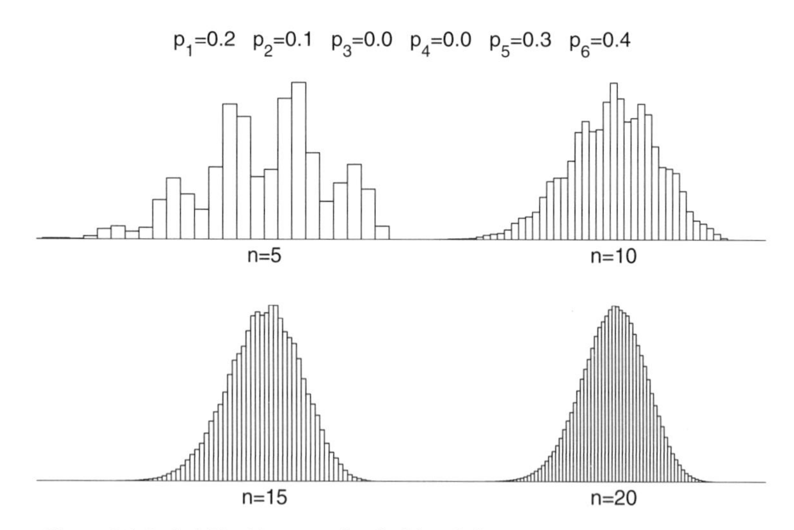

Figure 5.6 Probability histogram for the biased die.

5.6 Statistical applications

The central limit theorem has numerous applications in probability theory and statistics. In this section, we discuss a few illustrative applications.

5.6.1 Normal approximation of the binomial distribution

Suppose that X is a binomially distributed random variable with parameters n and p. The random variable X can be interpreted as the total number of successes in n independent repetitions of a Bernoulli experiment with a success probability of p (see Section 4.1). Consequently, the random variable X can be represented as

$$X = I_1 + I_2 + \cdots + I_n,$$

where the indicator variable I_k is defined by

$$I_k = \begin{cases} 1 & \text{if the } k\text{th trial leads to a success} \\ 0 & \text{otherwise.} \end{cases}$$

It is left to the reader to verify that the Bernoulli variable I_k has an expected value of $\mu = p$ and a standard deviation of $\sigma = \sqrt{p(1-p)}$. The random variables $I_1, \ldots, I_n$ are independent. Using Property 2 from Section 5.2, it now follows that the expected value and the standard deviation of the binomially distributed random variable X are given by

$$E(X) = np \qquad \text{and} \qquad \sigma(X) = [p(1-p)]^{1/2}\sqrt{n}.$$

The central limit theorem now tells us that the (discrete) binomial distribution can be approximated by the (continuous) normal distribution when n is large.[5] A guideline is to use the normal approximation when $np > 5$ and $n(1-p) > 5$, but of course this would depend on the accuracy required. The normal probability density function then has sufficient space to "unfurl" around the expected value np without too much of the probability mass falling into the negative axis.

Example 5.1 A student has passed a final exam by supplying correct answers for 26 out of 50 multiple-choice questions. For each question, there was a choice of three possible answers, of which only one was correct. The student claims not to have learned anything in the course and not to have studied for the exam,

[5] This approximation can be improved by using the so-called *continuity correction*: approximate $P(X > k)$ by $P\left(U \geq k + \frac{1}{2}\right)$ and $P(X < k)$ by $P\left(U \leq k - \frac{1}{2}\right)$, where U is a normal random variable with expected value np and standard deviation $[p(1-p)]^{1/2}\sqrt{n}$.

and says that his correct answers are the product of guesswork. Do you believe him?

This problem can be approached as follows: assume that the student did guess at all the answers and calculate the probability of identifying 26 or more correct answers through guesswork. If this probability is below a threshold value you have chosen in advance, you judge that the student is bluffing. If all the answers are guessed at, then the number of correct answers can be seen as the number of successes in $n = 50$ independent trials of a Bernoulli experiment having a success probability of $p = \frac{1}{3}$. The binomial probability model is thus applicable. A quick and generally useful method of determining whether 26 correct answers is exceptional goes as follows: Check to find out how many standard deviations lie between the observed number of correct answers achieved and the expected number. Then, approximate, the binomial distribution with parameters $n = 50$ and $p = \frac{1}{3}$ by a normal distribution with expected value $np = 16\frac{2}{3}$ and standard deviation $\sqrt{np(1-p)} = 3\frac{1}{3}$ and use the rule of thumb stating that the probability of a normally distributed random variable taking on a value lying three or more standard deviations above the expected value is very small (the probability is 0.0013). The observed value of 26 correct answers lies $(26 - 16\frac{2}{3})/3\frac{1}{3} = 2.8$ standard deviations above the expected value. The probability of such a deviation occurring is quite small. There is very good reason, therefore, to suppose that the student is bluffing, and that he in fact did prepare for the exam.

5.6.2 The z-value

The key to finding the solution to the problem in Example 5.1 is to measure the number of standard deviations separating the observed value from the expected value. The normal distribution will allow you to establish whether the difference between the observed value and the expected value can be explained as a chance variation or not. The z-value is defined as

$$z = \frac{\text{observed value} - \text{expected value}}{\text{standard deviation}}.$$

It is often used in the testing of hypotheses. This is illustrated with the famous example of the Salk vaccine.

Example 5.2 The Salk vaccine against polio was tested in 1954 in a carefully designed field experiment. Approximately 400,000 children took part in this experiment. Using a randomization procedure, the children were randomly divided into two groups of equal size, a treatment group and a control group. The vaccine was given only to the children in the treatment group; the control group children received placebo injections. The children did not know which of the two groups they had been placed into. The diagnosticians also lacked this information (double-blind experiment). Fifty-seven children in the treatment group went on to contract polio, while 142 children in the control group contracted the illness. Based on these results, how reliable is the claim that the vaccine worked?

This famous experiment is commonly misperceived. It is often claimed that such an experiment including the participation of 400,000 children cannot deliver reliable conclusions when those conclusions are based on the fact that two relatively small groups of 142 and 57 children contracted polio. People subscribing to this train of thought are misled by the magnitude of the group at large; what they should really be focusing on is the difference between the number of polio occurrences in each of the two groups as compared to the total number of polio occurrences. The test group must be large because statistically founded conclusions can only be drawn when a sufficiently large number of cases have been observed. Incidentally, since the probability of contracting polio is very small and the group sizes are very large, the realized number of polio occurrences in each group can be seen as an outcome of a Poisson distribution.

In order to find out whether the difference in outcomes between the two groups is a significant difference and not merely the result of a chance fluctuation, the following reasoning is used. Suppose that assignment to treatment or control had absolutely no effect on the outcome. Under this hypothesis, each of the 199 children was doomed to contract polio regardless of which group he/she was in. Now we have to ask ourselves this question: what is the probability that of the 199 affected children only 57 or less will belong to the treatment group? This problem strongly resembles the problem of determining the probability of not more than 57 heads turning up in 199 tosses of a fair coin. This problem can be solved with the binomial model with parameters $n = 199$ and $p = \frac{1}{2}$. This binomial model can be approximated by the normal model. For the z-value we find

$$z = \frac{57 - 199 \times 0.5}{\sqrt{199 \times 0.5 \times 0.5}} = -6.03.$$

Thus, the observed number of polio cases in the treatment group registers at more than six standard deviations below the expected number. The probability of this occurring in a normal distribution is on the order of 10^{-9}. It is therefore extremely unlikely that the difference in outcomes between the two groups can be explained as a chance variation. This in turn makes clear that the hypothesis is incorrect and that the vaccine does, in fact, work.

5.6.3 The z-value and the Poisson distribution

For many of the everyday situations of a statistical nature that occur, we only have averages available from which to draw conclusions. For example, records are kept of the average number of traffic accidents per year, the average number of bank robberies per year, etc. When working with this kind of information, the Poisson model is often suitable. The Poisson distribution is completely determined by its expected value. In Chapter 9, it will be shown that a Poisson distributed random variable with expected value λ has a standard deviation of $\sqrt{\lambda}$. Also, for λ sufficiently large (say $\lambda \geq 25$), the Poisson distribution with expected value λ can be approximated by a normal distribution with expected value λ and standard deviation $\sqrt{\lambda}$. The explanation is that the Poisson distribution is a limiting case of the binomial distribution (see Chapter 4). In the beginning of this section, we saw that the binomial distribution can be approximated by the normal distribution. The normal approximation to the Poisson distribution and the concept of the z-value allow one to make statistical claims in situations such as those mentioned above. Suppose, for example, you read in the paper that, based on an average of 1,000 traffic deaths per year in previous years, the number of traffic deaths for last year rose 12%. How can you evaluate this? The number of traffic deaths over a period of one year can be modeled as a Poisson distributed random variable with expected value 1,000 (why is this model reasonable?). An increase of 12% on an average of 1,000 is an increase of 120, or rather an increase of $120/\sqrt{1{,}000} = 3.8$ standard deviations above the expected value 1,000. The probability that a normally distributed random variable will take on a value of more than three standard deviations above the expected value is quite small. In this way, we find justification for the conclusion that the increase in the number of traffic deaths is not coincidental, but that something for which concrete explanations can be found has occurred. What would your conclusions have been if, based on an average of 100 traffic deaths per year, a total of 112 traffic deaths occurred in the past year?

5.7 Confidence intervals for simulations

In the preceding chapters we encountered several examples of simulation studies for stochastic systems. In these studies, we obtained point estimates for unknown probabilities or unknown expected values. It will be seen in this more technical section that the central limit theorem enables us to give a probabilistic judgment about the accuracy of the point estimate. Simulation of a stochastic system is in fact a statistical experiment in which one or more unknown parameters of the system are estimated from a sequence of observations that are obtained from independent simulation runs of the system. Let's first consider the situation in which we wish to estimate the unknown expected value $\mu = E(X)$ of a random variable X defined for a given stochastic system (e.g., the expected value of the random time until a complex electronic system fails for the first time). Later on, when we encounter the estimating of probabilities, we will see that this turns out to be none other than a special case of estimating an expected value.

Let X be a random variable defined on the sample space of a random experiment. The goal is to estimate the unknown expected value $\mu = E(X)$. Suppose that n independent repetitions of a random experiment are performed. The kth performance of the experiment yields the representative X_k of the random variable X. An estimator for the unknown expected value $\mu = E(X)$ is given by the sample mean

$$\overline{X}(n) = \frac{1}{n} \sum_{k=1}^{n} X_k.$$

It should be noted that this statistic, being the arithmetic mean of the random sample $X_1, \ldots, X_n$, is a random variable. The central limit theorem tells us that for n large,

$$\frac{X_1 + \cdots + X_n - n\mu}{\sigma\sqrt{n}}$$

has an approximately standard normal distribution, where $\sigma = \sigma(X)$ is the standard deviation of the random variable X. Dividing the numerator and the denominator of the above expression by n, we find that

$$\frac{\overline{X}(n) - \mu}{\sigma/\sqrt{n}}$$

has an approximately standard normal distribution. For any number α with $0 < \alpha < 1$ the percentile $z_{1-\frac{1}{2}\alpha}$ is defined as the unique number for which the area under the standard normal curve between the points $-z_{1-\frac{1}{2}\alpha}$ and $z_{1-\frac{1}{2}\alpha}$ equals $100(1-\alpha)\%$. The percentile $z_{1-\frac{1}{2}\alpha}$ has the values 1.960 and 2.324 for the often used values 0.05 and 0.01 for α. Since $\left[\overline{X}(n) - \mu\right] / \left(\sigma/\sqrt{n}\right)$ is an approximately standard normal random variable, it follows that

$$P\left(-z_{1-\frac{1}{2}\alpha} \leq \frac{\overline{X}(n) - \mu}{\sigma/\sqrt{n}} \leq z_{1-\frac{1}{2}\alpha}\right) \approx 1 - \alpha$$

or, stated differently,

$$P\left(\overline{X}(n) - z_{1-\frac{1}{2}\alpha}\frac{\sigma}{\sqrt{n}} \leq \mu \leq \overline{X}(n) + z_{1-\frac{1}{2}\alpha}\frac{\sigma}{\sqrt{n}}\right) \approx 1 - \alpha.$$

Voila! You have now delimited the unknown expected value μ on two ends. Both endpoints involve the standard deviation σ of the random variable X. In most situations σ will be unknown when the expected value μ is unknown, but fortunately, this problem is easily circumvented by replacing σ by an estimator based on the sample values $X_1, \ldots, X_n$. Just as the unknown expected value $\mu = E(X)$ is estimated by the sample mean $\overline{X}(n) = (1/n)\sum_{k=1}^{n} X_k$, the standard deviation σ is estimated by the square root of the *sample variance*. This statistic is denoted and defined by

$$S^2(n) = \frac{1}{n}\sum_{k=1}^{n}\left[X_k - \overline{X}(n)\right]^2.$$

The definition of the statistic $S^2(n)$ resembles the definition of the variance $\sigma^2(X) = E[(X - \mu)^2]$ (usually one defines $S^2(n)$ by dividing through $n - 1$ rather than n, but for large n the two definitions of $S^2(n)$ boil down to the same thing). Using the law of large numbers, it can be shown that the statistic $S^2(n)$ converges to σ^2 as n tends to infinity.

The sample variance enables us to give a probability judgment about the quality or accuracy of the estimate $\overline{X}(n)$ for the unknown expected value $\mu = E(X)$. It can be proved that the central limit theorem remains valid when σ is replaced by its estimator $S(n)$. That is, for n large,

$$P\left(-z_{1-\frac{1}{2}\alpha} \leq \frac{\overline{X}(n) - \mu}{S(n)/\sqrt{n}} \leq z_{1-\frac{1}{2}\alpha}\right) \approx 1 - \alpha$$

or, stated differently,

$$P\left(\overline{X}(n) - z_{1-\frac{1}{2}\alpha}\frac{S(n)}{\sqrt{n}} \leq \mu \leq \overline{X}(n) + z_{1-\frac{1}{2}\alpha}\frac{S(n)}{\sqrt{n}}\right) \approx 1 - \alpha.$$

This result is the basis for an interval estimate of the unknown parameter μ rather than a point estimate. Such an interval estimate is called a *confidence interval*. The following important result holds:

for n large, an approximate $100(1 - \alpha)\%$ confidence interval for the unknown expected value $\mu = E(X)$ is given by

$$\overline{X}(n) \pm z_{1-\frac{1}{2}\alpha}\frac{S(n)}{\sqrt{n}}.$$

When speaking about large n, it is better to think in terms of values of n on the order of tens of thousands than on the order of hundreds.[6] In practice, one often chooses $\alpha = 0.05$ and thus constructs a 95% confidence interval. The percentile $z_{1-\frac{1}{2}\alpha}$ is 1.960 for $\alpha = 0.05$.

When n independent simulation runs are performed to estimate the unknown expected value μ of the random variable X, then

$$\begin{aligned}&\text{the width of the approximate } 100(1-\alpha)\% \text{ confidence interval}\\ &= 2z_{1-\alpha/2}\frac{S(n)}{\sqrt{n}}\\ &= \frac{2z_{1-\alpha/2}}{\sqrt{n}} \times (\text{estimate for the unknown standard deviation of } X).\end{aligned}$$

The estimator $S(n)$ of the unknown standard deviation σ of X will not change much after some initial period of the simulation. This means that the width of the confidence interval is nearly proportional to $1/\sqrt{n}$ for n sufficiently large. This conclusion leads to a practically important rule of thumb:

to reduce the width of a confidence interval by a factor of two, about four times as many observations are needed.

5.7.1 Interpretation of the confidence interval

Let's say we have determined by simulation a 95% confidence interval (25.5, 27.8) for an unknown parameter μ. In this case, we cannot actually say that there is a 95% probability of μ falling within the interval (25.5, 27.8). Why not? The reason is simply that the unknown μ is a constant and not a random variable. Thus, either the constant μ falls within the interval (25.5, 27.8) or it does not. In other words, the probability of μ falling within the interval (25.5, 27.8) is 1 or 0. If the values of the sample data $X_1, \ldots, X_n$ had been different, the confidence interval would also have been different. Some simulation studies will produce confidence intervals that cover the value of μ and others will not. Before the simulation runs are done, it can be said that the 95% confidence interval that will result will cover the true value of μ with a probability of 95%. After the data are obtained, it can only be said that "we are 95% confident that the resultant interval covers the true value of μ." A more concrete interpretation of the $100(1-\alpha)\%$ confidence interval is provided by the frequentist approach. If you construct a large number of $100(1-\alpha)\%$ confidence intervals, each based on

[6]In the special case of the random variables X_i themselves being normally distributed, it is possible to give a confidence interval that is not only exact, but also applies to small values of n. This exact confidence interval is based on the so-called student-t distribution instead of the standard normal distribution (see Chapter 10).

the same number of simulation runs, then the proportion of intervals covering the unknown value of μ is approximately equal to $1-\alpha$. To illustrate this, consider Figure 5.7. This figure relates to a problem known as the newsboy problem and displays one hundred 95% confidence intervals for the expected value of the daily net profit of the newsboy. In this well-known inventory problem, a newsboy decides at the beginning of each day how many newspapers he will purchase for resale. Let's assume that the daily demand for newspapers is uniformly distributed between 150 and 250 papers. Demand on any given day is independent of demand on any other day. The purchase price per newspaper is one dollar. The resale price per newspaper is two dollars; the agency will buy back any unsold newspapers for fifty cents apiece. The performance measure we are interested in is the expected value μ of the daily net profit when at the beginning of each day the newsboy purchases 217 newspapers. We constructed

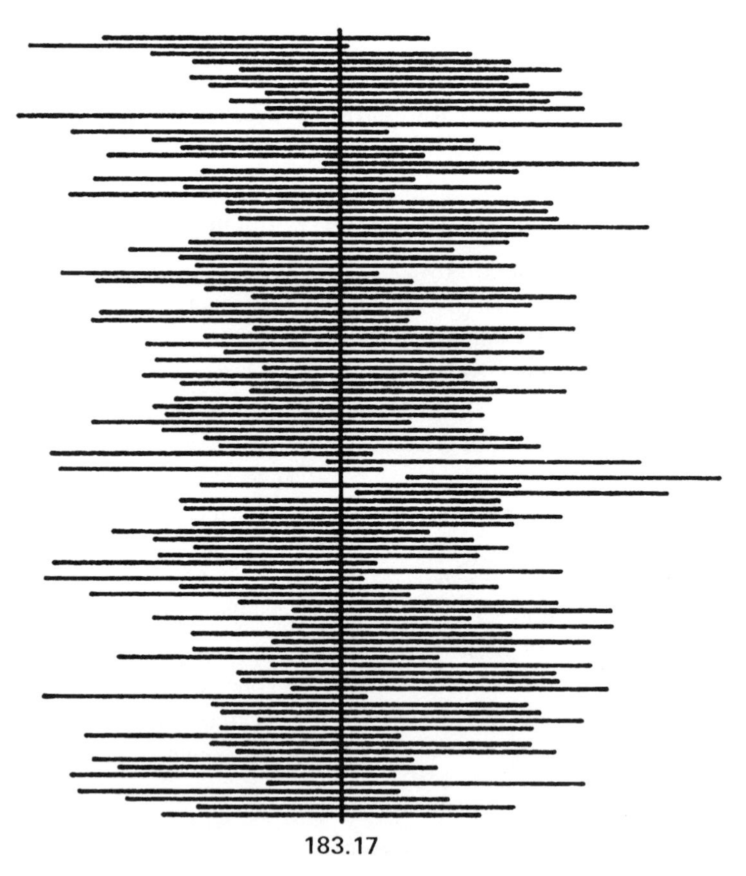

Figure 5.7 One hundred 95% confidence intervals.

one hundred approximate 95% confidence intervals for μ by simulating the sales one hundred times over $n = 2{,}000$ days. The resulting 95% confidence intervals for the expected value μ are given in Figure 5.7. It is instructive to have a look at the figure. Indeed, in approximately 95 of the 100 cases, the true value of μ is contained within the confidence interval (the true value of μ can analytically be shown to be equal to \$183.17).

5.7.2 Confidence interval for a probability

The goal of many simulation studies is to estimate an unknown probability $P(E)$ for a given event E in a random experiment. We shall demonstrate that the probability $P(E)$ can be seen as an expected value of an indicator variable. This means that we can apply this theory in order to find a confidence interval to correspond with the estimate for $P(E)$. Suppose that n independent repetitions of the experiment are simulated. Define the random variable X_i as

$$X_i = \begin{cases} 1 & \text{if event } E \text{ occurs in the } i\text{th trial} \\ 0 & \text{otherwise.} \end{cases}$$

The indicator variables $X_1, \ldots, X_n$ are independent Bernoulli variables each having the same distribution. Note that

$$E(X_i) = 0 \times P(X_i = 0) + 1 \times P(X_i = 1) = P(X_i = 1).$$

Since $P(X_i = 1) = P(E)$, it follows that the probability $P(E)$ is equal to the expected value of the indicator variables X_i. Thus, the sample mean

$$\overline{X}(n) = \frac{1}{n}\sum_{i=1}^{n} X_i$$

provides a point estimate for the unknown probability $P(E)$. The corresponding $100(1-\alpha)\%$ confidence interval $\overline{X}(n) \pm z_{1-\frac{1}{2}\alpha}\, S(n)/\sqrt{n}$ takes the insightful and simple form of

$$\overline{X}(n) \pm z_{1-\frac{1}{2}\alpha} \frac{\sqrt{\overline{X}(n)\left[1 - \overline{X}(n)\right]}}{\sqrt{n}}.$$

The explanation is that $S^2(n) = \overline{X}(n)[1 - \overline{X}(n)]$ for variables X_i that take on only the values 0 and 1. It is a matter of simple algebra to verify this fact. The simplified expression for $S^2(n)$ is in agreement with the fact that a Bernoulli variable X has variance $E(X)[1 - E(X)]$. It follows from the structure of the confidence interval for an unknown probability that it suffices to know the sample mean in order to construct the confidence interval.

As an illustration, suppose that someone tells you that he/she simulated the so-called game of ace-jack-two 2,500 times and found the point estimate of 0.8092 for the probability of the player winning. Then you know enough to conclude that the half-width of a 95% confidence interval for the probability of winning equals $1.96\sqrt{0.8092(1-0.8092}/\sqrt{2{,}500} = 0.015$. The game of ace-jack-two is played this way: 17 times in a row, a player chooses three cards from a deck of 52 thoroughly shuffled cards. Every time the group of three cards contains an ace, jack, or two, the player accrues one point; otherwise the bank wins a point. An analytical calculation of the player's probability of accruing the most points and winning the game is far from easily achieved. That's why simulation has been used for this problem.

The confidence interval for an unknown probability also gives insight into the necessary simulation efforts for *extremely small* probabilities. Let's say that the unknown probability $p = P(E)$ is on the order of 10^{-6}. How large must the number of simulation runs be before the half width of the 95% confidence interval is smaller than $f \times 10^{-6}$ for a given value of f between 0 and 1? To answer this question, note that, for n large,

$$\sqrt{\overline{X}(n)[1-\overline{X}(n)]} \approx \sqrt{p(1-p)} \approx \sqrt{p}$$

because $1 - p \approx 1$. The formula for the confidence interval for p now gives that the required number n of runs must satisfy $\frac{1.96\sqrt{p}}{\sqrt{n}} \approx f \times p$, or

$$n \approx \left(\frac{1.96}{f}\right)^2 \times \frac{1}{p}.$$

For $p = 10^{-6}$ and $f = 0.1$, this means approximately 400 million simulation runs. This shows how careful one must be when estimating extremely small probabilities with precision.

Example 5.1 A random sample of 1,000 voters is garnered from a population of 250,000 voters inhabiting a particular area. Interviews of the 1,000 voters in the sample group were conducted after which it was apparent that 520 of them voted Democrat in the last election. What is the 95% confidence interval for the fraction of Democrats among the voters in that area?

We can conceive of the process of interviewing 1,000 randomly chosen individuals as a simulation study with $n = 1{,}000$ independent trials of a Bernoulli experiment. Define the random variable X_i as

$$X_i = \begin{cases} 1 & \text{if the } i\text{th interviewee is Democrat} \\ 0 & \text{otherwise.} \end{cases}$$

The observed value of the sample mean $\overline{X}(n) = (1/n)\sum_{k=1}^{n} X_k$ is 520/1,000. Letting p represent the unknown fraction of Democrats among the voters, this fraction is estimated by the value

$$\overline{X}(n) = 0.52$$

with corresponding 95% confidence interval

$$\overline{X}(n) \pm 1.96\frac{\sqrt{\overline{X}(n)[1-\overline{X}(n)]}}{\sqrt{n}} = 0.52 \pm 0.03.$$

Imagine that you are asked how large the sample size must be in order to get a 95% confidence interval with a margin of ± 0.01. The answer to this question tells us that you would then need a sample size of about 9,000 people. Increasing the sample size by a factor of 9 reduces the margin of the confidence interval with a factor of about 3.

Example 5.2 It is commonly presumed that an unborn child has a 50% probability of being female. But is this really the case? Let's take a look at birth statistics for the Netherlands for the years 1989, 1990, and 1991. According to the Central Bureau of Statistics, there were, in total, 585,609 children born during the span of those years, of which 286,114 were girls. What is the estimate for the probability that a newborn child will be a girl and what is the corresponding 95% confidence interval?

We can model this problem as $n = 585{,}609$ independent trials of a Bernoulli experiment with an unknown success probability of p, where success is defined as the birth of a girl. Let the random variable X_i be equal to 1 if the ith trial of the experiment delivers a success and let X_i be otherwise equal to 0. Then the unknown probability p is estimated by the value

$$\overline{X}(n) = \frac{286{,}114}{585{,}609} = 0.4886.$$

The corresponding 95% confidence interval is

$$\overline{X}(n) \pm 1.96\,\frac{\sqrt{\overline{X}(n)\left[1-\overline{X}(n)\right]}}{\sqrt{n}} = 0.4886 \pm 0.0013.$$

In reality, then, the probability of a child being born female is slightly under 50% (the value 0.5 is also well outside the 99.99% confidence interval 0.4886 ± 0.0025). This probability appears to alter very little over time and applies to other countries as well. The celebrated French probability theorist Laplace, who also did much empirical research, investigated births over a long period in the eighteenth century and found that the probability of a newborn child being a girl had the value $\frac{21}{43} = 0.4884$ in each of the cities of Paris, London, Naples, and St. Petersburg. Interestingly, Laplace initially found a slightly deviating value for Paris, but in the end, after adjusting for the relatively large number of provincial girls placed in Parisian foundling homes, that probability was also reckoned at approximately $\frac{21}{43}$.

5.8 The central limit theorem and random walks

Random walks are among the most useful models in probability theory. They find applications in all parts of science. In Chapter 2, we introduced the random walk model based on the simple coin-tossing experiment and the random walk model of a gambler's fortune under the Kelly betting system. These elementary models uncover interesting, and occasionally profound, insights into the study of more complicated models. In this section, the central limit theorem will be used to reveal further properties of random walk models and to establish a link between random walks and the Brownian motion process. The Brownian motion process is widely applied to the modelling of financial markets. In particular, the famous Black-Scholes formula for the pricing of options will be discussed.

5.8.1 Fluctuations in a random walk

The central limit theorem allows us to make a mathematical statement about the behavior of the random walk in Figure 2.1. This random walk describes the evolution of the actual number of heads tossed minus the expected number when a fair coin is repeatedly tossed. The actual number of heads minus the expected number can be represented as $Z_n = X_1 + X_2 + \cdots + X_n - \frac{1}{2}n$, where the random variable X_i is equal to 1 if the ith toss of the coin shows heads and X_i is otherwise equal to 0. The random variable Z_n has approximately a normal distribution with expected value 0 and standard deviation $\sigma\sqrt{n}$ when n is large, where $\sigma = \frac{1}{2}$ is the standard deviation of the X_i. This fact explains the phenomenon that the range of the difference between the number of heads and the number of tails tossed in n fair coin tosses shows a tendency to grow proportionally with $\sqrt{n}$ as n increases (this difference is given by $X_1 + \cdots + X_n - (1 - X_1 + \cdots + 1 - X_n) = 2(X_1 + \cdots + X_n - \frac{1}{2}n)$). The proportion of heads in n coin tosses is $(X_1 + \cdots + X_n)/n$. The probability distribution of $(X_1 + \cdots + X_n)/n$ becomes more and more concentrated around the value 0.5 as n increases, where the deviations from the expected value of 0.5 are on the order of $1/\sqrt{n}$. A similar phenomenon appears in lotto drawings. The difference between the number of times the most frequently drawn number comes up and the number of times the least frequently drawn number comes up in n drawings shows a tendency to increase proportionally with $\sqrt{n}$ as n increases. This phenomenon can be explained using the multivariate central limit theorem (see Chapter 12).

5.8.2 Casino profits

The square-root law and the central limit theorem give further mathematical support to an earlier claim that operating a casino is in fact a risk-free undertaking. However small the house advantage may be, it is fairly well assured of large and stable profits if it spreads its risk over a very large number of gamblers. To illustrate this, let's consider the casino game of red-and-black. The plays are independent and at each play the gambler wins with probability p and loses with probability $1 - p$, where the win probability p satisfies $p < \frac{1}{2}$. The payoff odds are 1 to 1. That is, in case of a win the player gets paid out twice the stake; otherwise, the player loses the stake. Suppose that the player places n bets and stakes the same amount of cash on each bet. The total number of bets won by the player can be represented as the sum $X_1 + \cdots + X_n$, where the random variable X_i is equal to 1 if the player wins the ith bet and X_i is otherwise equal to 0. The expected value and the standard deviation of the Bernoulli variable X_i are given by

$$E(X_i) = 0 \times (1 - p) + 1 \times p = p$$

and

$$\sigma(X_i) = \sqrt{(0 - p)^2 \times (1 - p) + (1 - p)^2 \times p} = \sqrt{p(1 - p)}.$$

The central limit theorem tells us that the random variable $X_1 + \cdots + X_n$ has approximately a normal distribution with expected value np and standard deviation $[p(1 - p)]^{\frac{1}{2}}\sqrt{n}$ if n is sufficiently large. The casino loses money to the player only if the player wins $\frac{1}{2}n + 1$ or more bets (assume that n is even). In other words, the casino only loses money to the player if the number of bets the player wins exceeds the expected value np by

$$\beta_n = \frac{\frac{1}{2}n + 1 - np}{[p(1 - p)]^{1/2}\sqrt{n}}$$

or more standard deviations. The probability of this is approximately equal to $1 - \Phi(\beta_n)$. Because β_n increases proportionally to $\sqrt{n}$ as n increases, the probability $1 - \Phi(\beta_n)$ tends very rapidly to zero as n gets larger. In other words, it is practically impossible for the casino to lose money to the gambler when the gambler continues to play. The persistent gambler will always lose in the long run. The gambler's chances are the same as those of a lamb in the slaughterhouse. Assuming that the player stakes one dollar on each bet, then for n plays the profit of the casino over the gambler is equal to

$$W_n = n - 2(X_1 + \cdots + X_n).$$

Using Property 1 from Section 5.2 and the fact that $X_1 + \cdots + X_n$ has expected value np and standard deviation $[p(1-p)]^{\frac{1}{2}}\sqrt{n}$, it follows that

$$E(W_n) = n(1-2p) \qquad \text{and} \qquad \sigma(W_n) = 2[p(1-p)]^{\frac{1}{2}}\sqrt{n}.$$

The random variable W_n is approximately normally distributed for large n, because $X_1 + \cdots + X_n$ is approximately normally distributed and a linear transformation of a normally distributed random variable is again normally distributed. The fact that W_n is normally distributed allows us to give an insightful formula for the profit that the casino will grab with, say, 99% certainty. The standard normal density has 99% of its probability mass to the right of point -2.326. This means that, with a probability of approximately 99%, the profit of the casino over the player is greater than

$$n(1-2p) - 2.326 \times 2[p(1-p)]^{\frac{1}{2}}\sqrt{n}$$

dollars if the player places n bets of one dollar a piece.

In European roulette, the player has a win probability of $p = \frac{18}{37}$ when betting on red. It is interesting to see how quickly the probability of the casino losing money to the player tends to zero as the number (n) of bets placed by the player increases. For European roulette, the casino's loss probability has the values:

loss probability = 0.3553 if $n = 100$
loss probability = 0.1876 if $n = 1{,}000$
loss probability = 0.0033 if $n = 10{,}000$
loss probability = 6.1×10^{-18} if $n = 100{,}000$
loss probability = 3.0×10^{-161} if $n = 1{,}000{,}000$.

This clearly illustrates that the casino will not lose over the long run, notwithstanding the fact that, in the short run, an individual player has a reasonable chance of leaving the casino with a profit. Casinos are naturally more interested in long-run findings because over the long run a great many players will be encountered. The above calculations show that, in European roulette, the casino has a 99% probability of winning an amount of more than $0.02703n - 2.325\sqrt{n}$ dollars from a player when that player bets on n spins of the wheel and stakes one dollar on red each time. This is a steadily growing riskless profit!

5.8.3 Drunkard's walk

In "walking the line," a drunkard repeatedly takes a step to the right with a probability of $\frac{1}{2}$ or a step to the left with a probability of $\frac{1}{2}$. Each step the drunkard takes is of unit length. The consecutive steps are made independently from one another. Let random variable D_n represent the distance of the drunkard from the starting point after n steps. In Section 2.4, it was claimed that

$$E(D_n) \approx \sqrt{\frac{2}{\pi} n}$$

for n large. This claim can easily be proven correct with the help of the central limit theorem. Toward that end, D_n is represented as

$$D_n = |X_1 + \cdots + X_n|,$$

where the random variable X_i is equal to 1 if the ith step of the drunkard goes to the right and is otherwise equal to -1. The random variables $X_1, \ldots, X_n$ are independent and have the same distribution with expected value $\mu = 0$ and standard deviation $\sigma = 1$ (verify!). The central limit theorem now tells us that $X_1 + \cdots + X_n$ is approximately normally distributed with expected value 0 and standard deviation $\sqrt{n}$ for n large. Next, it is a question of integral calculus to arrive at the approximate expression for $E(D_n)$. The necessary tools to compute the expected value of a random variable $V = |W|$ with W having a normal distribution are provided in Chapter 10. Applying those tools gives the approximate expression for $E(D_n)$ (see Exercise 10.13).

5.8.4 Kelly betting

In Chapter 2, we saw that the Kelly betting system is an attractive system in a repeated sequence of favorable games. This system prescribes betting the same fixed fraction of your current bankroll each time. It maximizes the long-run rate of growth of your bankroll, and it has the property of minimizing the expected time needed to reach a specified but large size of your bankroll. In Section 2.7, the long-run rate of growth was found with the help of the law of large numbers. The central limit theorem enables you to make statements about the number of bets needed to increase your bankroll with a specified factor. Let's recapitulate the Kelly model. You face a sequence of favorable betting opportunities. Each time you can bet any amount up to your current bankroll. The payoff odds are $f - 1$ to 1. That is, in case of a win, the player gets paid out f times the amount staked; otherwise, the player loses the amount staked. The win probability p of the player is typically less than $\frac{1}{2}$, but it is assumed that the product pf is larger than 1 (a favorable bet). Under the Kelly system you bet the same fixed fraction α of your current bankroll each time. Assuming an initial capital of V_0, define the random variable V_n as

$$V_n = \text{ the size of your bankroll after } n \text{ bets.}$$

We ask ourselves the following two questions:

(a) What is the smallest value of n such that

$$E(V_n) \geq aV_0$$

for a given value of $a > 1$?

(b) What is the smallest value of n such that

$$P(V_n \geq aV_0) \geq 0.95$$

for a given value of $a > 1$?

The key to the answers to these questions is the relation

$$V_n = (1 - \alpha + \alpha R_1) \times \cdots \times (1 - \alpha + \alpha R_n)V_0,$$

where $R_1, \ldots, R_n$ are independent random variables with

$$P(R_i = f) = p \qquad \text{and} \qquad P(R_i = 0) = 1 - p.$$

This relation was obtained in Section 2.7. Next, note that

$$\ln(V_n) = \ln(1 - \alpha + \alpha R_1) + \cdots + \ln(1 - \alpha + \alpha R_n) + \ln(V_0).$$

Hence, except for the term $\ln(V_0)$, the random variable $\ln(V_n)$ is the sum of n independent random variables each having the same distribution. Denoting by μ and σ^2 the expected value and the variance of the random variables $\ln(1 - \alpha + \alpha R_i)$, then

$$\mu = p\ln(1 - \alpha + \alpha f) + (1 - p)\ln(1 - \alpha)$$

and

$$\sigma^2 = p[\ln(1 - \alpha + \alpha f) - \mu]^2 + (1 - p)[\ln(1 - \alpha) - \mu]^2.$$

The central limit theorem tells us that $\ln(V_n)$ is approximately $N(n\mu + \ln(V_0), n\sigma^2)$ distributed for n large. Next, we invoke a basic result that will be proved in Chapter 10. If the random variable U has a $N(\nu, \tau^2)$ distribution, then the random variable e^U has a so-called lognormal distribution with expected value $e^{\nu + \tau^2/2}$. This means that the random variable V_n is approximately lognormally distributed with expected value $e^{n\mu+\ln(V_0)+n\sigma^2/2}$ for n large. Thus, Question (a) reduces to finding the value of n for which

$$e^{n\mu+\ln(V_0)+n\sigma^2/2} \approx a V_0,$$

or, $n\mu + n\sigma^2/2 \approx \ln(a)$. For the data $V_0 = 1, a = 2, f = 3$, and $p = 0.4$, the optimal Kelly fraction is $\alpha = 0.1$. After some calculations, we find the answer $n = 36$ bets for Question (a). In order to answer Question (b), note that

$$\begin{aligned} P(V_n \geq a V_0) &= P\big(\ln(V_n) \geq \ln(a) + \ln(V_0)\big) \\ &= P\left(\frac{\ln(V_n) - n\mu - \ln(V_0)}{\sigma\sqrt{n}} \geq \frac{\ln(a) - n\mu}{\sigma\sqrt{n}}\right). \end{aligned}$$

The standardized variable $[\ln(V_n) - n\mu - \ln(V_0)]/[\sigma\sqrt{n}]$ has approximately a standard normal distribution for n large. Thus, the answer to Question (b) reduces to find the value of n for which

$$1 - \Phi\left(\frac{\ln(a) - n\mu}{\sigma\sqrt{n}}\right) \approx 0.95.$$

For the data $V_0 = 1, a = 2, f = 3$, and $p = 0.4$, the optimal Kelly fraction is $\alpha = 0.1$. After some calculations, we find the answer $n = 708$ bets for question (b).

5.8.5 Brownian motion[7]

Random movements are abundant in nature: butterfly movement, smoke particles in the air, or pollen particles in a water droplet. In 1828, the British botanist Robert Brown noticed that while studying tiny particles of plant pollen in water under a microscope, these pieces of pollen traveled about randomly. This apparently obscure phenomenon played a key role in the revolution that occurred in the field of physics in the first decade of the twentieth century. In a landmark 1905 paper, Einstein explained the motion of a tiny particle of pollen was the result of its collisions with water molecules. The rules describing this random motion are pretty similar to the rules describing the random walk of a drunkard. The random walk model and the Brownian motion model are among the most useful probability models in science. Brownian motion appears in an extraordinary number of places. It plays not only a crucial role in physics, but it is also widely applied to the modeling of financial markets. Think of a stock price as a small particle which is "hit" by buyers and sellers. The first mathematical description of stock prices utilizing Brownian motion was given in 1900 by the

[7]This section contains advanced material.

French mathematician Louis Bachelier (1870–1946), who can be considered as the founding father of modern option pricing theory. His innovativeness, however, was not fully appreciated by his contemporaries, and his work was largely ignored until the 1950s.

This section is aimed at giving readers a better perception of Brownian motion. We present an intuitive approach showing how Brownian motion can be seen as a limiting process of random walks. The central limit theorem is the link between the random walk model and the Brownian motion model. Let's assume a particle that makes every Δ time units either an upward jump or a downward jump of size δ with probabilities p and $1-p$, where δ and p depend on Δ. The idea is to choose smaller and smaller step sizes for the time and to make the displacements of the random walk smaller as well. As the step sizes get closer and closer to zero, the discrete-time random walk looks more and more like a continuous-time process, called the Brownian motion. To make this more precise, let us fix for the moment the step size Δ. For any $t > 0$, let us define the random variable $X^{\Delta}(t)$ as

$$X^{\Delta}(t) = \text{the position of the particle at time } t.$$

It is assumed that the initial position of the particle is at the origin. The random variable $X^{\Delta}(t)$ can be represented as the sum of independent random variables X_i with

$$X_i = \begin{cases} \delta & \text{with probability } p \\ -\delta & \text{with probability } 1-p. \end{cases}$$

Letting $\lfloor u \rfloor$ denote the integer that results by rounding down the number u, it holds for any $t > 0$ that

$$X^{\Delta}(t) = X_1 + \cdots + X_{\lfloor t/\Delta \rfloor}.$$

Invoking the central limit theorem, it follows that the random variable $X^{\Delta}(t)$ is approximately normally distributed for t large. Using the fact that

$$E(X_i) = (2p-1)\delta \qquad \text{and} \qquad \text{Var}(X_i) = 4p(1-p)\delta^2$$

(verify!), the expected value and the variance of $X^{\Delta}(t)$ are given by

$$E[X^{\Delta}(t)] = \lfloor t/\Delta \rfloor (2p-1)\delta \qquad \text{and} \qquad \text{Var}[X^{\Delta}(t)] = \lfloor t/\Delta \rfloor 4p(1-p)\delta^2.$$

By choosing the displacement size δ and the displacement probability p in a proper way as function of the time-step size Δ and letting Δ tend to zero, we can achieve for any $t > 0$ that

$$\lim_{\Delta \to 0} E[X^{\Delta}(t)] = \mu t \qquad \text{and} \qquad \lim_{\Delta \to 0} \text{Var}[X^{\Delta}(t)] = \sigma^2 t$$

for given numbers μ and σ with $\sigma > 0$. These limiting relations are obtained by taking

$$\delta = \sigma\sqrt{\Delta} \quad \text{and} \quad p = \frac{1}{2}\left\{1 + \frac{\mu}{\sigma}\sqrt{\Delta}\right\}.$$

It is a matter of simple algebra to verify this result. The details are left to the reader.

We now have made plausible that the random variable $X^{\Delta}(t)$ converges in a probabilistic sense to an $N(\mu t, \sigma^2 t)$ distributed random variable $X(t)$ when the time-step size Δ tends to zero and δ and p are chosen according to $\delta = \sigma\sqrt{\Delta}$ and $p = \frac{1}{2}\{1 + \frac{\mu}{\sigma}\sqrt{\Delta}\}$. The random variable $X(t)$ describes the position of the particle in a continuous-time process at time t. Using deep mathematics, it can be shown that the random process $\{X(t)\}$ has the following properties:

(a) the sample paths of the process are continuous functions of t
(b) the increments $X(t_1) - X(t_0)$, $X(t_2) - X(t_1), \ldots, X(t_n) - X(t_{n-1})$ are independent for all $0 \leq t_0 < t_1 < \cdots < t_{n-1} < t_n$ and $n > 1$
(c) $X(s + t) - X(s)$ is $N(\mu t, \sigma^2 t)$ distributed for all $s \geq 0$ and $t > 0$.

A random process $\{X(t)\}$ having these properties is called a *Brownian motion* with drift parameter μ and variance parameter σ^2. The Brownian motion process is often referred to as the Wiener process after the American mathematician Norman Wiener who laid the mathematical foundation of Brownian motion and showed the existence of a random process $X(t)$ satisfying properties (a)–(c). A peculiar feature of Brownian motion is that the probability of occurrence of a sample path being either decreasing or increasing on any finite time interval is zero, no matter how short the interval is. In other words, the sample paths are very kinky and nowhere differentiable, although they are continuous functions of the time t. An intuitive explanation of this remarkable property is as follows. Divide any given small time interval of length L in many smaller disjoint subintervals of length Δ and note that the increments of the Brownian motion in the disjoint subintervals are independent. Each increment is normally distributed with mean $\mu\Delta$ and thus takes on a positive or a negative value each with an approximate probability of 0.5 as Δ tends to zero. The probability of having increments of the same sign in all of the L/Δ subintervals is thus of the order $0.5^{L/\Delta}$ and tends to zero as Δ approaches zero. This explains why a typical Brownian path is nowhere differentiable, in agreement with the phenomenon that a Brownian particle jiggles about randomly. It is instructive to simulate Brownian motion on the computer. In Monte Carlo simulation, the position of the particle is numerically advanced with the update equation $X(t + \Delta) = X(t) + I(\Delta)$ for a small time-step Δ, where $I(\Delta)$ is $N(\mu\Delta, \sigma^2\Delta)$ distributed. An effective method

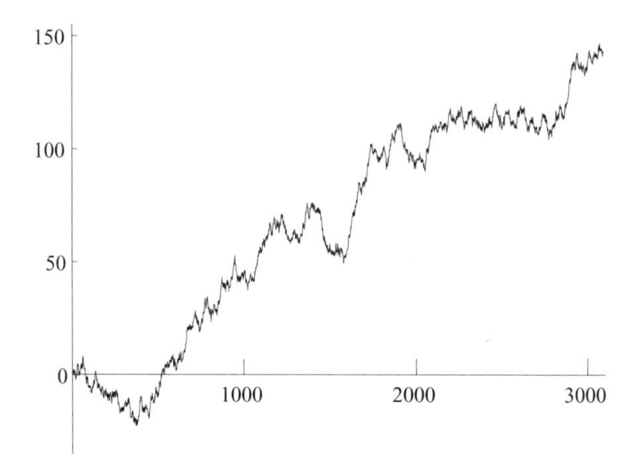

Figure 5.8 A realization of Brownian motion.

to simulate random observations from a normal distribution is given in Section 11.3. Figure 5.8 displays a simulated realization of Brownian motion.

As pointed out before, Brownian motion has applications in a wide variety of fields. In particular, the application of Brownian motion to the field of finance received a great deal of attention. It was found that the logarithms of common-stock prices can often be very well modeled as Brownian motions. In agreement with this important finding is the result we found in the previous paragraph for the wealth process in the situation of Kelly betting. Other important applications of Brownian motion arise by combining the theory of fractals and Brownian motion. Fractals refer to images in the real world that tend to consist of many complex patterns that recur at various sizes. The fractional Brownian motion model regards naturally occurring rough surfaces like mountains and clouds as the end result of random walks.

5.8.6 Stock prices and Brownian motion

Let's assume a stock whose price changes every Δ time units, where Δ denotes a small increment of time. Each time the price of the stock goes up by the factor δ with probability p or goes down by the same factor δ with probability $1-p$, where δ and p are given by

$$\delta = \sigma\sqrt{\Delta} \qquad \text{and} \qquad p = \frac{1}{2}\left\{1 + \frac{\mu}{\sigma}\sqrt{\Delta}\right\}$$

for given values of μ and σ with $\sigma > 0$. It is assumed that Δ is small enough such that $0 < \delta < 1$ and $0 < p < 1$. The initial price of the stock is S_0. If we let the time step Δ tend to zero, what happens to the random process describing the stock price? Letting S_t denote the stock price at time t in the limiting process, the answer is that the random process describing $\ln(S_t/S_0)$ is a Brownian motion with drift parameter $\mu - \frac{1}{2}\sigma^2$ and variance parameter σ^2. An intuitive explanation is as follows. Denote by S_t^Δ the stock price at time t when the stock price changes every Δ time units. Then, for any $t > 0$,

$$S_t^\Delta = (1 + X_1) \times \cdots \times (1 + X_{\lfloor t/\Delta \rfloor})S_0,$$

where $X_1, X_2, \ldots$ are independent random variables with

$$P(X_i = \delta) = p \quad \text{and} \quad P(X_i = -\delta) = 1 - p.$$

Hence

$$\ln\left(\frac{S_t^\Delta}{S_0}\right) = \sum_{i=1}^{\lfloor t/\Delta \rfloor} \ln(1 + X_i).$$

The next step is to use the power series expansion

$$\ln(1 + x) = x - \frac{x^2}{2} + \frac{x^3}{3} - \cdots, \qquad |x| < 1.$$

Letting Δ tend to zero, we may assume that t/Δ is an integer. Noting that $|X_i| = \sigma\sqrt{\Delta}$, we obtain

$$\sum_{i=1}^{t/\Delta} \frac{X_i^2}{2} = \frac{1}{2}\sigma^2 t.$$

Also, for Δ small, $\sum_{i=1}^{t/\Delta} \frac{1}{3}X_i^3$ is on the order of $\sqrt{\Delta}$, $\sum_{i=1}^{t/\Delta} \frac{1}{4}X_i^4$ is on the order of Δ, and so on. Hence, the contribution of these terms become negligible as $\Delta \to 0$. Thus, using the expansion of $\ln(1 + X_i)$, we find that $\ln\left(S_t^\Delta/S_0\right)$ is approximately distributed as

$$\sum_{i=1}^{t/\Delta} X_i - \frac{1}{2}\sigma^2 t$$

for Δ small. It was argued earlier that the random walk process $\sum_{i=1}^{t/\Delta} X_i$ becomes a Brownian motion with drift parameter μ and variance parameter σ^2 when the time-step Δ tends to zero. The sum of an $N(\nu, \tau^2)$ random variable and a constant a has an $N(a + \nu, \tau^2)$ distribution. This is the last step in the intuitive explanation that the process describing $\ln(S_t/S_0)$ is a Brownian motion with drift parameter $\mu - \frac{1}{2}\sigma^2$ and variance parameter σ^2.

5.8.7 Black-Scholes formula

The Black-Scholes formula is the most often-used formula with probabilities in finance. It shows how to determine the value of an option. An option is a financial product written on another financial product. The latter is typically referred to as the "underlying." A call option gives the holder the right, but not the obligation, to buy some underlying stock at a given price, called the exercise price, on a given date. The buyer pays the seller a premium for this right. The premium is the value of the option. Taking $t = 0$ as the current date, let

$$\begin{aligned} T &= \text{time to maturity of the option (in years)} \\ S_0 &= \text{current stock price (in dollars)} \\ K &= \text{exercise price of the option (in dollars)} \\ r &= \text{risk-free interest rate (annualized)} \\ \sigma &= \text{underlying stock volatility (annualized).} \end{aligned}$$

The risk-free interest rate is assumed to be continuously compounded. Thus, if $r = 0.07$, this means that in one year \$1 will grow to $e^{0.07}$ dollars. The volatility parameter σ is nothing else than the standard deviation parameter of the Brownian motion process that is supposed to describe the process $\ln(S_t/S_0)$ with S_t denoting the price of the underlying stock at time t. On the basis of economical considerations, the drift parameter η of this Brownian motion process is chosen as

$$\eta = r - \frac{1}{2}\sigma^2.$$

An intuitive explanation for this choice is as follows. In an efficient market, it is reasonable to assume that betting on the price change of the stock in a short time interval is a fair bet. That is, the condition $E(S_\Delta) - e^{r\Delta}S_0 = 0$ is imposed for Δ small. Letting $W = \ln(S_\Delta/S_0)$ and using the fact that $e^{\ln(a)} = a$, we have $E(S_\Delta/S_0) = E\left(e^W\right)$. Since W is $N(\eta\Delta, \sigma^2\Delta)$ distributed, a basic result for the normal distribution tells us that $E\left(e^W\right) = e^{\eta\Delta + \frac{1}{2}\sigma^2\Delta}$ (see Example 14.4 in Chapter 14). Thus, the condition $E(S_\Delta) - e^{r\Delta}S_0 = 0$ is equivalent to $e^{\eta\Delta+\frac{1}{2}\sigma^2\Delta} = e^{r\Delta}$, yielding $\eta = r - \frac{1}{2}\sigma^2$. In the real world, the volatility parameter σ is estimated from the sample variance of the observations $\ln(S_{i\Delta}/S_{(i-1)\Delta})$ for $i = 1, 2, \ldots, h$ (say, $h = 250$) over the last h trading days of the stock.

We now turn to the determination of the price of the option. To do so, it is assumed that the option can only be exercised at the maturity date T. Furthermore, it is assumed that the stock will pay no dividend before the maturity date. The option will be exercised at the maturity date T only if $S_T > K$. Hence, at maturity, the option is worth

$$C_T = \max(0, S_T - K).$$

The net present value of the worth of the option is $e^{-rT} C_T$. Using the fact that $\ln(S_T/S_0)$ has an $N\big((r - \frac{1}{2}\sigma^2)T, \sigma^2 T\big)$ distribution, it is matter of integral calculus and standard formulas to evaluate the expression for the option price $e^{-rT} E(C_T)$. This gives the *Black-Scholes formula*

$$e^{-rT} E(C_T) = \Phi(d_1) S_0 - \Phi(d_2) K e^{-rT}$$

with

$$d_1 = \frac{\ln(S_0/K) + (r + \frac{1}{2}\sigma^2)T}{\sigma\sqrt{T}} \qquad \text{and} \qquad d_2 = d_1 - \sigma\sqrt{T},$$

where $\Phi(x)$ is the standard normal distribution function. This beautiful mathematical formula was developed by Fisher Black, Robert Merton, and Myron Scholes. Its publication in 1973 removed the guesswork and reliance on individual brokerage firms from options pricing and brought it under a theoretical framework that is applicable to other derivative products as well. The Black-Scholes formula changed the world, financial markets, and indeed capitalism as well. It helped give rise to a standardized options industry dealing in the hundreds of billions of dollars.

As a numerical illustration, consider a European call option on 100 shares of a nondividend-paying stock ABC. The option is struck at \$50 and expires in 0.3 years. ABC is trading at \$51.25 and has 30% implied volatility. The risk-free interest is 7%. What is the value of the option? Applying the Black-Scholes formula with $S_0 = 51.25$, $K = 50$, $T = 0.3$, $r = 0.07$, and $\sigma = 0.3$, the value of the option per share of ABC is \$4.5511. The call option is for 100 shares and so it is worth \$455.11. In doing the calculations, the values of the standard normal distribution function $\Phi(x)$ were calculated from the polynomial-type of approximation

$$\Phi(x) \approx 1 - \frac{1}{2}\left(1 + d_1 x + d_2 x^2 + d_3 x^3 + d_4 x^4 + d_5 x^5 + d_6 x^6\right)^{-16}, \qquad x \geq 0,$$

where the constants $d_1, \ldots, d_6$ are given by

$$d_1 = 0.0498673470,\ d_2 = 0.0211410061,\ d_3 = 0.0032776263,$$
$$d_4 = 0.0000380036,\ d_5 = 0.0000488906,\ d_6 = 0.0000053830.$$

The absolute error of this approximation is less than 1.5×10^{-7} for all $x \geq 0$.

5.9 Falsified data and Benford's law

Most people have preconceived notions of randomness that often differ substantially from true randomness. Truly random datasets often have unexpected properties that go against intuitive thinking. These properties can be used to test whether datasets have been tampered with when suspicion arises. To illustrate, suppose that two people are separately asked to toss a fair coin 120 times and take note of the results. Heads is noted as a "one" and tails as a "zero." The following two lists of compiled zeros and ones result:

1 1 0 0 1 0 0 1 0 1 1 0 0 1 0 0 0 1 1 0
1 0 1 0 0 1 1 0 1 0 0 1 0 1 0 1 1 0 1 1
0 0 1 1 0 1 1 1 0 1 0 0 1 0 0 1 1 0 1 0
0 1 1 0 1 0 0 1 1 0 1 0 1 1 0 0 1 1 1 0
0 1 0 1 0 1 0 0 0 1 0 1 0 1 0 1 0 1 0 1
1 0 0 1 0 0 1 0 1 1 0 0 1 0 0 1 1 0 1 1

and

1 1 1 0 0 0 1 1 1 0 1 0 1 1 1 1 1 1 0 1
0 0 0 1 1 0 0 1 1 0 1 0 1 0 0 0 1 1 0 1
0 0 1 1 1 0 1 0 0 0 0 1 0 1 1 1 0 1 1 0
0 1 1 1 0 1 1 0 0 1 1 1 1 1 1 0 1 1 0 1
0 1 1 1 0 0 0 0 0 0 0 0 1 1 0 1 1 1 0 1
1 1 1 0 1 1 1 1 0 1 0 1 1 0 1 1 0 1 0 1.

One of the two individuals has cheated and has fabricated a list of numbers without having tossed the coin. Which is the fabricated list? The key to solving this dilemma lays in the fact that in 120 tosses of a fair coin, there is a very large probability that *at some point* during the tossing process, a sequence of five or more heads or five or more tails will naturally occur. The probability of this is 0.9865. In contrast to the second list, the first list shows no such sequence of five heads in a row or five tails in a row. In the first list, the longest sequence of either heads or tails consists of three in a row. In 120 tosses of a fair coin, the probability of the longest sequence consisting of three or less in a row is equal to 0.000053, which is extremely small indeed. Thus, the first list is almost certainly a fake. Most people tend to avoid noting long sequences of consecutive heads or tails. Truly random sequences do not share this human tendency!

5.9.1 Success runs[8]

How can we calculate the probability of the occurrence of a success run of a certain length in a given number of coin tosses? Among other things, this probability comes in handy when tackling questions such as the one posed in Chapter 1: what is the probability of a basketball player with a 50% success rate shooting five or more baskets in a row in twenty attempts? We learned in Section 2.1.3, with the help of computer simulation, that the player has approximately a 25% probability of achieving such a lengthy success run. However, this probability can also be exactly calculated. In this paragraph, we give an exact method to use in answering the following question: what is the probability of getting a run of r heads in n fair coin tosses? To answer this question, let's say that the tossing process is in state (i, k) when there are still k tosses to go and heads came up in the last i tosses but so far a run of r heads has not occurred. Define

$$u_k(i) = \text{the probability of getting a run of } r \text{ heads during } n \text{ tosses when the current state of the tossing process is } (i, k).$$

The index k runs through $0, 1, \ldots, n$ and the index i through $0, 1, \ldots, r$. The probability $u_n(0)$ is being sought. To set up a recursion equation for the probability $u_k(i)$, we condition on the outcome of the next toss after state (i, k). Heads comes up in the next toss with probability $\frac{1}{2}$. If this happens, the next state of the tossing process is $(i + 1, k - 1)$; otherwise, the next state is $(0, k - 1)$. Thus, by the law of conditional probabilities, we find the following recursion for $k = 1, 2, \ldots, n$:

$$u_k(i) = \frac{1}{2}u_{k-1}(i + 1) + \frac{1}{2}u_{k-1}(0) \quad \text{for} \quad i = 0, 1, \ldots, r - 1.$$

This recursion equation has the boundary conditions

$$u_0(i) = 0 \quad \text{for} \quad 0 \leq i \leq r - 1 \qquad \text{and} \qquad u_k(r) = 1 \quad \text{for} \quad 0 \leq k \leq n - 1.$$

The recursion equation leads to a simple method in order to calculate the probability $u_n(0)$ exactly. Beginning with $u_0(i) = 0$ for $0 \leq i \leq r - 1$ and $u_0(i) = 1$ for $i = r$, we first calculate $u_1(i)$ for $0 \leq i \leq r$, then $u_2(i)$ for $0 \leq i \leq r$ and going on recursively, we eventually arrive at the desired probability $u_n(0)$.

Applying the recursion with $n = 20$ and $r = 5$ leads to the value 0.2499 for the probability that in twenty shots a basketball player with a successful shot rate of 50% will shoot five or more baskets in a row (Problem 2 from Chapter 1). This is the same value as was found earlier with computer simulation in Chapter

[8] This section contains advanced material.

2. Isn't it fascinating to see how two fundamentally different approaches lead to the same answer?

A similar recursion can be given to calculate the probability that in n fair coin tosses a run of r heads or r tails occurs. In this case, we say that the tossing process is in state (i, k) when there are k tosses still to go and the last i tosses all showed the same outcome but so far no run of r heads or r tails has occurred. The probability $v_k(i)$ is defined as

$$v_k(i) \;=\; \text{the probability of getting a run of } r \text{ heads or } r \text{ tails during } n \text{ tosses when the current state of the tossing process is } (i, k).$$

The probability $v_{n-1}(1)$ is being sought (why?). Verify for yourself that the following recursion applies for $k = 1, 2, \ldots, n$:

$$v_k(i) = \frac{1}{2}v_{k-1}(i+1) + \frac{1}{2}v_{k-1}(1) \quad \text{for} \quad i = 1, \ldots, r-1.$$

The boundary conditions are $v_0(i) = 0$ for $1 \le i \le r-1$ and $v_j(r) = 1$ for $0 \le j \le n-1$. If you apply the recursion with $n = 120$ and $r = 5$, then you arrive at the earlier found value $v_{n-1}(1) = 0.9865$ for the probability of tossing five heads or five tails in a row in 120 fair coin tosses. The recursion with $n = 120$ and $r = 4$ gives the value $1 - v_{n-1}(1) = 0.000053$ for the probability that in 120 fair coin tosses the longest run of either heads or tails has a length of no more than three.

5.9.2 Benford's law

In 1881, the astronomer/mathematician Simon Newcomb published a short article in which he noticed that the pages of logarithm tables with small initial digits were dirtier than those with larger initial digits. Apparently, numbers beginning with 1 were more often looked up than numbers beginning with 2, and numbers beginning with 2 more often than numbers beginning with 3, etc. Newcomb quantified this surprising observation in a logarithmic law giving the frequencies of occurrence of numbers with given initial digits. This law became well known for the first time, many years later, as Benford's law. In 1938, physicist Frank Benford rediscovered the "law of anomalous numbers," and published an impressive collection of empirical evidence supporting it. Benford's law says that in many naturally occurring sets of numerical data, the first significant (nonzero) digit of an arbitrarily chosen number is not equally likely to be any one of the digits $1, \ldots, 9$, as one might expect, but instead is

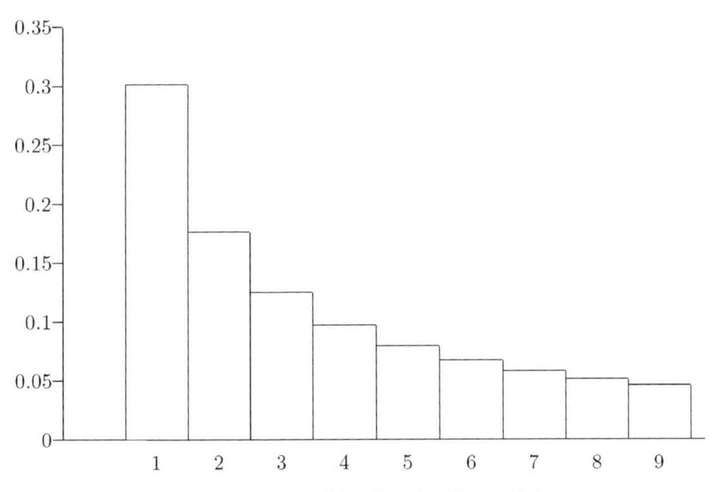

Figure 5.9 Probability distribution of the first significant digit.

closely approximated by the logarithmic law

$$P(\text{first significant digit} = d) = \log_{10}\left(1 + \frac{1}{d}\right) \quad \text{for} \quad d = 1, 2, \ldots, 9.$$

Figure 5.9 shows the values of these probabilities. Benford's empirical evidence showed that this logarithmic law was fairly accurate for the numbers on the front pages of newspapers, the lengths of rivers, universal constants in physics and chemistry, numbers of inhabitants of large cities, and many other tables of numerical data. It appeared that the logarithmic law was a nearly perfect approximation if all these different data sets were combined. Of course, not every dataset follows Benford's law. For example, consider the times for the Olympic 400 meter race. Very few of those times will begin with a 1! The same is true for the telephone numbers in New York City.

This led to the question of what properties "natural" datasets must satisfy in order to follow Benford's law. It can be proven mathematically that, if a collection of numbers satisfies Benford's law, then the same collection still satisfies this law if every number in the collection is multiplied by the same positive constant. This shows, for example, that, for Benford's law, it does not matter whether the lengths of rivers are expressed in miles or in kilometers. Moreover, the logarithmic distribution is the only distribution that is scale invariant. This still does not explain why so many "natural" datasets satisfy Benford's law. An explanation for this phenomenon was recently given by the American mathematician Ted Hill. Roughly speaking, Hill showed the following: if numbers are selected at *random* from *different* arbitrarily chosen collections of data, then the numbers in the *combined* sample will tend to follow Benford's law. The larger and more varied the sample from the different datasets, the more likely it is that the relative frequency of the first significant digits will tend to obey Benford's law. This result offers a plausible theoretical explanation, for example, of the fact that the numbers from the front pages of newspapers are a very good fit to Benford's law. Those numbers typically arise from many sources, and are influenced by many factors.

Benford's law, which at first glance appears bizarre, does have practical applications. The article by Ted Hill, "The difficulty of faking data," *Chance Magazine,* 1999, 12(3), 27–31, discusses an interesting application of Benford's law to help detect possible fraud in tax returns. Empirical research in the United States has shown, for example, that in actual tax returns that correctly reported income, the entries for interest paid, and interest received are a very good fit to Benford's law. Companies' returns that deviate from this law, over the course of many years, appear to be fraudulent in many cases.

5.10 The normal distribution strikes again

How to pick a winning lottery number is the subject of many a book about playing the lottery. The advice extended in these entertaining books is usually based on the so-called secret of balanced numbers. Let's take Lotto 6/49 as an example. In the Lotto 6/49, the player must choose six different numbers from the numbers $1, \ldots, 49$. For this lottery, players are advised to choose six numbers whose sum add up to a number between 117 and 183. The basic idea here is that the sum of six randomly picked numbers in the lottery is approximately normally distributed. Indeed, this is the case. In Lotto 6/49, the sum of the six winning numbers is approximately normally distributed with expected value 150 and a spread of 32.8. It is known that a sample from the normal distribution with expected value μ and spread σ will be situated between $\mu - \sigma$ and $\mu + \sigma$, with a probability of approximately 68%. This is the basis for advising players to choose six numbers that add up to a number between 117 and 183. The reasoning is that playing such a number combination raises the probability of winning a 6/49 Lottery prize. This is nothing but poppycock. It is true enough that we can predict which stretch the sum of the six winning numbers will fall into, but the combination of six numbers that adds up to a given sum can in no way be predicted. In Lotto 6/49, there are in total 165,772 combinations of six numbers that add up to the sum of 150. The probability of the winning numbers adding up to a sum of 150 is equal to 0.0118546. If you divide this probability by 165,772, then you get the exact value of $1/\binom{49}{6}$ for the probability that a *given* combination of six numbers will be drawn!

We speak of the Lotto r/s when r different numbers are randomly drawn from the numbers $1, \ldots, s$. Let the random variable X_i represent the ith number drawn. The random variables $X_1, \ldots, X_r$ are dependent, but for reasons of symmetry, each of these random variables has the same distribution. From $P(X_1 = k) = 1/s$ for $k = 1, \ldots, s$, it follows that $E(X_1) = \frac{1}{2}(s+1)$. This leads to

$$E(X_1 + \cdots + X_r) = \frac{1}{2} r (s+1).$$

It is stated without proof that

$$\sigma^2(X_1 + \cdots + X_r) = \frac{1}{12} r (s-r)(s+1).$$

Also, for r and $s - r$ both sufficiently large, it can proved that the sum $X_1 + \cdots + X_r$ is approximately normally distributed with expected value $\frac{1}{2} r (s+1)$ and variance $\frac{1}{12} r (s-r)(s+1)$. Figure 5.10 displays the simulated frequency diagram of $X_1 + \cdots + X_r$ for $r = 6$ and $s = 49$. The simulation consisted of one million runs. A glance at Figure 5.10 confirms that the probability histogram of $X_1 + \cdots + X_r$ can indeed be approximated by the normal density function.

Statistics and probability theory

In this chapter, we have already introduced several statistical problems. Statistics and probability theory are distinct disciplines. Probability theory is a branch of mathematics. In mathematics, we reason from the general to the specific.

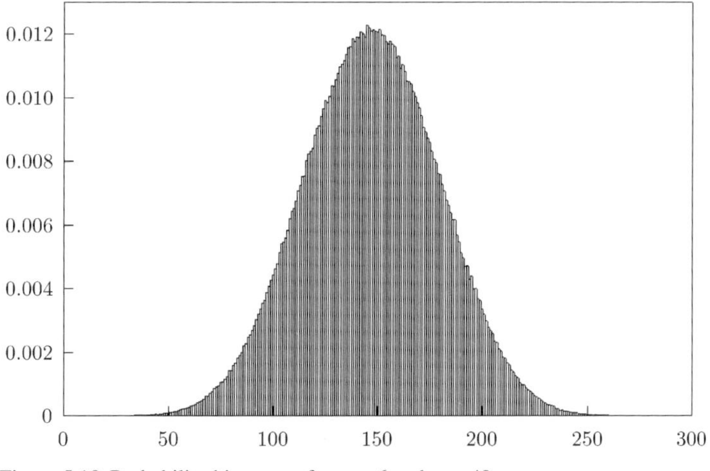

Figure 5.10 Probability histogram for $r = 6$ and $s = 49$.

Given a number of axioms, we can derive general propositions that we can then apply to specific situations. This is called *deductive* reasoning. The deductive nature of probability theory is clearly demonstrated in Chapter 7. Statistics, on the other hand, works the other way around by reasoning from the specific to the general. Statistics is therefore a science based on *inductive* reasoning. In statistics, we attempt to draw more generally valid conclusions based on data obtained from a specific situation. For example, statisticians attempt to discern the general effectiveness of new medicines based on their effectiveness in treating limited groups of test patients. To do so, statisticians must select a method based on one of two schools of thought. Most statisticians base their methods on the *classical* approach, whereas others base their methods on the *Bayesian* approach. In the classical approach, the test of the null hypothesis is based on the idea that any observed deviation from what the null hypothesis predicts is solely the product of chance. If something that is unusual under the null hypothesis happens, then the null hypothesis is rejected. It is common to use a significance level of 5% or 1% as a benchmark for the probability to be judged. Note that in the classical approach, the probability of rejecting the null hypothesis is not the same as the probability that the null hypothesis is false.

The Bayesian approach assumes an *a priori* probability distribution (called the *prior*) as to whether or not the null hypothesis is true. The prior distribution is then updated in the light of the new observations. Simply put, the classical approach is based on $P(\text{data}|H_0)$, whereas the Bayesian approach makes use of $P(H_0|\text{data})$. The need to specify a prior distribution before analyzing the data introduces a subjective element into the analysis and this is often regarded as a weakness of the Bayesian approach. It is, however, important to keep in mind that the classical approach is not entirely objective either. The choice whether to reject the null hypothesis at the 5% significance level instead of at, for example, the 0.1% level is also subjective. The fundamental difference between the classical and Bayesian approach can be best illustrated via Example 5.1. This example deals with a multiple-choice exam consisting of 50 questions, each of which has three possible answers. A student receives a passing grade if he/she correctly answers more than half of the questions. Take the case of a student who manages to answer 26 of the 50 questions correctly and claims not to have studied, but rather to have obtained 26 correct answers merely by guessing. In Section 5.6, we see that the classical approach is based on the probability that 26 or more of the 50 questions could be answered by luck alone. The Bayesian approach is based on a different probability: the probability that the student could have guessed each answer given that he/she answered 26 of the 50 questions correctly. The Bayesian approach to determining this probability requires that we first specify a prior distribution for the various ways the student may have prepared for the exam. This distribution concerns the situation *before* the exam and can be a purely subjective assessment (although it can also be based on information of the student's earlier academic performance on homework or previous exams). Let us assume for simplicity's sake that there are only two possibilities: either the student was totally unprepared (hypothesis H) or that the student was well prepared (the complementary hypothesis $\overline{H}$). We furthermore assume that the assessment before the exam was that with a probability of 50%, the student was well prepared. In other words, $P(H) = P(\overline{H}) = \frac{1}{2}$. Using the binomial distribution with $n = 50$ and $p = \frac{1}{3}$, it then follows that with a probability of 0.00492 the student could pass the exam if the student did not study (and therefore guessed the answer to all of the questions). Let us now make the additional assumption that based on experience it is known that a well-prepared student passes an exam 70% of the time. The Bayesian approach then enables us to conclude that with a probability of 0.7% the student did not study and could only complete the exam by guessing, given the fact that he/she passed the exam (see Problem 8.8 in Chapter 8). Although this represents at least partially a subjective estimate, it is in any case based on a "reasonable" choice for the prior probabilities. If we had instead assumed the prior probabilities $P(H) = 0.8$ and $P(\overline{H}) = 0.2$, then our estimated Bayesian probability would have been 2.7%. Generally speaking, we come to the same conclusion we found using the classical approach: it is very likely that the student is bluffing if he/she claims to have passed the exam without studying.

The following example also clearly demonstrates the differences between classical and Bayesian statistics.[9] Imagine that you participate in a game requiring you to guess the number of heads resulting from 50 coin tosses. We would expect approximately 25 heads, but imagine that the actual result is 18 heads. Is this result the product of chance, or is the coin not a fair one? The Bayesian approach makes it possible to estimate the probability that heads results less than 50% of the time given the observation of 18 heads. The approach requires the specification of the prior probability of obtaining heads based only on information available before the game begins. In the classical approach, on the other hand, we determine the probability of 18 or fewer heads given the hypothesis that the coin is fair. This, however, does not result in a statement about the probability of the coin being fair. For such a statement, we need Bayesian

[9] An especially readable book on Bayesian statistics with emphasis on medical applications is D. A. Berry's *Statistics: A Bayesian Perspective*, Duxbury Press, Belmont, CA, 1996. Of particular relevance in the case of medical applications, and contrary to the classical approach, the Bayesian approach permits us to draw intermediate conclusions based on partial results from an ongoing experiment and, as a result, to modify the future course of the experiment in light of these conclusions.

analysis. Bayesian statistics is discussed in more detail in Chapter 8. The spirit of Reverend Bayes (1702–1761) is still very much alive!

5.11 Problems

5.1 You draw twelve random numbers from (0,1) and average these twelve random numbers. Which of the following statements is then correct?
(a) the average has the same uniform distribution as each of the random numbers
(b) the distribution of the average becomes more concentrated in the middle and less at the ends.

5.2 Someone has written a simulation program in an attempt to estimate a particular probability. Five hundred simulation runs result in an estimate of 0.451 for the unknown probability with 0.451 ± 0.021 as the corresponding 95% confidence interval. One thousand simulation runs give an estimate of 0.453 with a corresponding 95% confidence interval of 0.453 ± 0.010. Give your opinion:
(a) there is no reason to question the programming
(b) there is an error in the simulation program.

5.3 The annual rainfall in Amsterdam is normally distributed with an expected value of 799.5 mm and a standard deviation of 121.4 mm. Over many years, what is the proportion of years that the annual rainfall in Amsterdam is below 550 mm?

5.4 The cholesterol level for an adult male of a specific racial group is normally distributed with an expected value of 5.2 mmol/L and a standard deviation of 0.65 mmol/L. Which cholesterol level is exceeded by 5% of the population?

5.5 Gestation periods of humans are normally distributed with an expected value of 266 days and a standard deviation of 16 days. What is the percentage of births that are more than 20 days overdue?

5.6 In a single-product inventory system a replenishment order will be placed as soon as the inventory on hand drops to the level s. You want to choose the reorder point s such that the probability of a stockout during the replenishment lead time is no more than 5%. Verify that s should be taken equal to $\mu + 1.645\sigma$ when the total demand during the replenishment lead time is $N(\mu, \sigma^2)$ distributed.

5.7 Suppose that the rate of return on stock A takes on the values 30%, 10%, and -10% with respective probabilities 0.25, 0.50, and 0.25 and on stock B the values 50%, 10%, and -30% with the same probabilities 0.25, 0.50, and 0.25. Each stock, then, has an expected rate of return of 10%. Without calculating the actual values of the standard deviation, can you argue why the standard deviation of the rate of return on stock B, is twice as large as that on stock A?

5.8 You wish to invest in two funds, A and B, both having the same expected return. The returns of the funds are negatively correlated with correlation coefficient ρ_{AB}. The standard deviations of the returns on funds A and B are given by σ_A and σ_B. Demonstrate that you can achieve a portfolio with the lowest standard deviation by investing a fraction f of your money in fund A and a fraction $1-f$ in fund B, where the optimal fraction f is given by $\left(\sigma_B^2 - \sigma_A\sigma_B\rho_{AB}\right) / \left(\sigma_A^2 + \sigma_B^2 - 2\sigma_A\sigma_B\rho_{AB}\right)$.

5.9 You want to invest in two stocks A and B. The rates of return on these stocks in the coming year depend on the development of the economy. The economic prospects for the coming year consist of three equally likely case scenarios: a strong economy, a normal economy, and a weak economy. If the economy is strong, the rate of return on stock A will be equal to 34% and the rate of return on stock B will be equal to -20%. If the economy is normal, the rate of return on stock A will be equal to 9.5% and the rate of return on stock B will be equal to 4.5%. If the economy is weak, stocks A and B will have rates of return of -15% and 29%, respectively.

(a) Is the correlation coefficient of the rates of return on the stocks A and B positive or negative? Calculate this correlation coefficient.

(b) How can you divide the investment amount between two stocks if you desire a portfolio with a minimum variance? What are the expected value and standard deviation of the rate of return on this portfolio?

5.10 Suppose that the random variables $X_1, X_2, \ldots, X_n$ are defined on a same probability space. In Chapter 11, it will be seen that

$$\sigma^2\left(\sum_{i=1}^{n} X_i\right) = \sum_{i=1}^{n} \sigma^2(X_i) + 2\sum_{i=1}^{n-1}\sum_{j=i+1}^{n} \operatorname{cov}\left(X_i, X_j\right).$$

For the case that $X_1, \ldots, X_n$ all have the same variance σ^2 and $\operatorname{cov}(X_i, X_j)$ is equal to a constant $c \neq 0$ for all i, j with $i \neq j$, verify that the variance of $\overline{X}(n) = (1/n)\sum_{k=1}^{n} X_k$ is given by

$$\sigma^2\left(\overline{X}(n)\right) = \frac{\sigma^2}{n} + \left(1 - \frac{1}{n}\right)c.$$

In investment theory, the first term σ^2/n is referred to as the nonsystematic risk, and the second term $(1-1/n)c$ is referred to as the systematic risk. The nonsystematic risk can be significantly reduced by diversifying to a large number of stocks, but a bottom-line risk cannot be altogether eliminated. Can you explain this in economic terms?

5.11 Consider the investment example from Section 5.2 in which a retiree invests $100,000 in a fund in order to reap the benefits for twenty years. The rate of return on the fund for the past year was 14%, and the retiree hopes for a yearly profit of $15,098 over the coming twenty years. If the rate of return remained at 14%, for each year, then at the end of the xth year, the invested capital would be equal to $f(x) = (1+r)^x A - \sum_{k=0}^{x-1}(1+r)^k b$ for $x = 1, \ldots, 20$, where $A = 100,000$, $r = 0.14$, and $b = 15,098$. However, the yearly rate of return fluctuates with an average value of 14%. If last year the rate of return was r%, then next year the rate of return will be r%, $(1+f)r$%, or $(1-f)r$% with respective probabilities p, $\frac{1}{2}(1-p)$, and $\frac{1}{2}(1-p)$. For each of the cases, $p = 0.8$, $f = 0.1$ and $p = 0.5$, $f = 0.2$, simulate a histogram of the distribution of the number of years during the twenty-year period that the invested capital at the end of the year will fall below or on the curve of the function $f(x)$.

5.12 The Argus Investment Fund's Spiderweb Plan is a 60-month-long contract according to which the customer agrees to deposit a fixed amount at the beginning of each month. The customer chooses beforehand for a fixed deposit of $100, $250, or $500. Argus then immediately deposits 150 times that monthly amount to remain in the fund over the five-year period (i.e., Argus deposits $15,000 of capital in the fund if the customer opts for the $100 fixed monthly deposit). The monthly amount deposited by the customer is actually the interest payment (8%) on the capital invested by the fund. Five years later, the customer receives the value of the investment minus the initial capital investment. Let's assume that the yearly rate of return on the Argus investment fund fluctuates according to the following probability model: if the return was r% for the previous year, then for the coming year the return will remain at r% with a probability of p_s, will change to $(1-f_d)r$% with a probability of p_d, and will change to $(1+f_u)r$% with a probability of p_u, where $p_u + p_d + p_s = 1$ and $0 < f_d, f_u < 1$. Choose reasonable values for the parameters p_s, p_d, p_u, f_d, and f_u. Simulate a histogram for the probability distribution of the customer's capital after five years. Also, use simulation to estimate the expected value and the standard deviation of the customer's rate of return on the monthly deposits.

5.13 An investor decides to place $2,500 in an investment fund at the beginning of each year for a period of twenty years. The rate of return on the fund was 14% for the previous year. If the yearly rate of return remained at 14% for each year, then, at the end of twenty years, the investor will have an amount of $\sum_{k=1}^{20}(1+0.14)^k 2{,}500 = 259{,}421$ dollars. Suppose now that the yearly rate of return fluctuates according to the following probability model: if last year the rate of return was $r\%$, then during the coming year the rate of return will be $r\%$, $(1 + f)r\%$, or $(1 - f)r\%$ with respective probabilities p, $\frac{1}{2}(1 - p)$, and $\frac{1}{2}(1 - p)$. Using computer simulation, determine the probability distribution of the investor's capital after twenty years.

5.14 Women spend on average about twice as much time in the restroom as men, but why is the queue for the women's restroom on average four or more times as long as the one for the men's? This intriguing question was answered in the article "Ladies in waiting" by Robert Matthews in *New Scientist*, 167 (July 29, 2000): 2249, Explain the answer using the Pollaczek-Khintchine formula discussed in Section 5.2. Assume that there is one restroom for women only and one restroom for men only, the arrival processes of women and men are Poisson processes with equal intensities, and the coefficient of variation of the time people spend in the restroom is the same for women as for men.

5.15 What happens to the value of the probability of getting at least r sixes in one throw of $6r$ dice as $r \to \infty$? Explain your answer.

5.16 The owner of a casino in Las Vegas claims to have a perfectly balanced roulette wheel. A spin of a perfectly balanced wheel stops on red an average of 18 out of 38 times. A test consisting of 2,500 trials delivers 1,105 red finishes. If the wheel is perfectly balanced, is this result plausible? Answer this question by using the normal distribution only.

5.17 Each year in Houndsville, an average of 81 letter carriers are bitten by dogs. In the past year, 117 such incidents were reported. Is this number exceptionally high?

5.18 In a particular area, the number of traffic accidents hovers around an average of 1,050. Last year, however, the number of accidents plunged drastically to 920. Authorities suggest that the decrease is the result of new traffic safety measures that have been in effect for one year. Statistically speaking, is there cause to doubt this explanation? What would your answer be if, based on a yearly average of 105 traffic accidents, the record for the last year decreased to 92 accidents?

5.19 A national information line gets approximately 100 telephone calls per day. On a particular day, only 70 calls come in. Is this extraordinary?

5.20 A large table is marked with parallel and equidistant lines a distance D apart. A needle of length $L\,(\leq D)$ is tossed in the air and falls at random onto the table. The eighteenth century French scientist Georges-Louis Buffon proved that the probability of the needle falling across one of the lines is $\frac{2L}{\pi D}$. The Italian mathematician M. Lazzarini carried out an actual experiment in 1901, where the ratio L/D was taken equal to $5/6$. He made 3,408 needle tosses and observed that 1,808 of them intersected one of the lines. This resulted in a remarkably accurate estimate of 3.14159292 for $\pi = 3.14159265\ldots$ (an error of about 2.7×10^{-7}). Do you believe that Lazzarini performed the experiment in a statistically sound way?

5.21 A gambler claims to have rolled an average of 3.25 points per roll in 1,000 rolls of a fair die. Do you believe this?

5.22 A standard medication cures a particular infection 75% of the time, on average. A new product appears to have cured 85 of the first 100 tested patients. It is suggested that this performance convincingly shows the superiority of the new medication. What is your opinion?

5.23 The Dutch lotto formerly consisted of drawing six numbers from the numbers $1, \ldots, 45$, but the rules were changed. In addition to six numbers from $1, \ldots, 45$, a colored ball is drawn from six distinct colored balls. A statistical analysis of the lotto drawings in the first two years of the new lotto revealed that the blue ball was drawn 33 times in the 107 drawings. The lottery officials hurriedly announced that the painted balls are all of the same weight and that this outcome must have been due to chance. What do you think about this statement?

5.24 In a particular small hospital, approximately 25 babies per week are born, while in a large hospital approximately 75 babies per week are born. Which hospital, do you think, has a higher percentage of weeks during which more than 60% of the newborn babies are boys? Argue your answer without making any calculations. Using the continuity correction, calculate an approximation for each hospital for the probability that in a given week more than 60% of the newborn babies will be boys and compare this approximation with the exact value of the binomial probability.

5.25 A damage claims insurance company has 20,000 policyholders. The amount claimed yearly by policyholders has an expected value of \$150 and a standard deviation of \$750. Give an approximation for the probability that the total amount claimed in the coming year will be larger than 3.3 million dollars.

5.26 The Nero Palace casino has a new, exciting gambling machine: the multiplying bandit. How does it work? The bandit has a lever or "arm" that the player may depress up to ten times. After each pull, an H (heads) or a T (tails) appears, each with probability $\frac{1}{2}$. The game is over when heads appears for the first time, or when the player has pulled the arm ten times.

The player wins $\$2^k$ if heads appears after k pulls ($1 \leq k \leq 10$), and wins $\$2^{11} = \$2{,}048$ if after ten pulls heads has not appeared. In other words, the payoff doubles every time the arm is pulled and heads does not appear. The initial stake for this game is \$15. What is the house advantage? Assume there are 2,000 games played each day. Give an approximation for the probability that the casino will lose money on a given day.

5.27 The Dutch Ministry of Education has taken a random sampling of the student population of four hundred. The students in the sample group were asked if they were in favor of the introduction of a weekend pass for public transportation. Suppose that 208 students were in favor of the pass. Give a 95% confidence interval for the estimate of the percentage of students from the entire student population that would be in favor of the pass. How large must the sample be to ensure that the 95% confidence interval has a margin of no more than 2%?

5.28 Six million voters are expected to vote in the upcoming presidential election. There are two candidates, A and B. The voters cast their ballots independently of one another and each voter will vote for candidate A with probability p and for candidate B with probability $1 - p$. Calculate for both $p = 0.5$ and $p = 0.501$ the probability that the difference in number of votes cast for each of the two candidates will be less than 300.

5.29 In 1986, an article appeared on the front page of *the New York Times* about the results of a research project on the effect of a light dose of aspirin on the incidence of heart attacks. By means of a carefully selected randomization method, a group of 22,000 healthy middle-aged males was randomly sorted into two groups of the same size: an aspirin group and a placebo group. In the aspirin group, 104 heart attacks occurred, while 209 heart attacks occurred in the placebo group. How can you argue, on the grounds of these results, that it is beyond a reasonable doubt that aspirin contributes to the prevention of heart attacks?

5.30 You are interested in assembling a random sample of young people that occasionally use soft drugs. To prevent people from falsely claiming not to use soft drugs, you have thought of the following procedure. The interviewer asks each young person to toss a coin, keeping the result of the toss a secret. The young person is then instructed that if he/she tosses heads he/she must answer "yes" to the question asked even if the true answer to the question is "no" and that if he/she tosses tails, he/she must simply answer the question with the truth. Suppose that the random sample consists of n young people. Let X_i equal 1 if the ith person answers "yes" and otherwise let X_i be equal to 0. Verify that the unknown value of the fraction of young people that use soft drugs can be estimated by $2\overline{X}(n) - 1$ with the corresponding 95% confidence interval $2\,\overline{X}(n) - 1 \,\pm 1.96 \times 2\sqrt{\overline{X}(n)\left[1 - \overline{X}(n)\right]}/\sqrt{n}$.

5.31 In order to test a new pseudo-random number generator, we let it generate 100,000 random numbers. From this result, we go on to form a binary sequence in which the ith element will be equal to 0 if the ith randomly generated number is smaller than $\frac{1}{2}$, and will otherwise be equal to 1. The binary sequence turns out to consist of 49,487 runs. A run begins each time a number in the binary sequence differs from its direct predecessor. Do you trust the new random number generator on the basis of this test outcome?

5.32 You have an economy with a risky asset and a riskless asset. Your strategy is to hold always a constant proportion α of your wealth in the risky asset and the remaining proportion of your wealth in the riskless asset, where $0 < \alpha < 1$. The initial value of your wealth is V_0. The rate of return on the risky asset is described by a Brownian motion with a drift of 15% and a standard deviation of 30%. The instantaneous rate of return on the riskless asset is 7%. Let V_t denote your wealth at time t. Use a random-walk discretization of the process of rate of return on the risky asset in order to give an intuitive explanation of the result that $\ln(V_t/V_0)$ is a Brownian motion with drift parameter $r + \alpha(\mu - r) - \frac{1}{2}\alpha^2\sigma^2$ and variance parameter σ^2, where $r = 0.07$, $\mu = 0.15$, and $\sigma = 0.3$. Next, argue that the long-run rate of growth of your wealth is maximal for the Kelly fraction $\alpha^* = 0.89$. How much time is required in order to double your initial wealth with a probability of 90%?

5.33 Set up a recursion equation to calculate the probability of rolling a six three or more times in a row in 25 rolls of a fair die.

5.34 What is the probability that after n spins of the wheel in European roulette the ball will have fallen on red r times in a row, or on black r times in a row? Argue that this probability is given by $v_n(0)$, where $v_n(0)$ is found by applying the following recursion formula for $k = 1, \ldots, n$:

$$v_k(0) = \tfrac{1}{37} v_{k-1}(0) + \tfrac{36}{37} v_{k-1}(1)$$

$$v_k(i) = \tfrac{1}{37} v_{k-1}(0) + \tfrac{18}{37} v_{k-1}(1) + \tfrac{18}{37} v_{k-1}(i+1), \quad 1 \le i \le r-1.$$

The boundary conditions are $v_0(i) = 0$ for $1 \le i \le r-1$ and $v_j(r) = 1$ for $0 \le j \le n-1$. Verify experimentally that the probability $v_n(0)$ is well approximated by the Poisson probability $1 - e^{-2nqp^r}$ with $p = \frac{18}{37}$ and $q = 1 - p$ when n and r are sufficiently large and r is relatively small in comparison with n.

5.35 In Problem 3.33, we defined the increase factor of a permutation $(k_1, \ldots, k_n)$ of the integers $1, \ldots, n$. Using computer simulation, verify that the increase factor of a random permutation of the integers $1, \ldots, n$ is approximately normally distributed with expected value $\frac{1}{4}n(n-1)$ and variance $\frac{1}{72}n(n-1)(2n+5)$ if n is sufficiently large ($n > 10$).

5.36 For a random permutation $\sigma = (\sigma_1, \ldots, \sigma_n)$ of the integers $1, \ldots, n$, define the random variable X as $X = \frac{1}{2}\sum_{i=1}^{n} |\sigma_i - i|$. Using computer simulation, verify that the distribution of the (discrete) random variable X is well approximated by a normal distribution with expected value $\frac{1}{6}(n^2 - 1)$ and variance $\frac{1}{180}(n+1)(2n^2+7)$ if n is sufficiently large (say, $n > 10$).

5.37 Suppose that $U_1, \ldots, U_{n+m}$ are independent random observations from the uniform distribution on the interval (0,1). Rank these $n + m$ data values. Give the smallest data value rank 1, the second smallest rank 2, and so on. Let R_i denote the rank of the data value X_i for $i = 1, \ldots, n$. Define the test statistic T by $T = \sum_{i=1}^{n} R_i$ Use computer simulation to verify experimentally that T has approximately a normal distribution with mean $\frac{1}{2}n(n+m+1)$ and variance $\frac{1}{12}nm(n+m+1)$ provided that $n > 8$ and $m > 8$. *Remark*: this result can be proved to hold for random observations from any continuous probability distribution and underlies the classical Mann-Whitney test, or Wilcoxon test, for testing whether two independent series of independent observations come from a same probability distribution.

6

Chance trees and Bayes' rule

–NO WONDER I GOT IT SO CHEAP!

Chance trees provide a useful tool for a better understanding of uncertainty and risk. A lot of people have difficulties assessing risks. Many physicians, for example, when performing medical screening tests, overstate the risk of actually having the disease in question to patients testing positive for the disease. They underestimate the false-positives of the test. Likewise, prosecutors often misunderstand the uncertainties involved in DNA evidence. They confuse the not-guilty probability of a suspect matching the trace evidence with the probability of a person randomly selected from a population matching the trace evidence. Incorrect reasoning with conditional probabilities is often the source of erroneous conclusions. A chance tree is useful in such once-only decision situations containing a degree of uncertainty. It depicts the uncertainty in an insightful way and it clarifies conditional probabilities by decomposing a compound event into its simpler components. We begin our discussion of chance trees with some entertaining problems, such as the three-doors problem and the related three prisoners problem. A lot of time and energy have been expended in the solving of these two problems; numerous people have racked their brains in search of their solutions, alas to no avail. There are several productive ways of analyzing the two problems, but by using a chance tree we run the least amount of risk of falling into traps. This chapter also provides an illustration of how the concept of the chance tree is used for analyzing uncertainties in medical screening tests. Bayes' rule provides an alternative approach in the analysis of situations in which probabilities must be revised in light of new information. This rule will also be discussed in this chapter.

6.1 The Monty Hall dilemma

Seldom has a probability problem so captured the imagination as the one we refer to as the Monty Hall dilemma. This problem, named after the popular 1970s game show host, attracted worldwide attention in 1990 when American columnist Marilyn vos Savant took it on in her weekly column in the Sunday *Parade* magazine. It goes like this. The contestant in a television game show must choose between three doors. An expensive automobile awaits the contestant behind one of the three doors, and gag prizes await him behind the other two. The contestant must try to pick the door leading to the automobile. He chooses a door randomly, appealing to Lady Luck. Then, as promised beforehand, the host opens one of the other two doors concealing one of the gag prizes. With two doors remaining unopened, the host now asks the contestant whether he wants to remain with his choice of door, or whether he wishes to switch to the other remaining door. The candidate is faced with a dilemma. What to do? In her weekly *Parade* column, Marilyn vos Savant advised the contestant to switch to the other remaining door, thereby raising his odds of winning the automobile to a $\frac{2}{3}$ probability. In the weeks that followed, vos Savant was inundated with thousands of letters, some rather pointed to say the least, from readers who disagreed with her solution to the problem. Ninety percent of the letter writers, including some professional mathematicians, insisted that it made no difference whether the player switched doors or not. Their argument was that each of the two remaining unopened doors had a $\frac{1}{2}$ probability of concealing the automobile. The matter quickly transcended the borders of the United States, gathering emotional impact along the way. Note the reaction in this letter to the editor published in a Dutch newspaper: "The unmitigated gall! Only sheer insolence would allow someone who failed mathematics to make the claim that the win probability is raised to $\frac{2}{3}$ by switching doors. Allow me to expose the columnist's error: Suppose there are one hundred doors, and the contestant chooses for door number one. He then has a 1% probability of having chosen the correct door, and there is a 99% probability that the automobile is concealed behind one of the other ninety-nine doors. The host then proceeds to open all of the doors from 2 through 99. The automobile does not appear behind any of them, and it then becomes apparent that it must be behind either door number one or door number one hundred. According to the columnist's reasoning, door number one hundred now acquires a 99% probability of concealing the automobile. This, of course, is pure balderdash. What we actually have here is a new situation consisting of only two possibilities, each one being equally probable." Now, not only is this writer completely wrong, he also provides, unintentionally, an ironclad case in favor of changing doors. Another writer claims in his letter to vos Savant: "As a professional mathematician it concerns me to see a growing lack of mathematical proficiency among the general public. The probability in question must be $\frac{1}{2}$; I caution you in future to steer clear of issues of which you have no understanding."[1] Martin Gardner, the spiritual father of the Monty Hall problem, writes: "There is no other branch of mathematics in which experts can so easily blunder as in probability theory."

The fact that there was so much dissension over the correct solution to the Monty Hall dilemma can be explained by the psychological given that many people naturally tend to assign equal probabilities to the two remaining doors at the last stage of the game. Some readers of vos Savant's column may have thought that when the game show host promised to open a door, he meant that he would pick a door at random. Were this actually the case, it would not be to the contestant's advantage to switch doors later. But the quizmaster had promised to open a door concealing one of the gag prizes. This changes the situation and brings relevant, previously unknown information to light. At the beginning of the game when there are three doors to choose from, the contestant has a $\frac{1}{3}$ probability that the automobile will be hidden behind his chosen door, and a

[1] Many other reactions and a psychological analysis of those reactions can be found in Marilyn vos Savant's *The Power of Logical Thinking*, St. Martin's Press, New York, 1997.

$\frac{2}{3}$ probability that it will not be behind his door. At this point, the host opens a door. The promise he makes at the outset is that after the contestant indicates his choice of door, and regardless of what that choice is (this is an essential given), the host will open one of the remaining doors without an automobile behind it. If the automobile is not behind the contestant's door, then it is in all certainty behind the door remaining after the host opens a door. In other words, there is a $\frac{2}{3}$ probability that switching doors at this stage will lead to the contestant's winning the automobile, while there is a $\frac{1}{3}$ probability that switching doors will not lead to the contestant's winning the automobile. Of course, it does rankle when contestants switch doors only to find that their original choice was the correct one.

6.1.1 Chance tree

The reasoning that leads to the correct answer of $\frac{2}{3}$ is simple, but you do have to get started along the right pathway. How can you reach the correct answer in a more systematic way without stumbling into a pattern of faulty intuitive thinking? One answer lies in computer simulation, another in the playing of a streamlined version of the game in which a ten-dollar bill is hidden under one of three coasters. A systematic approach using nothing but pencil and paper is also a possibility. This last option is carried out with the help of a chance tree. Chance trees make very clear that probabilities depend on available information. Figure 6.1 shows the chance tree for the Monty Hall problem. It shows all possible events with their corresponding probabilities. To make it as straightforward as possible, we have labeled the door first chosen by the contestant as door 1. The host's promise to open a door behind which there will be no automobile can also be seen in the chance tree. He will either open

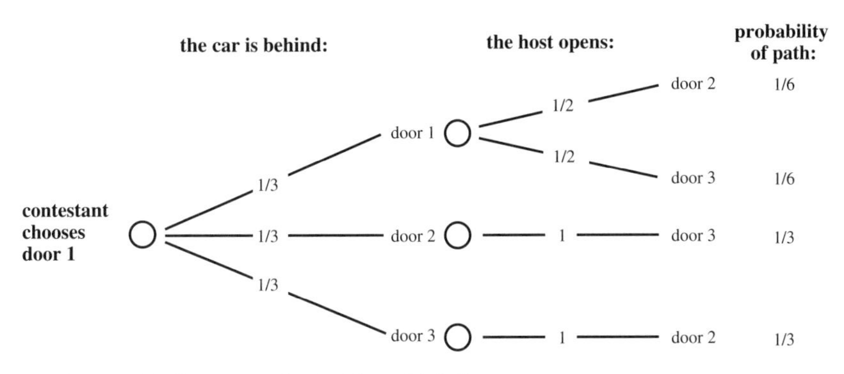

Figure 6.1 Chance tree for the Monty Hall dilemma.

door 2 (if the automobile is behind door 3) or door 3 (if the automobile is behind door 2), and will open either door 2 or 3 randomly if the automobile is behind door 1. The numbers associated with the lines branching out from a node show the probability of the feasible events that may occur at that particular node. The probabilities of the possible pathways are calculated by multiplying the probabilities located at the various branches along the pathway. We can see by looking at the chance tree in Figure 6.1 that the two last pathways lead to the winning of the automobile. The probability of winning the automobile by switching doors is given by the sum of the probabilities of these two paths and is thus equal to $\frac{1}{3} + \frac{1}{3} = \frac{2}{3}$. And the correct answer is, indeed, $\frac{2}{3}$.

The Monty Hall dilemma clearly demonstrates how easy it is to succumb to faulty intuitive reasoning when trying to solve some probability problems. The same can be said of the following, closely related problem.

6.1.2 The problem of the three prisoners

Each of three prisoners A, B, and C is eligible for early release due to good behavior. The prison warden has decided to grant an early release to one of the three prisoners and is willing to let fate determine which of the three it will be. The three prisoners eventually learn that one of them is to be released, but do not know who the lucky one is. The prison guard does know. Arguing that it makes no difference to the odds of his being released, prisoner A asks the guard to tell him the name of one co-prisoner that will not be released. The guard refuses on the grounds that such information will raise prisoner A's release probability to $\frac{1}{2}$. Is the guard correct in his thinking, or is prisoner A correct? The answer is prisoner A. If the guard provides the information sought by prisoner A, the latter's release probability does not change, but remains equal to $\frac{1}{3}$. This is readily seen with a glance at the chance tree in Figure 6.2. Another way of arriving at the conclusion that the answer must be $\frac{1}{3}$ is to see this problem in the light of the Monty Hall problem. The to-be-freed prisoner is none other than the door with the automobile behind it. The essential difference between the two problems is that, in the prisoner's problem, there is no switching of doors/prisoners. If the contestant in the Monty Hall problem does not switch doors, the probability of his winning the automobile remains at $\frac{1}{3}$, even after the host has opened a door revealing no automobile!

6.1.3 Sushi delight

One fish is contained within the confines of an opaque fishbowl. The fish is equally likely to be a piranha or a goldfish. A sushi lover throws a piranha into

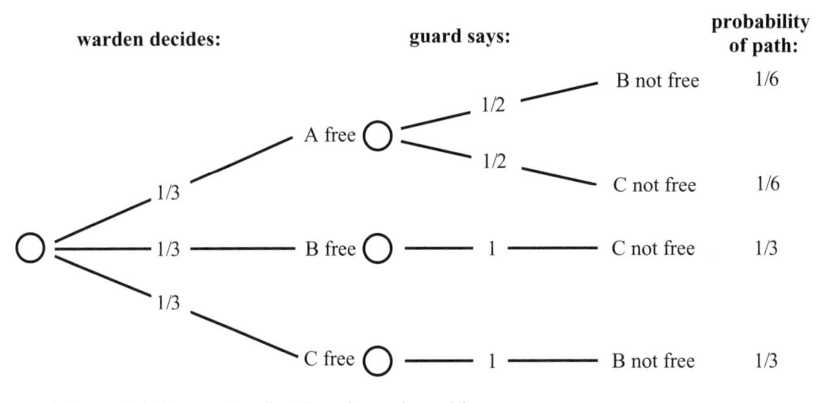

Figure 6.2 Chance tree for the prisoner's problem.

the fish bowl alongside the other fish. Then, immediately, before either fish can devour the other, one of the fish is blindly removed from the fishbowl. The fish that has been removed from the bowl turns out to be a piranha. What is the probability that the fish that was originally in the bowl by itself was a piranha? This is another problem that can instigate heated discussions.

The correct answer to the question posed is $\frac{2}{3}$. This is easily seen from the chance tree in Figure 6.3. The first and second paths in the tree lead to the removal of a piranha from the bowl. The probability of occurrence of the first is $\frac{1}{2} \times 1 \times 1 = \frac{1}{2}$, and the probability of occurrence of the second path is $\frac{1}{2} \times 1 \times \frac{1}{2} = \frac{1}{4}$. The desired probability of the fishbowl originally holding a piranha is the probability of occurrence of the first path given that the first or

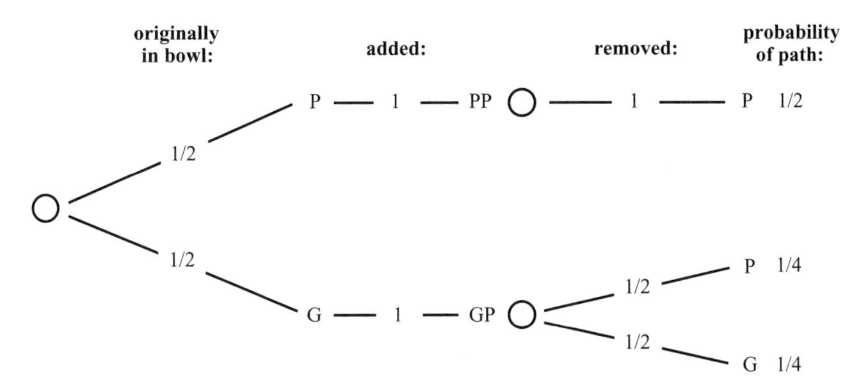

Figure 6.3 Chance tree for the sushi delight problem.

the second path has occurred. The definition of conditional probability gives

$$P(\text{path } 1 \mid \text{path } 1 \text{ or path } 2) = \frac{P(\text{path } 1)}{P(\text{path } 1) + P(\text{path } 2)}$$
$$= \frac{1/2}{1/2 + 1/4} = \frac{2}{3}.$$

In the same way, it can be verified that the probability we are seeking is equal to

$$\frac{p}{p + \frac{1}{2}(1 - p)}$$

if the fishbowl originally held a piranha with probability p and a goldfish with probability $1 - p$.

The sushi delight problem was originated by American scientist/writer Clifford Pickover and is a variant of the classic problem we will discuss at the end of this chapter in Problem 6.8.

6.1.4 Daughter-son problem

You are told that a family, completely unknown to you, has two children and that one of these children is a daughter. Is the chance of the other child being a daughter equal to $\frac{1}{2}$ or $\frac{1}{3}$? Are the chances altered if, aware of the fact that the family has two children only, you ring their doorbell and a daughter opens the door? In Section 2.9, computer simulation was used to obtain the answers $\frac{1}{3}$ and $\frac{1}{2}$ to the first and second questions. The assumption was made that each newborn child is equally likely to be a boy or a girl. In answering the second question, we also made the assumption that, randomly, one of the children will open the door. The answers $\frac{1}{3}$ and $\frac{1}{2}$ to the first and second questions can also be verified using a chance tree. We leave it to the reader to do so.

6.2 The test paradox

An inexpensive diagnostic test is available for a certain disease. Although the test is very reliable, it is not 100% reliable. If the test result for a given patient turns out to be positive, then further, more in-depth testing is called for to determine with absolute certainty whether or not the patient actually does suffer from the particular disease. Among persons who actually do have the disease, the test gives positive results in an average of 99% of the cases. For patients who do not have the disease, there is a 2% probability that the test will give a false-positive result. In one particular situation, the test is used at a policlinic to test a subgroup of persons among whom it is known that 1 out of 2 has the disease. For a given person out of this subgroup, what is the probability that he will turn out to have the disease after having tested positively? To arrive at an answer, we must create a chance tree like the one shown in Figure 6.4. If we take the product of the probabilities along each pathway in the chance tree, we see that the first and third pathways lead to positive test results with probabilities 0.495 and 0.01, respectively. The probability that the person has the disease after testing positively is equal to the probability of the appearance of the first pathway given that the first or the third pathway has appeared. Next, the definition of conditional probability leads to:

$$P(\text{path 1} \mid \text{path 1 or path 3}) = \frac{P(\text{path 1})}{P(\text{path 1}) + P(\text{path 3})}.$$

The probability we are seeking, then, is equal to

$$\frac{0.495}{0.495 + 0.01} = 0.9802.$$

In other words, in the subgroup, an average 98.0% of the positive test results are correct.

Let's now suppose that, based on the success of the test, it is suggested that the entire population be tested for this disease on a yearly basis. Among the general population, an average of 1 out of 1,000 persons has this disease. Is it a good idea to test everyone on a yearly basis? In order to answer this, we will calculate the probability of a randomly chosen person turning out to have the disease given that the person tests positively. To do this, we will refer to the chance tree in Figure 6.5. This figure shows that the probability of a randomly chosen person having the disease given that the person tests positively is

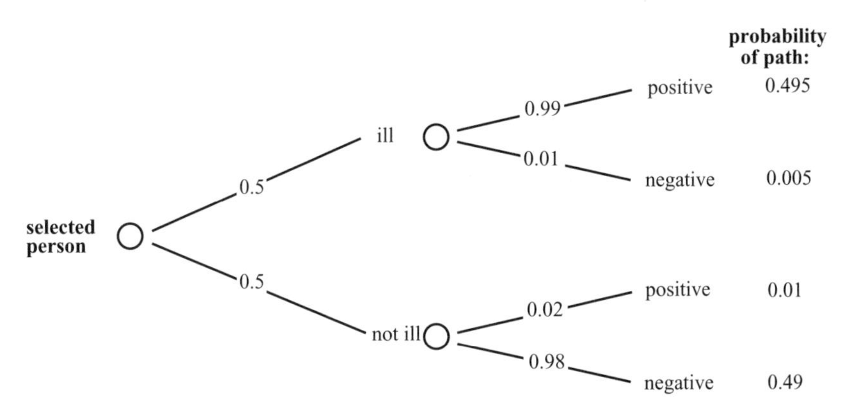

Figure 6.4 Chance tree for the subgroup.

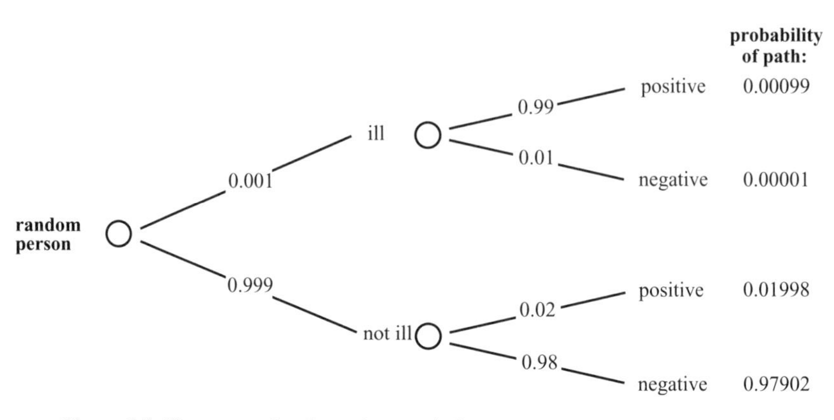

Figure 6.5 Chance tree for the entire population.

equal to

$$\frac{0.00099}{0.00099 + 0.01998} = 0.0472.$$

This leads to the seemingly paradoxical result that, in an average of more than 95% of the cases that test positively, the persons in question do not actually have the disease. Considering this fact, many people would be unnecessarily distressed if the entire population were to be tested. An explanation for the fact that a reasonably reliable test works so unsatisfactorily for the entire population lies in the fact that the vast majority of the population does not have the disease. The result is that even though there is only a small probability of receiving a positive test result when one does not have the disease, the people among the general population who do not have the disease, by virtue of their sheer numbers, will nevertheless get a much larger number of positive results than the small group of people who are actually ill. In other words, the number of false-positives far outstrips the number of correct diagnoses when the entire population undergoes the test. This is underlined by the following reasoning. Suppose you test 10,000 randomly chosen people. There will be on average 9,990 people who do not have the illness, and ten people who do. This means, on average, $0.02 \times 9{,}990 = 199.8$ false-positives and $0.99 \times 10 = 9.9$ true-positives.

In the example above, we can see how important it is to keep an eye on the *basic proportions* between the various categories of people. If we ignore these proportions, we can end up coming to weird conclusions such as: "Statistics show that 10% of traffic accidents are caused by drunken drivers, which means that the other 90% are caused by sober drivers . . . is it then not sensible to allow only drunken drivers onto the roads?" This statement is attributed to M. Samford and should give politicians pause to refrain from making similar statements.

6.2.1 Bayes' rule[2]

The chance tree in Figure 6.5 describes uncertainties in a process that evolves over time. Initially, before the test is done, you have an estimate of 0.001 of the probability of the disease. After the test is done, you have a revised estimate of this probability. The former estimate is called the *prior* probability, and the latter estimate is called the *posterior* probability. An alternative method to calculate the posterior probability is Bayes' rule. The reasoning of this rule is based on a (subtle) use of conditional probabilities. Bayes' rule will be illustrated for the situation that the test is used for the whole population. In order to find the posterior probability of the disease given a positive test result, we first list the data

$$P(\text{disease}) = 0.001, \qquad P(\text{no disease}) = 0.999,$$

$$P(\text{positive} \mid \text{disease}) = 0.99, \qquad P(\text{negative} \mid \text{disease}) = 0.01,$$

$$P(\text{positive} \mid \text{no disease}) = 0.02, \qquad P(\text{negative} \mid \text{no disease}) = 0.98.$$

The posterior probability $P(\text{disease} \mid \text{positive})$ satisfies the relation

$$P(\text{disease} \mid \text{positive}) = \frac{P(\text{positive and disease})}{P(\text{positive})}.$$

A repeated application of the definition of conditional probability gives

$$P(\text{positive and disease}) = P(\text{positive} \mid \text{disease})P(\text{disease})$$

and

$$\begin{aligned} P(\text{positive}) &= P(\text{positive and disease}) + P(\text{positive and no disease}) \\ &= P(\text{positive} \mid \text{disease})P(\text{disease}) \\ &\quad + P(\text{positive} \mid \text{no disease})P(\text{no disease}). \end{aligned}$$

Consequently, the desired probability $P(\text{disease} \mid \text{positive})$ satisfies the formula

$$P(\text{disease} \mid \text{positive}) = \frac{P(\text{positive} \mid \text{disease})P(\text{disease})}{P(\text{positive} \mid \text{disease})P(\text{disease}) + P(\text{positive} \mid \text{no disease})P(\text{no disease})}.$$

[2]This rule is named after British parson Thomas Bayes (1702–1761), in whose posthumously published *Essay Toward Solving a Problem in the Doctrine of Chance*, an early attempt is made at establishing what we now refer to as Bayes' rule.

By filling in the above data, we find that

$$P(\text{disease} \mid \text{positive}) = \frac{0.99 \times 0.001}{0.99 \times 0.001 + 0.02 \times 0.999} = 0.0472.$$

This is the same value as the one we found earlier.

The above derivation of the conditional probability $P(\text{disease} \mid \text{positive})$ is an illustration of Bayes' rule. It is possible to give a general mathematical formula for Bayes' rule. However, in specific applications, one better calculate the posterior probability according to Bayes' rule by using first principles as done in the above example.

Doctors should be more knowledgeable about chance trees and Bayes formula. Consider the following situation. A doctor discovers a lump in a woman's breast during a routine physical exam. The lump could be a cancer. Without performing any further tests, the probability that the woman has breast cancer is 0.01. A mammogram is a test that, on average, is correctly able to establish whether a tumor is benign or cancerous 90% of the time. A positive test result indicates that a tumor is cancerous. What is the probability that the woman has breast cancer if the test result from a mammogram is positive? Results from a psychological study indicate that many doctors think that the probability $P(\text{cancer} \mid \text{positive})$ is slightly lower than the probability $P(\text{positive} \mid \text{cancer})$ and estimate the former probability as being about 80%. The actual value for the probability $P(\text{cancer} \mid \text{positive})$, however, is only 8.3% (verify)! A similar misconception sometimes occurs in court cases when the probability of innocence in an accused person with the same physical characteristics as the perpetrator is confused with the probability that a randomly selected member of the public looks like the perpetrator. Most such mistakes can be prevented by presenting the relevant information in terms of frequencies instead of probabilities. In the example of the mammogram test, the information might then consist of the fact that, of 1,000 women examined, there were ten who had cancer. Of these 10, eight had a positive mammogram, whereas of the 990 healthy women, 99 had a positive mammogram. Based on the information presented in this way, most doctors would then be able to correctly estimate the probability of breast cancer given a positive mammogram as being equal to $9/(9+99) \approx 8.3\%$. Changing risk representations from probabilities into natural frequencies can turn the innumeracy of nonstatisticians into insight.[3]

[3]In his book *Calculated Risks* (Simon & Schuster, 2002), Gerd Gigerenzer advocates that doctors and lawyers be educated in more understandable representations of risk.

6.3 Problems

6.1 The roads are safer at nonrush hour times than during rush hour because fewer accidents occur outside of rush hour than during the rush hour crunch. Do you agree or do you disagree?

6.2 On a table before you are two bowls containing red and white marbles; the first bowl contains seven red and three white marbles, and the second bowl contains 70 red and 30 white marbles. You are asked to select one of the two bowls, from which you will blindly draw two marbles (with no replacing of the marbles). You will receive a prize if at least one of the marbles you picked is white. In order to maximize your probability of winning the prize, do you choose the first bowl or the second bowl?

6.3 You are one of fifty thousand spectators at a baseball game. Upon entering the ballpark, each spectator has received a ticket bearing an individual number. A winning number will be drawn from all of these fifty thousand numbers. At a certain point, five numbers are called out over the loudspeaker. These numbers are randomly drawn and include the winning number. Your number is among the five numbers called. What is the probability of your ticket bearing the winning number?

6.4 Now consider the Monty Hall dilemma from Section 6.1 with the following difference: you learned beforehand that there is a 0.2 probability of the automobile being behind door 1, a 0.3 probability of its being behind door 2, and a 0.5 probability of its being behind door 3. Your strategy is to choose the door with the lowest probability (door 1) in the first round of the game, and then to switch doors to one with a higher probability after the host has opened a gag prize door. Set up a chance tree to determine your probability of winning the automobile.

6.5 Consider the Monty Hall dilemma with the following twists: there are five doors, and the host promises to open two of the gag prize doors after the contestant has chosen a door. Set up a chance tree to calculate the probability of the contestant winning the automobile by switching doors.

6.6 Consider the following variant of the Monty Hall dilemma. There are now four doors, behind one of which there is an automobile. You first indicate a door. Then the host opens another door behind which a gag prize is to be found. You are now given the opportunity to switch doors. Regardless of whether or not you switch, the host then opens another door (not the door of your current choice) behind which no automobile is to be found. You are now given a final opportunity to switch doors. What is the best strategy for playing this version of the game?

6.7 The final match of world championship soccer is to be played between England and the Netherlands. The star player for the Dutch team, Dennis Nightmare, has been injured. The probability of his being fit enough to play in the final is being estimated at 75%. Pre-game predictions have estimated that, without Nightmare, the probability of a Dutch win is 30% and with Nightmare, 50%. Later, you hear that the Dutch team has won the match. Without having any other information about events that occurred, what would you say was the probability that Dennis Nightmare played in the final?

6.8 Passers-by are invited to take part in the following sidewalk betting game. Three cards are placed into a hat. One card is red on both sides, one is black on both sides, and one is red on one side and black on the other side. A participant is asked to pick a card out of the hat at random, taking care to keep just one side of the card visible. After having picked the card and having seen the color of the visible side of the card, the owner of the hat bets the participant equal odds that the other side of the card will be the same color as the one shown. Is this a fair bet?

6.9 Alcohol checks are regularly conducted among drivers in a particular region. Drivers are first subjected to a breath test. Only after a positive breath test result is a driver taken for a blood test. This test will determine whether the driver has been driving under the influence of alcohol. The breath test yields a positive result among 90% of drunken drivers and yields a positive result among only 5% of sober drivers. As it stands at present, a driver can only be required to do a breath test after having exhibited suspicious driving behavior. It has been suggested that it might be a good idea to subject drivers to breath tests randomly. Current statistics show that one out of every twenty drivers on the roads in the region in question is driving under the influence. Calculate the probability of a randomly tested driver being unnecessarily subjected to a blood test after a positive breath test.

6.10 You know that bowl A has three red and two white balls inside and that bowl B has four red and three white balls. Without your being aware of which one it is, one of the bowls is randomly chosen and presented to you. Blindfolded, you must pick two balls out of the bowl. You may proceed according to one of the following strategies:

(a) you will choose and replace (i.e., you will replace your first ball into the bowl before choosing your second ball).

(b) you will choose two balls without replacing any (i.e., you will not replace the first ball before choosing a second).

The blindfold is then removed and the colors of both of the balls you chose are revealed to you. Thereafter you must make a guess as to which bowl your two balls came from. For each of the two possible strategies, determine how you can make your guess depending on the colors you have been shown. Which strategy offers the higher probability for a correct guess as to which bowl the balls came from? Does the answer to this question contradict your intuitive thinking?

6.11 There are two taxicab companies in a particular city, "Yellow Cabs" and "White Cabs." Of all the cabs in the city, 85% are "Yellow Cabs" and 15% are "White Cabs." The issue of cab color has become relevant in a hit-and-run case before the courts in this city, in which witness testimony will be essential in determining the guilt or innocence of the cab driver in question. In order to test witness reliability, the courts have set up a test situation similar to the one occurring on the night of the hit-and-run accident. Results showed that 80% of the participants in the test case correctly identified the cab color, whereas 20% of the participants identified the wrong company. What is the probability that the accused hit-and-run cabbie is a "White Cabs" employee? (This problem is taken from the book of Kahneman et al.; see the footnote in Chapter 1.)

6.12 A doctor finds evidence of a serious illness in a particular patient and must make a determination about whether or not to advise the patient to undergo a dangerous operation. If the patient does suffer from the illness in question, there is a 95% probability that he will die if he does not undergo the operation. If he does undergo the operation, he has a 50% probability of survival. If the operation is conducted and it is discovered that the patient does not suffer from the illness, there is a 10% probability that the patient will die due to complications resulting from the operation. If it has been estimated that there is a 20% to 30% probability of the patient actually having the illness in question, how should the doctor advise her patient?

6.13 Consider the sushi delight problem from Section 6.1. Suppose now that both a piranha and a goldfish are added to the fishbowl alongside the original fish. What is the probability that the original fish is a piranha if a piranha is taken out of the bowl?

6.14 Suppose that there is a DNA test that determines with 100% accuracy whether or not a particular gene for a certain disease is present. A woman would like to do the DNA test, but wants to have the option of holding out hope that the gene is not present in her DNA even if it is determined that the gene for the illness is, indeed, present. She makes the following arrangement with her doctor. After the test, the doctor will toss a fair coin into the air, and will tell the woman the test results only if those results are negative and the coin has turned up heads. In every other case, the doctor will not tell her the test results. Suppose that there is a one out of one hundred probability that the woman does have the gene for the disease in question before she is tested. What is the revised value of this probability if the woman's doctor does not inform her over the test results? (Marilyn vos Savant, *Parade* magazine, February 7, 1999)

6.15 At a particular airport, each passenger must pass through a special fire arms detector. An average of 1 out of every 100,000 passengers is carrying a fire arm. The detector is 100% accurate in the detection of fire arms, but in an average of 1 in 10,000 cases, it results in a false alarm while the passenger is not carrying a fire arm. In cases when the alarm goes off, what is the probability that the passenger in question is carrying a fire arm?

6.16 A sum of money is placed in each of two envelopes. The amounts differ from one another, but you do not know what the values of the two amounts are. You do know that the values lie between two boundaries m and M with $0 < m < M$. You choose an envelope randomly. After inspecting its contents, you may switch envelopes. Set up a chance tree to verify that the following procedure will give you a probability of greater than $\frac{1}{2}$ of winding up with the envelope holding the most cash.

1. Choose an envelope and look to see how much cash is inside.
2. Pick a random number between m and M.
3. If the number you drew is greater than the amount of cash in your envelope, you exchange the envelope. Otherwise, you keep the envelope you have.

6.17 In a television game show, you can win 10,000 dollars by guessing the composition of red and white marbles contained in a nontransparent vase. The vase contains a very large number of marbles. You must guess whether the vase has twice as many red marbles as white ones, or whether it has twice as many white ones as red ones. Beforehand, both possibilities are equally likely to you. To help you guess, you are given a one-time opportunity of picking one, two, or three marbles out of the vase. This action, however, comes at the expense of the 10,000 dollar prize money. If you opt to choose one marble out of the vase, \$750 will be subtracted from the 10,000 should you win. Two marbles will cost you \$1,000 and three marbles will cost you \$1,500. Set up a chance tree to determine which strategy will help you maximize your winnings.

TWO

Essentials of probability

"Little help?"

7
Foundations of probability theory

Constructing the mathematical foundations of probability theory has proven to be a long-lasting process of trial and error. The approach consisting of defining probabilities as relative frequencies in cases of repeatable experiments leads to an unsatisfactory theory. The frequency view of probability has a long history that goes back to Aristotle. It was not until 1933 that the great Russian mathematician Andrej Nikolajewitsch Kolmogorov (1903–1987) laid a satisfactory mathematical foundation of probability theory. He did this by taking a number of axioms as his starting point, as had been done in other fields of mathematics. Axioms state a number of minimal requirements that the mathematical objects in question (such as points and lines in geometry) must satisfy. In the axiomatic approach of Kolmogorov, probability figures as a function on subsets of a sample space. The axioms are the basis for the mathematical theory of probability. As a milestone, the law of large numbers can be deduced from the axioms by logical reasoning. The law of large numbers confirms our intuition that the probability of an event in a repeatable experiment can be estimated by the relative frequency of its occurrence in many repetitions of the experiment. This law, which was already discussed in Chapter 2, is the fundamental link between theory and the real world. The purpose of this chapter is to discuss the axioms of probability theory in more detail and to derive from the axioms the most basic rules for the calculation of probabilities. These rules include the addition rule and the more general inclusion-exclusion rule. Various examples will be given to illustrate the rules.

7.1 Axioms of probability

A probability model for a random experiment consists of a complete description of all possible outcomes (the sample space) and an assignment of probability to the possible outcomes of the experiment. Many examples were given in Section 2.2 in Chapter 2. The classic example is the experiment of tossing a coin. Then, the sample space consists of the two outcomes H and T, and each of the two outcomes gets assigned a probability of $\frac{1}{2}$ if the coin is fair. In Section 2.2, we formulated the axioms of probability for a probability measure on a finite sample space. The axioms of probability must be slightly altered to accommodate a probability measure defined on an infinite sample space. We begin, however, by introducing a number of concepts from set theory.

7.1.1 Countable and uncountable sets

The set of natural numbers (positive integers) is an infinite set and is the prototype of a countable set. In general, a set is called *countable* if a one to one function exists that maps the elements of the set to the set of natural numbers. In other words, every element of the set can be assigned to a unique natural number and conversely each natural number corresponds to a unique element of the set. For example, the set of squared numbers $1, 4, 9, 16, 25, \ldots$ is countable. Not all sets with an infinite number of elements are countable. The set of all points on a line and the set of all real numbers between 0 and 1 are examples of infinite sets that are not countable. The German mathematician Georg Cantor (1845–1918) proved this result in the nineteenth century. This discovery represented an important milestone in the development of mathematics and logic (the concept of infinity, to which even scholars from ancient Greece had devoted considerable energy, obtained a solid theoretical basis for the first time through Cantor's work). Sets that are neither finite nor countable are called *uncountable*. Random experiments with either countable or uncountable sample spaces are common.

Example 7.1 Consider the random experiment consisting of the number of tosses of a fair coin needed to obtain three heads in a row. The sample space of this experiment is the set of integers $3, 4, 5, \ldots,$ where the outcome k indicates that three heads in a row was first achieved after k coin tosses. This sample space is countable.

Example 7.2 Consider the experiment in which an arbitrary point in a circle with radius R is chosen by a blindfolded person throwing a dart at a dartboard. The sample space of this experiment consists of the set of pairs of real numbers (x, y) where $x^2 + y^2 \leq R^2$. This sample space is uncountable.

The axioms of probability for an experiment with a countable sample space are the same as those for one with an uncountable sample space. A distinction must be made, however, between the sorts of subsets to which probabilities can be assigned, whether these subsets occur in countable or uncountable sample spaces. In the case of a countable sample space, probabilities can be assigned to each subset of the sample space (the probability of a subset being given by the sum of the individual probabilities associated with each of the basic outcomes in the subset). In the case of uncountable sample spaces, not all subsets of the sample space are assigned probabilities. Instead, for more fundamental mathematical reasons, only sufficiently well-behaved subsets are assigned probabilities. For example, when the sample space is the set of real numbers, then essentially only those subsets consisting of a finite interval, the complement of each finite interval, and the union of each countable number of finite intervals is assigned a probability. If the probability measure on the sample space is denoted by P, then P must satisfy the following properties

Axiom 1. $P(A) \geq 0$ *for each subset* A.

Axiom 2. $P(A) = 1$ *when* A *is equal to the sample space.*

Axiom 3. $P\left(\bigcup_{i=1}^{\infty} A_i\right) = \sum_{i=1}^{\infty} P(A_i)$ *for every collection of pairwise disjoint subsets* $A_1, A_2, \ldots$.

The notation $\bigcup_{i=1}^{\infty} A_i$ indicates the set of all outcomes which belong to at least one of the subsets $A_1, A_2, \ldots$. The subsets $A_1, A_2, \ldots$ are said to be *pairwise disjoint* when any two subsets have no element in common. The first two axioms simply express a probability as a number between 0 and 1. The crucial axiom 7.3 states that, for any sequence of *mutually exclusive* events, the probability of at least one of these events occurring is the sum of their individual probabilities. In probability terms, any subset of the sample space is called an *event*. If the outcome of the random experiment belongs to A, the event A is said to *occur*. The events $A_1, A_2, \ldots$ are said to be *mutually exclusive* if the corresponding sets $A_1, A_2, \ldots$ are pairwise disjoint.

The entire theory of probability is built on the above three axioms. For a finite sample space, these axioms are equivalent to those given in Section 2.2. An immediate consequence of the above axioms is that

$$P\left(\bigcup_{i=1}^{n} A_i\right) = \sum_{i=1}^{n} P(A_i)$$

for any finite sequence of pairwise disjoint sets $A_1, \ldots, A_n$. A formal proof of this (obvious) result will be postponed to Section 7.3.

The standard notation for the sample space is the symbol Ω. An element in Ω (in other words an outcome or elementary event of the random experiment) is denoted by ω. As has already been pointed out in Chapter 2, for a finite or countable sample space Ω, it is sufficient to assign a probability $p(\omega)$ to each element $\omega \in \Omega$ where $p(\omega) \geq 0$ and $\sum_{\omega \in \Omega} p(\omega) = 1$. A probability measure P on Ω is then defined by specifying the probability of each subset A of Ω as

$$P(A) = \sum_{\omega \in A} p(\omega)$$

(the notation $\omega \in A$ means that ω belongs to the set A and $\sum_{\omega \in A} p(\omega)$ is the notation for the sum of all $p(\omega)$'s with $\omega \in A$). Elementary mathematics is sufficient in order to prove that P satisfies Axioms 7.1 to 7.3 (details of the proof are left to the reader).

To illustrate the choice of a probability measure for an uncountable sample space, consider Example 7.2 again. The assumption of the dart hitting the dartboard at a random point is translated by assigning the probability

$$P(A) = \frac{\text{the area of the region } A}{\pi R^2}$$

to each subset A of the sample space. If the bull's-eye of the dartboard has radius b, the probability of the dart hitting the bull's-eye is $\pi b^2/(\pi R^2) = b^2/R^2$. For a better understanding of the concept of continuous random variable to be discussed in Chapter 10, it might be helpful to point out that each subset consisting of a single point has probability zero. The probability of hitting a *specific* point is zero. It makes only sense to speak of the probability of hitting a given region of the dartboard. This observation expresses a fundamental difference between a probability model with a finite or countable sample space and a probability model with an uncountable sample space.

Many problems asking for the formulation of a probability space are contained in Section 2.2 and the problem section of Chapter 2. In each of these problems, the sample space is finite. Here, we give a problem involving an uncountable sample space.

Problem 7.1 The game of franc-carreau was a popular game in eighteenth-century France. In this game, a toin is tossed on a chessboard. The player wins if the coin does not fall on one of the lines of the board. Suppose now that a round coin with a diameter of d is blindly tossed on a large table. The surface of the table is divided into squares whose sides measure a in length, such that $a > d$. Define an appropriate probability space and calculate the probability of the coin falling entirely within the confines of a square. *Hint*: consider the position of the coin's middle point.

Remark 7.1 Probability is a continuous set function. To explain this property, consider a nondecreasing sequence of sets $E_1, E_2, \ldots$. The sequence $E_1, E_2, \ldots$ is said to be *nondecreasing* if the set E_{n+1} contains the set E_n

for all $n \geq 1$. Let's define the set E by $E = \bigcup_{i=1}^{\infty} E_i$ and denote this set by $E = \lim_{n\to\infty} E_n$. Then, the continuity property states that

$$\lim_{n\to\infty} P(E_n) = P(\lim_{n\to\infty} E_n).$$

The proof is instructive. Define $F_1 = E_1$ and let the set F_{n+1} consist of the points of E_{n+1} that are not in E_n for $n \geq 1$. It is readily seen that the sets $F_1, F_2, \ldots$ are pairwise disjoint and satisfy $\bigcup_{i=1}^{n} F_i = \bigcup_{i=1}^{n} E_i (= E_n)$ for all $n \geq 1$ and $\bigcup_{i=1}^{\infty} F_i = \bigcup_{i=1}^{\infty} E_i$. Thus,

$$\begin{aligned} P\big(\lim_{n\to\infty} E_n\big) &= P\Big(\bigcup_{i=1}^{\infty} E_i\Big) = P\Big(\bigcup_{i=1}^{\infty} F_i\Big) = \sum_{i=1}^{\infty} P(F_i) \\ &= \lim_{n\to\infty} \sum_{i=1}^{n} P(F_i) = \lim_{n\to\infty} P\Big(\bigcup_{i=1}^{n} F_i\Big) \\ &= \lim_{n\to\infty} P\Big(\bigcup_{i=1}^{n} E_i\Big) = \lim_{n\to\infty} P(E_n), \end{aligned}$$

proving the continuity property. The proof uses Axiom 7.3 in the third and fifth equalities (Axiom 7.3 implies that $P(\bigcup_{i=1}^{n} A_i) = \sum_{i=1}^{n} P(A_i)$ for any finite sequence of pairwise disjoint sets $A_1, \ldots, A_n$; see Rule 7.1 in Section 7.3).

The result $\lim_{n\to\infty} P(E_n) = P(\lim_{n\to\infty} E_n)$ holds also for a *nonincreasing* sequence of sets $E_1, E_2, \ldots$ (E_{n+1} is contained in E_n for all $n \geq 1$), in which case the set $\lim_{n\to\infty} E_n$ is defined as the intersection of all sets E_i for $i \geq 1$. The intersection is defined as the set of all outcomes that belong to each of the sets E_i.

Problem 7.2 Use the axioms to prove the following results:

(a) $P(A) \leq P(B)$ if the set A is contained in the set B.
(b) $P\left(\cup_{k=1}^{\infty} A_k\right) \leq \sum_{k=1}^{\infty} P\left(A_k\right)$ for any sequence of subsets $A_1, A_2, \ldots$ (this result is known as *Boole's inequality*).

Problem 7.3 Let $A_1, A_2, \ldots$ be an infinite sequence of subsets with the property that $\sum_{k=1}^{\infty} P(A_k) < \infty$. Define the set C as $C = \{\omega : \omega \in A_k \text{ for infinitely many } k\}$. Use the continuity property of probabilities to prove that $P(C) = 0$ (this result is known as the *Borel-Cantelli lemma*).

7.2 Compound random experiments

A random experiment is called a *compound* experiment if it consists of several elementary random experiments. In Section 2.2, several examples were given of compound experiments along with the corresponding probability spaces. The question arises as to how, in general, we define a probability space for a compound experiment in which the elementary experiments are physically independent of each other. By physically independent, we mean that the outcomes from any one of the elementary experiments have no influence on the functioning or outcomes of any of the other elementary experiments. We first answer the question for the case of a finite number of physically independent elementary experiments $\varepsilon_1, \ldots, \varepsilon_n$. Assume that each experiment ε_k has a finite or countable sample space Ω_k on which the probability measure P_k is defined such that the probability $p_k(\omega_k)$ is assigned to each element $\omega_k \in \Omega_k$. The sample space of the compound experiment is then given by the set Ω consisting of $\omega = (\omega_1, \ldots, \omega_n)$, where $\omega_k \in \Omega_k$ for $k = 1, \ldots, n$. A natural choice for the probability measure P on Ω arises by assigning the probability $p(\omega)$ to each element $\omega = (\omega_1, \ldots, \omega_n) \in \Omega$ obtained using the *product rule*:

$$p(\omega) = p_1(\omega_1) \times p_2(\omega_2) \times \cdots \times p_n(\omega_n).$$

This choice for the probability measure is not only intuitively the obvious one, but we can also prove that it is the only probability measure satisfying property $P(AB) = P(A)\,P(B)$ when the elementary experiments that generate event A are physically independent of those elementary experiments which give rise to event B. This important result of the uniqueness of the probability measure satisfying this property justifies the use of the product rule for compound random experiments.

Example 7.3 Two desperados play Russian roulette in which they take turns pulling the trigger of a six-cylinder revolver loaded with one bullet (after each pull of the trigger, the magazine is spun to randomly select a new cylinder to fire). What is the probability that the desperado who begins will be the one to shoot himself dead?

The sample space we use for this random experiment is the set

$$\Omega = \{F, MF, MMF, \ldots\} \cup \{MM\ldots\},$$

where the element $M \ldots MF$ with the first $n-1$ letters all being M represents the event that the first $n-1$ times that the trigger is pulled, no shot is fired, and that the fatal shot is fired on the nth attempt. The element $MM\ldots$ represents the event that no fatal shot is fired when the two desperados repeatedly pull the trigger without ever stopping. By representing the event $M \ldots MF$ with the first $n-1$ letters all being M by the integer n, and the event $MM\ldots$ by the symbol ∞, we can also express the sample space Ω as

$$\Omega = \{1, 2, \ldots\} \cup \{\infty\}.$$

This latter representation of the countable infinite sample space is the one we will use. Given the fact that the outcomes of the pulling of the trigger are independent of one another, and that with each pull of the trigger there is a probability of $\frac{1}{6}$ that the fatal shot will be fired, it is reasonable to assign the probability

$$p(n) = \left(\frac{5}{6}\right)^{n-1} \frac{1}{6} \quad \text{for} \qquad n = 1, 2, \ldots$$

to the event n from the sample space for $n \in \{1, 2, \ldots\}$. To complete the specification of the probability measure P, we have to assign a probability to the element ∞. The probabilities $p(n)$ satisfy $\sum_{n=1}^{\infty} p(n) = 1$ because the geometric series $\sum_{k=1}^{\infty} x^k$ sums to $\frac{1}{1-x}$ for all $0 < x < 1$. Axiom 7.2 then implies that P must assign the value 0 the element ∞. If we now define A as the event that the fatal shot is fired by the desperado who begins, then $P(A)$ is given by

$$P(A) = \sum_{n=0}^{\infty} p(2n+1) = \sum_{n=0}^{\infty} \left(\frac{5}{6}\right)^{2n} \frac{1}{6}$$

$$= \frac{1}{6} \sum_{n=0}^{\infty} \left(\frac{25}{36}\right)^{n} = \frac{1}{6} \left(\frac{1}{1 - \frac{25}{36}}\right),$$

which corresponds to a probability of 0.5436. We can obtain the same answer by performing a computer simulation of this random experiment a very large number of times. This confirms that the theoretical probability model we selected was the right one.

Problem 7.4 In a tennis tournament between three players A, B, and C, each player plays the others once. The strengths of the player are as follows: $P(A$ beats $B) = 0.5$, $P(A$ beats $C) = 0.7$, and $P(B$ beats $C) = 0.4$. Assuming independence of the match results, calculate the probability that player A wins at least as many games as any other player.

Problem 7.5 Two people take turns selecting a ball at random from a bowl containing 3 white balls and 7 red ones. The winner is the person who is the first to select a white ball. It is assumed that the balls are selected with replacement. Define an appropriate sample space and calculate the probability that the person who begins will win.

Problem 7.6 In repeatedly rolling two dice, what is the probability of getting a total of 6 before a total of 7? What about an 8 and a 7? What is the probability of getting a total of 6 and a total of 8 in any order before two 7's?

7.2.1 A coin-tossing experiment[1]

When a compound random experiment consists of an infinite number of independent elementary random experiments, it has an uncountable sample space and the choice of an appropriate probability measure is less obvious. We illustrate how we deal with such experiments by way of an illustration of a compound experiment consisting of an infinite number of tosses of a fair coin. We model the sample space of this experiment using all infinite sequences $\omega = (\omega_1, \omega_2, \dots)$, where ω_i is equal to H when the ith coin toss comes up heads, and is equal to T otherwise. It can be proved that this sample space is uncountable. In order to be able to define a probability measure on this sample space, we must begin by restricting our attention to a class of appropriately chosen subsets. The so-called cylinder sets form the basis of this class of subsets. In the case of our random experiment, a *cylinder set* is the set of all outcomes ω where the first n elements $\omega_1, \dots, \omega_n$ have specified outcomes for finite values of n. A natural choice for the probability measure on the sample space is to assign the probability $P^{(\infty)}(A) = \left(\frac{1}{2}\right)^n$ to each cylinder set A with n specified elements. In this way, the event that heads first occurs at the kth toss can be represented by the cylinder set A_k with the finite beginning $T, T, \dots, T, H$, and can be assigned a probability of $\left(\frac{1}{2}\right)^k$. The collection $\bigcup_{k=1}^{\infty} A_k$ represents the event that at some point heads occurs. The probability measure on the class of cylinder sets can be extended to one defined on a sufficiently general class of subsets capable of representing all possible events of this random experiment.

In Section 2.1, we stated that the fraction of coin tosses in which heads occurs converges to $\frac{1}{2}$ with probability 1 when the number of tosses increases without limit. We are now in a position to state this claim more rigorously with the help of the probability measure $P^{(\infty)}$. To do so, we adopt the notation $K_n(\omega)$ to represent the number of heads occurring in the first n elements of ω. Furthermore, let C be the collection of all outcomes ω for which $\lim_{n\to\infty} K_n(\omega)/n = \frac{1}{2}$. For very many sequences ω, the number $K_n(\omega)/n$ does not converge to $\frac{1}{2}$ as $n \to \infty$. However, "nature" chooses a sequence from the collection C according to $P^{(\infty)}$: the *theoretical (strong) law of large numbers* states that the probability

[1]This section can be skipped without loss of continuity.

$P^{(\infty)}$ measure assigns a probability of 1 to the collection C. In mathematical notation, the result is

$$P^{(\infty)}\left(\left\{\omega : \lim_{n\to\infty} \frac{K_n(\omega)}{n} = \frac{1}{2}\right\}\right) = 1.$$

This type of convergence is called *convergence with probability 1*. The strong law of large numbers is of enormous importance: it provides a direct link between theory and practice. It was a milestone in probability theory when around 1930 A. N. Kolmogorov proved this law from the simple axioms of probability theory. In general, the proof of the strong law of large numbers requires advanced mathematics beyond the scope of this book. However, for the special case of the coin-tossing experiment, an elementary proof can be given using the Borel-Cantelli lemma. We give no further details.

7.3 Some basic rules

The axioms of probability theory directly imply a number of basic rules that are useful for calculating probabilities. We first repeat some basic notation. The event that at least one of events A or B occurs is called the *union* of A and B and is written $A \cup B$. The event that both A and B occur is called the *intersection* of A and B and is written $A \cap B$, or simply AB. The notation AB for the intersection of events A and B will be used throughout this book. The notation for union and intersection of two events extends to finite sequences of events. Given events $A_1, \ldots, A_n$, the event that at least one occurs is written $A_1 \cup A_2 \cup \cdots \cup A_n$, and the event that all occur is written $A_1 A_2 \cdots A_n$.

Rule 7.1 *For any finite number of mutually exclusive events $A_1, \ldots, A_n$,*

$$P(A_1 \cup A_2 \cup \cdots \cup A_n) = P(A_1) + P(A_2) + \cdots + P(A_n).$$

Rule 7.2 *For any event A,*

$$P(A) = 1 - P(A^c),$$

where the event A^c consists of all outcomes that are not in A.

Rule 7.3 *For any two events A and B,*

$$P(A \cup B) = P(A) + P(B) - P(AB).$$

Rule 7.4 *For any finite number of events* $A_1, \ldots, A_n$,

$$P\left(\bigcup_{i=1}^{n} A_i\right) = \sum_{i=1}^{n} P(A_i) - \sum_{\substack{i,j:\\ i<j}} P(A_i A_j) + \sum_{\substack{i,j,k:\\ i<j<k}} P(A_i A_j A_k) - \cdots$$
$$+ (-1)^{n-1} P(A_1 A_2 \cdots A_n).$$

The proofs of these rules are simple and instructive and nicely demonstrate how useful propositions can be obtained from "minimal" axioms.

To prove Rule 7.1, denote by $\emptyset$ the empty set of outcomes (*null event*). We first show that

$$P(\emptyset) = 0.$$

Applying Axiom 7.3 with $A_i = \emptyset$ for $i = 1, 2, \ldots$ gives $P(\emptyset) = \sum_{i=1}^{\infty} a_i$, where $a_i = P(\emptyset)$ for each i. This implies that $P(\emptyset) = 0$. Let $A_1, \ldots, A_n$ be any finite sequence of pairwise disjoint sets. Augment this sequence with $A_{n+1} = \emptyset$, $A_{n+2} = \emptyset, \ldots$. Then, by Axiom 7.3,

$$P\left(\bigcup_{i=1}^{n} A_i\right) = P\left(\bigcup_{i=1}^{\infty} A_i\right) = \sum_{i=1}^{\infty} P(A_i) = \sum_{i=1}^{n} P(A_i).$$

It is noted that Rule 7.1 and the Property $P(\emptyset) = 0$ show that, for a finite sample space, Axioms 7.1 to 7.3 are equivalent to the axioms in Section 2.2. The added generality of Axiom 7.3 is necessary when the sample space is infinite.

The proof of Rule 7.2 is as follows. The set $A \cup A^c$ is by definition equal to the sample space. Hence, by Axiom 7.2, $P(A \cup A^c) = 1$. The sets A and A^c are disjoint. It now follows from Rule 7.1 that $P(A \cup A^c) = P(A) + P(A^c)$. This gives the complement rule $P(A) = 1 - P(A^c)$.

To prove Rule 7.3, denote by A_1 the set of outcomes that belong to A but not to B. Let B_1 be the set of outcomes that are in B but not in A, and let $C = AB$ be the set of outcomes that are both in A and B. The sets A_1, B_1, and C are pairwise disjoint. Moreover,

$$A \cup B = A_1 \cup B_1 \cup C, \qquad A = A_1 \cup C \qquad \text{and} \qquad B = B_1 \cup C.$$

Applying Rule 7.1 gives

$$P(A \cup B) = P(A_1) + P(B_1) + P(C).$$

Also, $P(A) = P(A_1) + P(C)$ and $P(B) = P(B_1) + P(C)$. By substituting the latter two relations into the expression for $P(A \cup B)$ and noting that $C = AB$, we find

$$P(A \cup B) = P(A) - P(C) + P(B) - P(C) + P(C)$$
$$= P(A) + P(B) - P(AB).$$

Rule 7.4 will only be proved for the special case that the sample space is finite or countably infinite. In this case, $P(A) = \sum_{\omega \in A} p(\omega)$, where $p(\omega)$ is the probability assigned to the individual element ω of the sample space. Fix ω. If $\omega \notin \cup_{i=1}^{n} A_i$, then ω does not belong to any of the sets A_i and $p(\omega)$ does not contribute to either the left-hand side or the right-hand side of the expression in Rule 7.4. Assume now that $\omega \in \cup_{i=1}^{n} A_i$. Then, there is at least one set A_i to which ω belongs. Let s be the number of sets A_i to which ω belongs. In the left-hand side of the expression in Rule 4, $p(\omega)$ contributes only once. In the first term of the right-hand side of this expression, $p(\omega)$ contributes s times, in the second term $\binom{s}{2}$ times, in the third term $\binom{s}{3}$ times, and so on. Thus, the coefficient of $p(\omega)$ in the right-hand side is

$$s - \binom{s}{2} + \binom{s}{3} - \cdots + (-1)^{s-1}\binom{s}{s}.$$

Rule 7.4 follows by proving that this coefficient is equal to 1. Since $\binom{s}{1} = s$ and $\binom{s}{0} = 1$, we find

$$\begin{aligned} s &- \binom{s}{2} + \binom{s}{3} - \cdots + (-1)^{s-1}\binom{s}{s} \\ &= 1 - \left[\binom{s}{0} - \binom{s}{1} + \binom{s}{2} - \binom{s}{3} + \cdots + (-1)^s\binom{s}{s}\right] \\ &= 1 - (-1+1)^s = 1, \end{aligned}$$

where the second equality uses Newton's binomium $(a+b)^n = \sum_{k=0}^{n}\binom{n}{k} a^k b^{n-k}$.

Next, we give several illustrative applications of the above properties. We first illustrate Rule 7.2, which is known as the *complement rule*. This rule states that the probability of an event occuring is one minus the probability that it does not occur. This simple property is extremely useful. It is often easier to compute the complementary probability than the probability itself.

Example 7.4 What is the probability of getting at least one ace in a poker hand of five cards dealt from 52 cards?

Let A be the event that you get at least one ace. It is easier to compute the probability of the complementary event A^c that you get no ace in a poker hand of five cards. For the sample space of the random experiment, we take all ordered five-tuples $(x_1, x_2, x_3, x_4, x_5)$, where x_i corresponds to the suit and value of the ith card you get dealt. The total number of possible outcomes equals $52 \times 51 \times 50 \times 49 \times 48$. The number of outcomes without ace equals $48 \times 47 \times 46 \times 45 \times 44$. Assuming that the cards are randomly dealt, all possible outcomes are equally likely. Then, the event A^c has the probability

$$P\left(A^c\right) = \frac{48 \times 47 \times 46 \times 45 \times 44}{52 \times 51 \times 50 \times 49 \times 48} = 0.6588.$$

Hence, the probability of getting at least one ace in a poker hand of five cards is $1 - P\left(A^c\right) = 0.3412$.

Rule 7.3 is often called the *addition rule* and is illustrated with the following example.

Example 7.5 A single card is randomly drawn from a thoroughly shuffled deck of 52 cards. What is the probability that the drawn card will be either a heart or an ace?

For the sample space of this random experiment, we take the set consisting of the 52 elements

$$\spadesuit A, \ldots, \spadesuit 2, \quad \heartsuit A, \ldots \heartsuit 2, \quad \clubsuit A, \ldots, \clubsuit 2, \quad \diamondsuit A, \ldots, \diamondsuit 2,$$

where, for example, the outcome $\clubsuit 7$ means that the seven of clubs is drawn. All possible outcomes are equally likely and thus each outcome gets assigned the same probability $\frac{1}{52}$. Let A be the event that the drawn card is a heart and B the event that the drawn card is an ace. These two events are not mutually exclusive. We are looking for the probability $P(A \cup B)$ that at least one of the events A and B occurs. This probability can be calculated by applying Rule 7.3:

$$P(A \cup B) = P(A) + P(B) - P(AB).$$

In this case, $P(AB)$ stands for the probability that the drawn card is the ace of hearts. The events A and B correspond to sets that contain 13 and 4 elements, respectively, and thus have respective probabilities $\frac{13}{52}$ and $\frac{4}{52}$. The event AB corresponds to a set that is a singleton and thus has probability $\frac{1}{52}$. Hence, the probability that the drawn card is either a heart or an ace equals

$$P(A \cup B) = \frac{13}{52} + \frac{4}{52} - \frac{1}{52} = \frac{16}{52}.$$

Example 7.6 In the Lotto 6/42, six different numbers are picked at random from the numbers $1, \ldots, 42$. What is the probability that number 10 is picked?

The answer is $\frac{6}{42}$. A formal derivation goes as follows. Take as sample space all possible permutations of the integers $1, \ldots, 42$. Imagine that the six numbers are picked by taking the numbers in the first six positions of a random permutation. For $i = 1, \ldots, 6$, let A_i be the event that number 10 is in the ith position. Then, $P(A_i) = 41!/42! = 1/42$ for $i = 1, \ldots, 6$. The events $A_1, \ldots, A_6$ are disjoint and so

$$\begin{aligned} P(\text{number 10 is picked}) &= P(A_1 \cup \cdots \cup A_6) \\ &= P(A_1) + \cdots + P(A_6) = \frac{6}{42}. \end{aligned}$$

Problem 7.7 The probability that the events A and B both occur is 0.3. The individual probabilities of the events A and B are 0.7 and 0.5. What is the probability that neither event A nor event B occurs?

Problem 7.8 In the casino game of Chuck-a-Luck, three dice are contained within an hourglass-shaped, rotating cage. You bet on one of the six possible numbers and the cage is rotated. You lose money only if your number does not come up on any of the three dice. Much to the pleasure of the casinos, people sometimes reason as follows: the probability of my number coming up on one die is 1/6 and so the probability of my number coming up on one of the three dice is three times 1/6, or 1/2. Why is this reasoning false? How do you calculate the correct value of the probability that your number will come up on any of the three dice? *Hint*: use a formula for $P(A \cup B \cup C)$.

Problem 7.9 An integer is chosen at random from the integers $1, \ldots, 1{,}000$. What is the probability that the integer chosen is divisible by 3 or 5? What is the probability that the integer chosen is divisible by 3, 5 or 7?

Problem 7.10 For the upcoming drawing of the Bingo Lottery, five extra prizes have been added to the pot. Each prize consists of an all-expenses paid vacation trip. Each prize winner may choose from among three possible destinations A, B, and C. The three destinations are equally popular. Calculate the probability that at least one of the destinations A and B will be chosen. Also, calculate the probability that not each of the three destinations will be chosen.

Rule 7.4 extends Rule 7.3 and is known as the *inclusion-exclusion rule*. This rule states that the probability of the union of n events equals the sum of the probabilities of these events taken one at a time, minus the sum of the probabilities of these events taken two at a time, plus the sum of the probabilities of these events taken three at a time, and so on. We illustrate this property with the following classic example.

Example 7.7 Letters to n different persons are randomly put into n pre-addressed envelopes. What is the probability that at least one person receives the correct letter?

For the formulation of the sample space for this random experiment, it is convenient to give the label i to the envelope with the address of person i for $i = 1, \ldots, n$. Then, we take the set of all possible orderings $(e_1, \ldots, e_n)$ of the integers $1, \ldots, n$ as our sample space. In the outcome $\omega = (e_1, \ldots, e_n)$, the letter to person i is put into the envelope with label e_i for $i = 1, \ldots, n$. The total number of possible outcomes is $n \times (n-1) \times \cdots \times 1 = n!$. Since the letters are put randomly into the envelopes, all possible orderings are equally likely and thus each outcome $(e_1, \ldots, e_n)$ gets assigned the same probability $\frac{1}{n!}$. For fixed i, let A_i be the event that the letter for person i is put into the envelope with label i. The probability that at least one person receives the correct letter is given by $P(A_1 \cup A_2 \cup \cdots \cup A_n)$. The probabilities on the right-hand side of the inclusion-exclusion formula in Rule 7.4 are easy to calculate. For fixed i, the total number of orderings $(e_1, \ldots, e_n)$ with $e_i = i$ is equal to $(n-1)!$. This gives

$$P(A_i) = \frac{(n-1)!}{n!} \quad \text{for} \quad i = 1, \ldots, n.$$

Next fix i and j with $i \neq j$. The number of orderings $(e_1, \ldots, e_n)$ with $e_i = i$ and $e_j = j$ is equal to $(n-2)!$. Hence

$$P(A_i A_j) = \frac{(n-2)!}{n!} \quad \text{for} \quad \text{all } i \quad \text{and} \quad j \text{ with } i \neq j.$$

Continuing in this way, we find

$$P(A_1 \cup A_2 \cup \cdots \cup A_n) = \binom{n}{1}\frac{(n-1)!}{n!} - \binom{n}{2}\frac{(n-2)!}{n!} + \binom{n}{3}\frac{(n-3)!}{n!} - \cdots + (-1)^{n-1}\binom{n}{n}\frac{1}{n!}.$$

Since $\binom{n}{k} = \frac{n!}{k!(n-k)!}$, this expression simplifies to

$$P(A_1 \cup A_2 \cup \cdots \cup A_n) = \frac{1}{1!} - \frac{1}{2!} + \frac{1}{3!} - \cdots + (-1)^{n-1}\frac{1}{n!} = 1 - \sum_{k=0}^{n} \frac{(-1)^k}{k!}.$$

A surprising conclusion can be drawn from this result. A basic result from calculus is that $\sum_{k=0}^{\infty} (-1)^k/k! = e^{-1}$ with $e = 2.718\ldots$ (see the Appendix). Thus, for large n, the probability that at least one person will receive the correct letter is approximately equal to $1 - e^{-1} = 0.632$, independently of how large n is.

Having obtained the probability of at least one person receiving a correct letter, it is not difficult to argue that

$$P(\text{exactly } j \text{ persons receive a correct letter}) = \frac{1}{j!}\sum_{k=0}^{n-j} \frac{(-1)^k}{k!}$$

for $j = 0, 1, \ldots, n$. To verify this, denote by N_m the number of permutations of the integers $1, \ldots, m$ so that no integer remains in its original position. Since the probability that a random permutation of $1, \ldots, m$ has this property is $\sum_{k=0}^{m} (-1)^k/k!$, it follows that $N_m/m! = \sum_{k=0}^{m} (-1)^k/k!$ The number of permutations of the integers $1, \ldots, n$ so that exactly j integers remain in their original positions equals $\binom{n}{j}N_{n-j}$. Thus,

$$P(\text{exactly } j \text{ persons receive a correct letter}) = \frac{\binom{n}{j}N_{n-j}}{n!}.$$

Noting that $\binom{n}{j}/n! = \frac{1}{j!(n-j)!}$ and inserting the expression for $N_{n-j}/(n-j)!$, the above formula for the probability of exactly j persons receiving a correct letter follows. This probability tends to the Poisson probability $e^{-1}/j!$ as the number of envelopes becomes large (see Section 4.2.3 for a discussion of the Poisson approximation).

Problem 7.11 What is the probability that in a player's hand of 13 cards at least one suit will be missing?

Problem 7.12 Consider the card game Jeu de Treize from Problem 3.36. Use the inclusion-exclusion rule to verify that the probability of the dealer winning the first round is 0.6431.

Many probability problems of a combinatorial nature can be solved by using the inclusion-exclusion rule. We conclude this chapter with two more examples.

Example 7.8 Fifteen tourists are stranded in a city with four hotels, all of which are located nearby each other in the city center. Each hotel has enough rooms available to accommodate all fifteen tourists. Each tourist randomly chooses a hotel independently of the choices made by the others. What is the probability that not all four hotels will be chosen? We leave it to the reader to verify that the desired probability is given by

$$\sum_{k=1}^{4}(-1)^{k+1}\binom{4}{k}\frac{(4-k)^{15}}{4^{15}} = 0.0533.$$

Example 7.9 Suppose $n = 10$ married couples are invited to a bridge party. Bridge partners are chosen at random, without regard to gender. What is the probability that no one will be paired with his or her spouse? Denote by A_i the event that couple i is paired as bridge partners. Take as sample space the set of all the possible permutations of the integers $1, \ldots, 2n$, where the integers $2i-1$ and $2i$ represent couple i. We leave it to the reader to verify that the complementary probability $P(A_1 \cup A_2 \cup \cdots \cup A_n)$ of at least one couple being paired as bridge partners is given by

$$\sum_{k=1}^{n}(-1)^{k+1}\binom{n}{k}\frac{n \times (n-1) \times \cdots \times (n-k+1) \times 2^k \times (2n-2k)!}{(2n)!}.$$

This probability has the value 0.4088 for $n = 10$ (the probability of at least one couple being paired as bridge partners tends to $1 - e^{-\frac{1}{2}} = 0.3935$ as the number of couples gets large).

8
Conditional probability and Bayes

In this chapter, we delve further into the law of conditional probabilities and Bayes' rule. The law of conditional probabilities provides a useful and natural tool for calculating probabilities. Several applications have already been discussed in Chapters 3 to 6. We touched on Bayes' rule in Chapter 6 in our discussion of the applications of chance trees. The purpose of this chapter is to give some further insights into the importance of the concept of conditional probability and its role in Bayes' rule.

8.1 Conditional probability

First, we will repeat the definition of conditional probability. The starting point is a random experiment for which a sample space and a probability measure P are defined. Let A be an event of the experiment. The probability $P(A)$ reflects our knowledge of the occurrence of event A *before* the experiment takes place. Therefore, the probability $P(A)$ is sometimes referred to as the *a priori* probability of A or the *unconditional* probability of A. Suppose now we are told that an event B has occurred in the experiment, but we still do not know the precise outcome in the set B. In light of this added information, the set B replaces the sample space as the set of possible outcomes and consequently the probability of the occurrence of event A changes. A conditional probability now reflects our knowledge of the occurrence of the event A given that event B has occurred. The notation for this new probability is $P(A \mid B)$.

Definition 8.1 *For any two events A and B with $P(B) > 0$, the conditional probability $P(A \mid B)$ is defined as*

$$P(A \mid B) = \frac{P(AB)}{P(B)},$$

where AB stands for the occurrence of both event A and event B. This is not an arbitrary definition. It can be intuitively reasoned through a comparable property of the relative frequency. In Section 2.1, we defined the relative frequency $f_n(E)$ of the occurrence of event E as $\frac{n(E)}{n}$, where $n(E)$ represents the number of times that E occurs in n repetitions of the experiment. Assume, now, that in n independent repetitions of the experiment, event B occurs r times simultaneously with event A and s times without event A. We can then say that $f_n(AB) = \frac{r}{n}$ and $f_n(B) = \frac{r+s}{n}$. If we divide $f_n(AB)$ by $f_n(B)$, then we find that

$$\frac{f_n(AB)}{f_n(B)} = \frac{r}{r+s}.$$

Now define $f_n(A \mid B)$ as the relative frequency of event A in those repetitions of the experiment in which event B has occurred. From $f_n(A \mid B) = \frac{r}{r+s}$ we now get the following relationship:

$$f_n(A \mid B) = \frac{f_n(AB)}{f_n(B)}.$$

This relationship accounts for the definition of conditional probability $P(A \mid B)$.

Example 8.1 Someone has rolled a fair die twice. You know that one of the rolls turned up a face value of six. What is the probability that the other roll turned up a six as well?

Take as sample space the set $\{(i, j) | i, j = 1, \ldots, 6\}$, where i and j denote the outcomes of the first and second rolls. A probability of $\frac{1}{36}$ is assigned to each element of the sample space. The event of two sixes is given by $A = \{(6, 6)\}$ and the event of at least one six is given by $B = \{(1, 6), \ldots, (5, 6), (6, 6), (6, 5), \ldots, (6, 1)\}$. Applying the definition of conditional probability gives

$$P(A \mid B) = \frac{P(AB)}{P(B)} = \frac{1/36}{11/36}.$$

Hence the desired probability is $\frac{1}{11}$ (not $\frac{1}{6}$).

This example illustrates once again how careful you have to be when you are interpreting the information a problem is conveying. The wording of the problem is crucial: you know that one of the dice turned up a six but you do not know which one. In the case where one of the dice had dropped on the floor and you had seen the outcome six for that die, the probability of the other die turning up a six would have been $\frac{1}{6}$.

Example 8.2 The number of jackpot winners in some lottery is Poisson distributed with an expected value of 1.23. The jackpot is won at the last drawing.

What is the probability of the jackpot being won by just one winner (i.e., that the jackpot will not have to be divided among multiple winners)?

The assumption of a Poisson distribution for the number of winners of the jackpot says that the probability of having exactly k winners is $e^{-\lambda}\lambda^k/k!$ for $k = 0, 1, \ldots,$ where $\lambda = 1.23$. Let A be the event that there is exactly one winner of the jackpot, and B the event that the jackpot is won. The unconditional probability of the event A is $P(A) = \lambda e^{-\lambda} = 0.35952$. However, the conditional probability of exactly one winner given that there is at least one winner equals

$$P(A \mid B) = \frac{P(AB)}{P(B)} = \frac{P(A)}{P(B)} = \frac{\lambda e^{-\lambda}}{1 - e^{-\lambda}}$$
$$= \frac{0.35952}{0.70771} = 0.5080.$$

If someone is lucky enough to be a jackpot winner in this lottery, the probability is 50.8% that this person will not have to share the jackpot with other winners.

Problem 8.1 Every evening, two weather stations issue a weather forecast for the next day. The weather forecasts of the two stations are independent of each other. On average, the weather forecast of station 1 is correct in 90% of the cases, irrespective of the weather type. This percentage is 80% for station 2. On a given day, station 1 predicts sunny weather for the next day, whereas station 2 predicts rain. What is the probability that the weather forecast of station 1 will be correct?

Problem 8.2 Someone has tossed a fair coin three times. You know that one of the tosses came up heads. What is the probability that at least one of the other two tosses came up heads as well?

Problem 8.3 Suppose a bridge player's hand of thirteen cards contains an ace. What is the probability that the player has only one ace? What is the answer to this question if you know that the player had the ace of hearts?

8.1.1 Assigning probabilities through conditional probabilities

The formula for the conditional probability $P(A \mid B)$ can be rewritten as

$$P(AB) = P(A \mid B)P(B).$$

This phrasing lines up more naturally with the intuitive way people think about probabilities. In many cases, $P(AB) = P(A \mid B)P(B)$ is used in attributing probabilities to elements of the sample space. In illustration of this, consider the experiment in which two marbles are randomly chosen without replacements from a receptacle holding seven red and three white marbles. One possible choice for the sample space of this experiment is the set consisting of four elements (R, R), (R, W), (W, W), and (W, R), where R stands for red and W for white. The first component of each element indicates the color of the first marble chosen and the second component the color of the second marble chosen. On grounds of the reasoning that $P\big(1^{\text{st}} \text{ marble is red }\big) = \frac{7}{10}$ and $P(2^{\text{nd}} \text{ marble is white } \big| 1^{\text{st}} \text{ marble is red}) = \frac{3}{9}$, we attribute the probability of $P(R, W) = \frac{7}{10} \times \frac{3}{9} = \frac{7}{30}$ to the element (R, W). In the same way, we attribute the probabilities $P(R, R) = \frac{7}{10} \times \frac{6}{9} = \frac{7}{15}$, $P(W, W) = \frac{3}{10} \times \frac{2}{9} = \frac{1}{15}$, and $P(W, R) = \frac{3}{10} \times \frac{7}{9} = \frac{7}{30}$ to the remaining elements. It is common practice in this type of problem to assign probabilities to the elements of the sample space as a product of probabilities, one marginal and the others conditional. To do so, one uses the formula

$$P(A_1 A_2 \cdots A_n) = P(A_1) \times P(A_2 \mid A_1) \times P(A_3 \mid A_1 A_2) \\ \times \cdots \times P(A_n \mid A_1 A_2 \cdots A_{n-1}),$$

this being an extension of the formula $P(A) = P(A \mid B)P(B)$.

Example 8.3 A group of fifteen tourists is stranded in a city with four hotels of the same class. Each of the hotels has enough room available to accommodate the fifteen tourists. The group's guide, who has a good working relationship with each of the four hotels, assigns the tourists to the hotels as follows. First, he randomly determines how many are to go to hotel A, then how many of the remaining tourists are to go to hotel B, and then how many are to go to hotel C. All remaining tourists are sent to hotel D. Note that each stage of the assignment the guide draws at random a number between zero and the number of tourists left. What is the probability that all four hotels receive guests from the group?

Let the outcome (i_A, i_B, i_C, i_D) correspond with the situation in which i_A tourists are sent to hotel A, i_B tourists to hotel B, i_C tourists to hotel C, and i_D tourists to hotel D. The probability

$$\frac{1}{16} \times \frac{1}{16 - i_A} \times \frac{1}{16 - i_A - i_B}$$

is assigned to the outcome (i_A, i_B, i_C, i_D) for $0 \leq i_A, i_B, i_C, i_D \leq 15$ and $i_A + i_B + i_C + i_D = 15$. The probability that all four hotels will receive guests is given by

$$\sum_{i_A=1}^{12} \sum_{i_B=1}^{13-i_A} \sum_{i_C=1}^{14-i_A-i_B} \frac{1}{16} \times \frac{1}{16 - i_A} \times \frac{1}{16 - i_A - i_B} = 0.2856.$$

Problem 8.4 Take another look at Problem 7.5 and solve it under the supposition that the balls are chosen without replacements.

Problem 8.5 Seven individuals have reserved tickets at the opera. The seats they have been assigned are all in the same row of seven seats. The row of seats is accessible from either end. Assume that the seven individuals arrive and take their seats in a random order. What is the probability of all seven individuals taking their seats without having to squeeze past an already seated individual? Use conditional probabilities to answer this question. *Hint*: assume that the individuals get up from their seats one by one and in a random order, and calculate the probability of the individuals leaving without having to squeeze past others in their row.

8.1.2 Independent events

In the special case of $P(A \mid B) = P(A)$, the occurrence of event A is not contingent on the occurrence or nonoccurrence of event B. Event A is then said to be independent of event B. In other words, if A is independent of B, then learning that event B has occurred does not change the probability that event A occurs. Since $P(A \mid B) = \frac{P(AB)}{P(B)}$, it follows that A is independent of B if the equation $P(AB) = P(A)P(B)$ holds true. This equation is symmetric in A and B: if A is independent of B, then B is also independent of A. Summarizing,

Definition 8.2 *Two events A and B are said to be independent if*

$$P(AB) = P(A)P(B).$$

The reader should be aware that independent events and disjoint events are completely different things. If events A and B are disjoint, you calculate the probability of the union $A \cup B$ by *adding* the probabilities of A and B. For independent events A and B, you calculate the probability of the intersection AB by *multiplying* the probabilities of A and B. Since $P(AB) = 0$ for disjoint events A and B, independent events are typically not disjoint.

Example 8.4 Suppose two fair dice are thrown. Let A be the event that the number shown by the first die is even, and B the event that the sum of the dice is odd. Do you think the events A and B are independent?

The experiment has 36 possible outcomes (i, j), where i is the number shown by the first die and j the number shown by the second die. All possible outcomes are equally likely. Simply, by counting, $P(A) = 18/36$, $P(B) = 18/36$, and $P(AB) = 9/36$. Since $P(AB) = P(A)P(B)$, events A and B are independent.

Remark 8.1 In the case that events A, B, and C are pairwise independent, it is not necessarily true that $P(ABC) = P(A)P(B)P(C)$. This can be shown using Example 8.3. In addition to the events A and B from Example 8.3, let C be the event that the number shown by the second die is even. Events A, B, and C are pairwise independent, but $P(ABC)(=0)$ is not equal to $P(A)P(B)P(C)$. In general events $A_1, \ldots, A_n$ are said to be independent if $P(A_{i_1} \ldots A_{i_k}) = P(A_{i_1}) \times \cdots \times P(A_{i_k})$ for every collection $A_{i_1}, \ldots, A_{i_k}$ and $2 \leq k \leq n$.

8.1.3 The law of conditional probabilities

It is often the case that the unconditional probability $P(A)$ of an event A is found most easily by expressing it in terms of conditional probabilities. The idea is to choose an appropriate sequence of *mutually exclusive* events $B_1, \ldots, B_n$ such that the event A can only occur when one of the disjoint events $B_1, \ldots, B_n$ occurs. Next, the probability $P(A)$ can be obtained by applying the following rule:

Rule 8.1 *Let A be an event that can only occur when one of mutually exclusive events $B_1, \ldots, B_n$ occurs. Then,*

$$P(A) = P(A \mid B_1)P(B_1) + P(A \mid B_2)P(B_2) + \cdots + P(A \mid B_n)P(B_n).$$

This rule is called the *law of conditional probabilities*. The proof of this law is simple and instructive. The assumption that event A can only occur if one of the events $B_1, \ldots, B_n$ also occurs means, in terms of sets, that the subset A of the sample space is contained in the union $B_1 \cup \cdots \cup B_n$ of the subsets $B_1, \ldots, B_n$. This implies

$$A = AB_1 \cup AB_2 \cup \cdots \cup AB_n,$$

where AB_i stands for the set of outcomes belonging both to set A and set B_i. The assumption that the sets $B_1, \ldots, B_n$ are disjoint implies that the sets $AB_1, \ldots, AB_n$ are also disjoint. By Rule 7.1 in Chapter 7, we then have

$$P(A) = P(AB_1) + P(AB_2) + \cdots + P(AB_n).$$

This relationship and the definition $P(A \mid B) = P(AB)/P(B)$ lead to the law of conditional probabilities. This law is naturally also applicable when the sample space is divided by a countable number of disjoint subsets $B_1, B_2, \ldots$ instead of by a finite number.

A nice illustration of the law of conditional probabilities is provided by the craps example in Section 3.3 of Chapter 3. Another nice illustrative example is the following one.

Example 8.5 The upcoming Tour de France bicycle tournament will take place from July 1 through July 23. One hundred eighty cyclists will participate in the event. What is the probability that two or more participating cyclists will have birthdays on the same day during the tournament?

Denoting by A the event that two or more participating cyclists will have birthdays on the same day during the tournament, event A can occur only if one of the mutually exclusive events $B_2, \ldots, B_{180}$ occurs. Event B_i occurs when exactly i participating cyclists have birthdays during the tournament. The conditional probability $P(A \mid B_i)$ is easy to calculate. It refers to the birthday problem with i persons coming from a "planet" where the year has 23 days. The birthday problem was studied in detail in Chapter 3. The reader may easily verify that

$$P(A \mid B_i) = \begin{cases} 1 - \frac{23 \times 22 \times \cdots \times (23-i+1)}{(23)^i}, & 2 \leq i \leq 23 \\ 1, & i \geq 24. \end{cases}$$

The probability $P(B_i)$ is given by the binomial probability

$$P(B_i) = \binom{180}{i} \left(\frac{23}{365}\right)^i \left(1 - \frac{23}{365}\right)^{180-i}, \qquad 0 \leq i \leq 180.$$

Putting the pieces together, we find

$$\begin{aligned} P(A) &= \sum_{i=2}^{180} P(A \mid B_i) P(B_i) \\ &= 1 - P(B_0) - P(B_1) - \sum_{i=2}^{23} \frac{23 \times 22 \times \cdots \times (23 - i + 1)}{(23)^i} P(B_i). \end{aligned}$$

This yields the value 0.8841 for the probability $P(A)$.

Problem 8.6 Let's return to the casino game Red Dog from Problem 3.27. Using the law of conditional probabilities, calculate the probability of the player winning. *Hint*: argue first that the probability of a spread of i points is given by $\frac{1}{52!}\left[(12 - i) \times 4 \times 4 \times 2\right]$.

Problem 8.7 Consider the scratch-lottery problem from Section 4.2.3. Each week one million scratch-lottery tickets are printed. Assume that in a particular week only one half of the tickets printed are sold. What is the probability of at least one winner in that week? *Hint*: use results from Example 7.7.

8.2 Bayes' rule in odds form

Bayes' rule specifies how probabilities must be updated in the light of new information. It can be best understood by considering the odds form of the rule for the situation where there is question of a hypothesis being either true or false. An example of such a situation is a court case where the defendant is either guilty or not guilty. Let H represent the event that the hypothesis is true, and $\overline{H}$ the event that the hypothesis is false. Before examining the evidence, a Bayesian analysis begins with assigning prior subjective probabilities $P(H)$ and $P(\overline{H}) = 1 - P(H)$ to the mutually exclusive events H and $\overline{H}$. How do the prior probabilities change once evidence in the form of the knowledge that the event E has occurred becomes available? In our example of the court case, event E could be the evidence that the accused has the same blood type as the perpetrator's, whose blood has been found at the scene of the crime. The updated value of the probability that the hypothesis is true given the fact that event E has occurred is denoted by $P(H \mid E)$. To calculate the posterior probability $P(H \mid E)$, we use Bayes' rule. This rule can be expressed in several different ways. A convenient form uses odds. Odds are often used to represent probabilities. Gamblers usually think in terms of "odds" instead of probabilities. For an event with probability of $\frac{2}{3}$, the odds are 2 to 1 (written 2:1), while for an event with a probability of $\frac{3}{10}$, the odds are 3:7. The odds form of Bayes' rule reads as follows:

Rule 8.2 *The posterior probabilities* $P(H \mid E)$ *and* $P\big(\overline{H} \mid E\big) = 1 - P(H \mid E)$ *satisfy*

$$\frac{P(H \mid E)}{P\big(\overline{H} \mid E\big)} = \frac{P(H)}{P\big(\overline{H}\big)} \frac{P(E \mid H)}{P\big(E \mid \overline{H}\big)}.$$

In words: *posterior odds* $=$ *prior odds* $\times$ *likelihood ratio*. This insightful formula follows by twice applying the definition of conditional probability. By doing so, we obtain

$$P(H \mid E) = \frac{P(HE)}{P(E)} = P(E \mid H)\frac{P(H)}{P(E)}.$$

The same expression holds for $P\big(\overline{H} \mid E\big)$ with H replaced by $\overline{H}$. Dividing the expression for $P(H \mid E)$ by the expression for $P\big(\overline{H} \mid E\big)$ results in the odds form of Bayes' rule.

Bayes' rule updates the prior odds of the hypothesis H by multiplying $P(H)/P(\overline{H})$ with the likelihood ratio $P(E \mid H)/P\big(E \mid \overline{H}\big)$. In the example of the court case, the probabilities $P(E \mid H)$ and $P\big(E \mid \overline{H}\big)$ are typically determined by an expert.[1] However it is not the expert's task to tell the court what the prior odds are. The prior probabilities $P(H)$ and $P(\overline{H})$ represent the personal opinion of the court before the evidence is taken into account.

Example 8.6 A murder is committed. The perpetrator is either one or the other of the two persons X and Y. Both persons are on the run from authorities, and after an initial investigation, both fugitives appear equally likely to be the perpetrator. Further investigation reveals that the actual perpetrator has blood type A. Ten percent of the population belongs to the group having this blood type. Additional inquiry reveals that person X has blood type A, but offers no information concerning the blood type of person Y. In light of this new information, what is the probability that person X is the perpetrator?

In answering this question, use H to denote the event that person X is the perpetrator. Let E represent the new evidence that person X has blood type A. The prior probabilities of H and $\overline{H}$ before the appearance of the new evidence E are given by

$$P(H) = P(\overline{H}) = \frac{1}{2}.$$

In addition, it is also true that

$$P(E \mid H) = 1 \qquad \text{and} \qquad P(E \mid \overline{H}) = \frac{1}{10}.$$

Applying Bayes' rule in odds form at this point, we find that

$$\frac{P(H \mid E)}{P(\overline{H} \mid E)} = \frac{1/2}{1/2} \times \frac{1}{1/10} = 10.$$

The odds in favor, then, are 10 to 1 that person X is the perpetrator given that this person has blood type A. Otherwise stated, from $P(H \mid E)/[1 - P(H \mid E)] = 10$, it follows that

$$P(H \mid E) = \frac{10}{11}.$$

[1] In both legal and medical cases, the conditional probabilities $P(H \mid E)$ and $P(E \mid H)$ are sometimes confused with each other. A classic example is the famous court case of People vs. Collins in Los Angeles in 1964. In this case, a couple matching the description of a couple that had committed an armed robbery was arrested. Based on expert testimony, the district attorney claimed that the frequency of couples matching the description was roughly 1 in 12 million. Although this was the estimate for $P(E \mid \overline{H})$, the district attorney treated this estimate as if it was $P(\overline{H} \mid E)$ and incorrectly concluded that the couple was guilty beyond reasonable doubt (see also the discussion concerning the court case from paragraph 4.2.2).

The probability of Y being the perpetrator is $1 - \frac{10}{11} = \frac{1}{11}$ and not, as may be thought, $\frac{1}{10} \times \frac{1}{2} = \frac{1}{20}$. The error in this reasoning is that the probability of person Y having blood type A is not $\frac{1}{10}$ because Y is not a randomly chosen person; rather, Y is first of all a person having a 50% probability of being the perpetrator, whether or not he is found at a later time to have blood type A.

Another nice illustration of Bayes' rule in odds form is provided by legal arguments used in the discussion of the O. J. Simpson trial.[2]

Example 8.7 Nicole Brown was murdered at her home in Los Angeles on the night of June 12, 1994. The prime suspect was her husband O. J. Simpson, at the time a well-known celebrity famous both as a TV actor as well as a retired professional football star. This murder lead to one of the most heavily publicized murder trials in the United States during the last century. The fact that the murder suspect had previously physically abused his wife played an important role in the trial. The famous defense lawyer Alan Dershowitz, a member of the team of lawyers defending the accused, tried to belittle the relevance of this fact by stating that only 0.1% of the men who physically abuse their wives actually end up murdering them. Was the fact that O. J. Simpson had previously physically abused his wife irrelevant to the case?

The answer to the question is no. In this particular court case, it is important to make use of the crucial fact that Nicole Brown was murdered. The question, therefore, is not what the probability is that abuse leads to murder, but the probability that the husband is guilty in light of the fact that he had previously abused his wife. This probability can be estimated with the help of Bayes formula and a few facts based on crime statistics. Define the following

E = the event that the husband has physically abused his wife in the past
M = the event that the wife has been murdered
G = the event that the husband is guilty of the murder of his wife.

The probability in question is the conditional probability $P(G \mid EM)$. We can use Bayes formula expressed in terms of the posterior odds to calculate this probability. In this example, Bayes formula is given by

$$\frac{P(G \mid EM)}{P(\overline{G} \mid EM)} = \frac{P(G \mid M)}{P(\overline{G} \mid M)} \frac{P(E \mid GM)}{P(E \mid \overline{G}M)},$$

where $\overline{G}$ represents the event that the husband is not guilty of the murder of his wife. How do we estimate the conditional probabilities on the right-hand side

[2]This example is based on the article J. F. Merz and J. P. Caulkins, "Propensity to abuse – propensity to murder?, *Chance Magazine,* 1995, 8(2), 14.

of this formula? In 1992, 4,936 women were murdered in the United States, of which roughly 1,430 were murdered by their (ex)husbands or boyfriends. This results in an estimate of $\frac{1,430}{4,936} = 0.29$ for the prior probability $P(G \mid M)$ and an estimate of 0.71 for the prior probability $P(\overline{G} \mid M)$. Furthermore, it is also known that roughly 5% of married women in the United States have at some point been physically abused by their husbands. If we assume that a woman who has been murdered by someone other than her husband had the same chance of being abused by her husband as a randomly selected woman, then the probability $P(E \mid \overline{G}M)$ is equal to 5%. The remaining probability on the right-hand side is $P(E \mid GM)$. We can base our estimate of this probability on the reported remarks made by Simpson's famous defense attorney, Alan Dershowitz, in a newspaper article. In the newspaper article, Dershowitz admitted that a substantial percentage of the husbands who murder their wives have, previous to the murders, also physically abused their wives. Given this statement, the probability $P(E \mid GM)$ will be taken to be 0.5. By substituting the various estimated values for the probabilities into the formula for the posterior odds, we see that the odds are

$$\frac{P(G \mid EM)}{P(\overline{G} \mid EM)} = \frac{0.29}{0.71}\frac{0.5}{0.05} = 4.08.$$

We can translate the odds into probabilities using the fact that $P(\overline{G} \mid EM) = 1 - P(G \mid EM)$. This results in a value for $P(G \mid EM)$ of 0.81. In other words, there is an estimated probability of 81% that the husband is the murderer of his wife in light of the knowledge that he had previously physically abused her. The fact that O. J. Simpson had physically abused his wife in the past was therefore certainly very relevant to the case.

Problem 8.8 In a certain region, it rains on average once in every ten days during the summer. Rain is predicted on average for 85% of the days when rainfall actually occurs, while rain is predicted on average for 25% of the days when it does not rain. Assume that rain is predicted for tomorrow. What is the probability of rainfall actually occurring on that day?

Problem 8.9 You have five coins colored red, blue, white, green, and yellow. Apart from the variation in color, the coins look identical. One of the coins is unfair and when tossed comes up heads with a probability of $\frac{3}{4}$; the other four are fair coins. You have no further information about the coins apart from having observed that the blue coin, tossed three times, came up heads on all three tosses. On the grounds of this observation, you indicate that the blue coin is the unfair one. What is the probability of your being correct in this assumption?

Problem 8.10 Verify the answers that were obtained by Bayesian analysis in the statistical problem discussed in Example 5.1.

8.3 Bayesian statistics[3]

In addition to its application in court cases, Bayesian statistics is often used by accountants and tax inspectors to perform audits. Bayesian statistics is also often used to predict election results based on the results of new opinion polls and to update the degree of belief in the effectiveness of medical treatments given new clinical data. One of the principal advantages of Bayesian statistics is the ability to perform the analysis sequentially, where new information can be incorporated into the analysis as soon as it becomes available.

Example 8.8 On January 1, 2002, the euro was introduced as the new coin in many European countries. Belgian students made the papers at the beginning of January 2002, with an experiment in which a one euro coin with the image of King Albert was tossed 250 times and came up heads 140 times. What can be said about this coin?

In classical statistics, the null hypothesis for this experiment would be that the coin is fair. The null hypothesis would then be tested by calculating the probability of 140 or more heads out of 250 tosses with a fair coin. This probability is 0.0332. The approach of classical statistics thus calculates the probability of the data occurring under the null hypothesis. In Bayesian statistics, by contrast, one computes the probability that the null hypothesis is true given the data. More precisely, in the Bayesian approach, one assumes a prior distribution of the probability that a toss of the coin comes up heads. This distribution is revised when data become available. To show how this process works, imagine that there are nine possible values 0.1, 0.2, ..., 0.9 for the probability θ that the toss of a coin will land heads up. To start with, you have assigned a prior probability $p_0(\theta_i)$ to each possible value $\theta_i = \frac{i}{10}$:

θ_i	$p_0(\theta_i)$	θ_i	$p_0(\theta_i)$	θ_i	$p_0(\theta_i)$
0.1	0.05	0.4	0.15	0.7	0.10
0.2	0.05	0.5	0.30	0.8	0.05
0.3	0.10	0.6	0.15	0.9	0.05

[3]This section can be skipped at first reading

That is, before running an experiment, you believe that with probability 0.05 you have a coin coming up heads on average once in ten tosses, with probability 0.05 you have a coin coming up heads on average twice in ten tosses, etc. Next, after having observed 140 heads in 250 tosses, you calculate for every value θ_i the probability

$$P(140 \text{ times heads in } 250 \text{ tosses} \mid \theta = \theta_i)$$
$$= \binom{250}{140} \theta_i^{140} (1 - \theta_i)^{250-140}.$$

If we denote this probability as $L(\theta_i)$, we find that:

θ_i	$L(\theta_i)$	θ_i	$L(\theta_i)$	θ_i	$L(\theta_i)$
0.1	0	0.4	1.16×10^{-7}	0.7	9.48×10^{-7}
0.2	0	0.5	0.008357	0.8	0
0.3	0	0.6	0.022250	0.9	0

Thereafter, you calculate the posterior probability

$$p(\theta_i) = P(\theta = \theta_i \mid 140 \text{ times heads in } 250 \text{ tosses}).$$

This is done by applying Bayes rule:

$$p(\theta_i) = \frac{L(\theta_i)p_0(\theta_i)}{\sum_{k=1}^{9} L(\theta_k)p_0(\theta_k)} \quad \text{for} \quad i = 1, \ldots, 9.$$

This rule follows from arguments that are familiar by now:

$$P(\theta = \theta_i \mid 140 \text{ times heads in } 250 \text{ tosses})$$
$$= \frac{P(\theta = \theta_i \text{ and } 140 \text{ times heads in } 250 \text{ tosses})}{P(140 \text{ heads in } 250 \text{ tosses})}$$
$$= \frac{P(140 \text{ heads in } 250 \text{ tosses} \mid \theta = \theta_i)p_0(\theta_i)}{\sum_{k=1}^{9} P(140 \text{ heads in } 250 \text{ tosses} \mid \theta = \theta_k)p_0(\theta_k)}.$$

Applying Bayes' rule gives the following numerical values for the posterior distribution:

θ_i	$p(\theta_i)$	θ_i	$p(\theta_i)$	θ_i	$p(\theta_i)$
0.1	0	0.4	2.98×10^{-6}	0.7	1.62×10^{-5}
0.2	0	0.5	0.42895	0.8	0
0.3	0	0.6	0.57102	0.9	0

Comparing the posterior distribution with the prior distribution, one can see the effect of the data. An attractive property of the Bayesian approach is that when extra data becomes available, previous data does not lose its value. When new data becomes available, one takes the current posterior distribution as the new prior distribution and adjusts it as shown above. For example, imagine that 250 additional tosses of a one-euro coin land heads up 127 times. The above noted posterior distribution would then be adjusted as follows (verify!):

θ_i	$p(\theta_i)$	θ_i	$p(\theta_i)$	θ_i	$p(\theta_i)$
0.1	0	0.4	0	0.7	0
0.2	0	0.5	0.9821	0.8	0
0.3	0	0.6	0.0179	0.9	0

One could also have arrived at this posterior distribution by adjusting the original prior distribution $p_0(\theta_i)$ on the basis of $140 + 127 = 267$ times heads in $250 + 250 = 500$ tosses of the coin! Finally, it is notable that the posterior distribution becomes increasingly less sensitive to the originally chosen prior distribution as the available data increase. For example, the prior distribution $p_0(\theta_i) = \frac{1}{9}$ for $i = 1, \ldots, 9$ leads to the following posterior distribution when 500 tosses turn up heads 267 times:

θ_i	$p(\theta_i)$	θ_i	$p(\theta_i)$	θ_i	$p(\theta_i)$
0.1	0	0.4	0	0.7	0
0.2	0	0.5	0.9649	0.8	0
0.3	0	0.6	0.0351	0.9	0

The posterior distributions in the last two tables are quite similar, even though the priors are far apart.

Example 8.9 Two candidates A and B are contesting the election of governor in a given state. The candidate who wins the popular vote becomes governor. A random sample of the voting population is undertaken to find out the preference of the voters. The sample size of the poll is 1,000 and 517 of the polled voters favor candidate A. What can be said about the probability of candidate A winning the election?

The number of respondents in the poll who favor candidate A has a binomial distribution whose success probability represents the fraction of the voting population in favor of candidate A. Let's assume that, prior to polling, this success probability has the following prior distribution $p_0(\theta_i)$ on the possible values $\theta = 0.30, 0.31, \ldots, 0.69, 0.70$:

$$p_0(\theta) = \begin{cases} \frac{\theta - 0.29}{4.41} & \text{for} \quad \theta = 0.30, \ldots, 0.50, \\ \frac{0.71-\theta}{4.41} & \text{for} \quad \theta = 0.51, \ldots, 0.70. \end{cases}$$

Hence, the prior probability of candidate A getting the majority of the votes at the election is $p_0(0.51) + \cdots + p_0(0.70) = 0.476$. However, 517 of the 1,000 polled voters favor candidate A. In light of this new information, what is the probability of candidate A getting the majority of the votes at the time of election? This probability is given by $p(0.51) + \cdots + p(0.70)$, where $p(\theta)$ is the posterior probability that the fraction of the voting population in favor of candidate A equals θ. This posterior probability is easily calculated from

$$p(\theta) = \frac{\binom{1000}{517}\theta^{517}(1-\theta)^{1000-517}p_0(\theta)}{\sum_{a=30}^{70}\binom{1000}{517}\left(\frac{a}{100}\right)^{517}\left(1-\frac{a}{100}\right)^{1000-517}p_0\left(\frac{a}{100}\right)}.$$

Performing the numerical calculations, we find that the posterior probability of candidate A getting the majority of the votes at the election equals

$$p(0.51) + \cdots + p(0.70) = 0.7632.$$

The posterior probability of a tie at the election equals $p(0.50) = 0.1558$.

9
Basic rules for discrete random variables

In the first part of this book, we worked many times with models of random variables. In performing a random experiment, one is often not interested in the particular outcome that occurs but in a specific numerical value associated with that outcome. Any function that assigns a real number to each outcome in the sample space of the experiment is called a *random variable*. Many examples of random variables have been seen in Chapters 2 to 5. Most of these examples dealt with so-called discrete random variables. A random variable is said to be discrete if its set of possible values is finite or countably infinite. An example of a discrete random variable is the sum of two dice when the experiment consists of throwing two dice. The purpose of this chapter is to familiarize the reader with a number of basic rules for calculating characteristics of random variables such as the expected value and the variance. These rules are easiest explained and understood in the context of discrete random variables. Therefore, the discussion in this chapter is restricted to the case of discrete random variables. However, the rules for discrete random variables apply with obvious modifications to other types of random variables as well. In Chapter 10, we discuss so-called continuous random variables. Such random variables have a continuous interval as the range of possible values.

9.1 Expected value

The most important characteristic of a random variable is its *expected value*. Synonyms for expected value are *expectation, mean*, and *first moment*. In Chapter 2, we informally introduced the concept of expected value. The expected value of a discrete random variable is a weighted mean of the values the random variable can take on, the weights being furnished by the probability mass function of the random variable. The nomenclature of expected value may be misleading. The expected value is in general not a typical value that the random variable can take on. It is often helpful to interpret the expected value of a random variable as the long-run average value of the variable over many independent repetitions of an experiment.

Let X be a random variable that is defined on the sample space of a random experiment, where P is the given probability measure on the sample space. The random variable X is a function that assigns a numerical value $X(\omega)$ to each outcome ω in the sample space. It is assumed that X is a discrete random variable with I as the set of possible values. The set I is finite or countably infinite and is called the *range* of X. The *probability mass function* of X is defined by $P(X = x), x \in I$, where the notation $P(X = x)$ is a shorthand for the probability mass assigned by P to the set of all outcomes ω for which $X(\omega) = x$.

Definition 9.1 *The expected value of the discrete random variable X is defined by*

$$E(X) = \sum_{x \in I} x\, P(X = x).$$

This definition is only meaningful if the sum is well-defined. The sum is always well-defined if the range I is finite. However, the sum over countably many terms is not always well-defined when both positive and negative terms are involved. For example, the infinite series $1 - 1 + 1 - 1 + \ldots$ has the sum 0 when you sum the terms according to $(1 - 1) + (1 - 1) + \ldots$, whereas you get the sum 1 when you sum the terms according to $1 + (-1 + 1) + (-1 + 1) + (-1 + 1) + \cdots$. Such abnormalities cannot happen when all terms in the infinite summation are nonnegative. The sum of countably many *nonnegative* terms is always well-defined, with ∞ as a possible value for the sum. For a sequence $a_1, a_2, \ldots$ consisting of both positive and negative terms, a basic result from the theory of series states that the infinite series $\sum_{k=1}^{\infty} a_k$ is always well-defined with a finite sum if the series is absolutely convergent, where absolute convergence means that $\sum_{k=1}^{\infty} |a_k| < \infty$. In case the series $\sum_{k=1}^{\infty} a_k$ is absolutely convergent, the sum is uniquely determined and does not depend on the order in which the individual terms are added. For a discrete random variable X with range I, it is said that the expected value $E(X)$ *exists* if X is nonnegative or if $\sum_{x \in I} |x|\, P(X = x) < \infty$. An example of a random variable X for which $E(X)$ does not exist is the random variable X with probability mass function $P(X = k) = \frac{3}{\pi^2 k^2}$ for $k = \pm 1, \pm 2, \ldots$ (a celebrated result from calculus is that $\sum_{k=1}^{\infty} \frac{1}{k^2} = \frac{\pi^2}{6}$). The reason that $E(X)$ does not exist is the well-known fact from calculus that $\sum_{k=1}^{\infty} 1/k = \infty$.

Many exercises on calculating the expected value of a random variable are given in Chapters 2 and 3. Here, we give three more problems.

Problem 9.1 Consider Problems 7.5 and 8.4 again. For each of these two problems, calculate the expected value of the number of drawings needed to draw a white ball.

Problem 9.2 Two-finger Morra is an old Italian game. It is played by two players A and B. At the same time the two players A and B show one or two fingers and simultaneously call out a guess for the number of fingers their opponent will show. If only one of the players guesses correctly, he wins an amount equal to the number of fingers shown by him and his opponent. If both players guess correctly or if neither guesses correctly, no money is exchanged. Denote by (i, j) the decision of any given player to show i fingers and to guess that his opponent will show j fingers for $i, j = 1, 2$. Suppose that player A chooses decision $(1, 2)$ with probability $\frac{4}{7}$ and decision (2, 1) with probability $\frac{3}{7}$. Independently of player A, player B chooses each of the four possible decisions with an equal probability of $\frac{1}{4}$. What is the expected value of the amount won by player A?

Problem 9.3 A stick is broken at random into two pieces. You bet on the ratio of the length of the longer piece to the length of the smaller piece. You receive \$$k$ if the ratio is between k and $k + 1$ for some k with $1 \leq k \leq m - 1$, while you receive \$$m$ if the ratio is larger than m. Here, m is a given positive integer. What should be your stake to make this a fair bet? How does your stake behave as a function of m when m becomes large?

9.2 Expected value of sums of random variables

Let X and Y be two random variables that are defined on the same sample space with probability measure P. The following basic rule is of utmost importance.

Rule 9.1 *For any two random variables X and Y,*

$$E(X + Y) = E(X) + E(Y),$$

provided that $E(X)$ and $E(Y)$ exist.

The proof is simple for the discrete case. Letting $Z = X + Y$, a key observation is

$$P(Z = z) = \sum_{x,y:\ x+y=z} P(X = x, Y = y),$$

where $P(X = x, Y = y)$ is the notation for the probability of the joint event that X takes on the value x and Y the value y. Also, we need the relations $\sum_y P(X = x, Y = y) = P(X = x)$ and $\sum_x P(X = x, Y = y) = P(Y = y)$. Thus

$$E(Z) = \sum_z zP(Z = z) = \sum_z z \sum_{x,y:\, x+y=z} P(X = x, Y = y)$$

$$= \sum_z \sum_{x,y:\, x+y=z} (x + y)P(X = x, Y = y) = \sum_{x,y} (x + y)P(X = x, Y = y)$$

and so

$$\begin{aligned} E(Z) &= \sum_{x,y} xP(X = x, Y = y) + \sum_{x,y} yP(X = x, Y = y) \\ &= \sum_x x \sum_y P(X = x, Y = y) + \sum_y y \sum_x P(X = x, Y = y) \\ &= \sum_x xP(X = x) + \sum_y yP(Y = y), \end{aligned}$$

which proves the desired result $E(Z) = E(X) + E(Y)$. The same result holds for any finite number of random variables, each having a finite expected value. That is,

$$E(X_1 + \ldots + X_n) = E(X_1) + \ldots + E(X_n)$$

if $E(X_i)$ exists for all $i = 1, \ldots, n$. The result that the expected value of a finite sum of random variables equals the sum of the expected values is extremely useful. It is only required that the relevant expected values exist, but dependencies between the random variables are allowed. The utility of this result has already been demonstrated by several examples in Chapters 2 and 3. A trick that is often applicable to calculate the expected value of a random variable is to represent the random variable as the sum of random variables that can take on only values 0 and 1.

Example 9.1 Suppose that n children of differing heights are placed in line at random. You then select the first child from the line and walk with her/him along the line until you encounter a child who is taller or until you have reached the end of the line. If you do encounter a taller child, you also have her/him to accompany you further along the line until you encounter yet again a taller child or reach the end of the line, etc. What is the expected value of the number of children selected from the line?

Letting the random variable X denote the number of children selected from the line, we can most easily compute $E(X)$ by writing

$$X = X_1 + \cdots + X_n,$$

where

$$X_i = \begin{cases} 1 & \text{if the } i\text{th child is selected from the line} \\ 0 & \text{otherwise.} \end{cases}$$

The probability that the ith child is the tallest among the first i children equals $\frac{1}{i}$. Hence,

$$E(X_i) = 0 \times (1 - \frac{1}{i}) + 1 \times \frac{1}{i} = \frac{1}{i}, \qquad i = 1, \ldots, n.$$

This gives

$$E(X) = 1 + \frac{1}{2} + \ldots + \frac{1}{n}.$$

An insightful approximation can be given to this expected value. It is known from calculus that $1 + \frac{1}{2} + \ldots + \frac{1}{n}$ can very accurately be approximated by $\ln(n) + \gamma + \frac{1}{2n}$, where $\gamma = 0.57722\ldots$ is Euler's constant.

The linearity property of the expected value is a special case of a general result in calculus for sums and integrals. This property holds not only for discrete random variables but holds for any type of random variables. Another important type of random variable is the continuous random variable with a continuous interval as its range of possible values. Continuous random variables are to be discussed in Chapter 10 and subsequent chapters. The models of discrete and continuous random variables are the most important ones, but are not exhaustive. Also, there are so-called *mixed* random variables having properties of both discrete and continuous random variables. Think of your delay in queue at a counter in a supermarket or the amount paid on an automobile insurance policy in a given year. These random variables take on either the discrete value zero with positive probability or a value in a continuous interval.

Problem 9.4 Consider Example 7.8 again. Calculate the expected number of hotels that remain empty. *Hint*: define the random variable X_i as equal to 1 if the ith hotel remains empty and 0 otherwise.

Problem 9.5 What is the expected number of distinct birthdays within a randomly formed group of 100 persons?

9.3 Substitution rule and variance

Suppose X is a discrete random variable with a given probability distribution. In many applications, we wish to compute the expected value of some function of X. Let $g(x)$ be a given real-valued function. Then, the quantity $g(X)$ is a discrete random variable as well. The expected value of $g(X)$ can directly be calculated from the probability distribution of X.

Rule 9.2 *For any function g of the random variable X,*

$$E\left[g(X)\right] = \sum_{x \in I} g(x)\, P(X = x)$$

provided that $\sum_{x \in I} |g(x)|\ P(X = x) < \infty$.

This rule is called the *substitution rule*. The proof of the rule is simple. If X takes on the values $x_1, x_2, \ldots$ with probabilities $p_1, p_2, \ldots$ and the assumption is made that $g(x_i) \neq g(x_j)$ for $x_i \neq x_j$, then the random variable $Z = g(X)$ takes on the values $z_1 = g(x_1), z_2 = g(x_2), \ldots$ with the same probabilities $p_1, p_2, \ldots$. Next apply the definition $E(Z) = \sum_k z_k P(Z = z_k)$ and substitute $z_k = g(x_k)$ and $P(Z = z_k) = P(X = x_k)$. The proof needs an obvious modification when the assumption $g(x_i) \neq g(x_j)$ for $x_i \neq x_j$ is dropped.

A frequently made mistake of beginning students is to set $E\left[g(X)\right]$ equal to $g\left(E(X)\right)$. In general, $E\left[g(X)\right] \neq g\left(E(X)\right)$! Stated differently, the average value of the input X does not determine in general the average value of the output $g(X)$. As a counterexample, take the random variable X with $P(X = 1) = P(X = -1) = 0.5$ and take the function $g(x) = x^2$. An exception is the case of a linear function $g(x) = ax + b$. Then, it always holds that

$$E(aX + b) = aE(X) + b.$$

9.3.1 Variance

An important case of a function of X is the random variable $g(X) = (X - \mu)^2$, where $\mu = E(X)$ denotes the expected value of X. The expected value of $(X - \mu)^2$ is called the *variance* of X and is denoted by

$$\text{var}\,(X) = E[(X - \mu)^2].$$

It is a measure of the spread of the possible values of X. Often, one uses the *standard deviation*, which is defined as the square root of the variance. It is useful to work with the standard deviation since it has the same units (e.g. dollar or cm) as $E(X)$. Since $(X - \mu)^2 = X^2 - 2\mu X + \mu^2$, it follows from the linearity of the expectation operator that $E[(X - \mu)^2] = E(X^2) - 2\mu E(X) + \mu^2$. Hence, $\text{var}(X)$ is also given by

$$\text{var}\,(X) = E(X^2) - \mu^2.$$

Example 9.2 Suppose the random variable X has the Poisson distribution $P(X = k) = e^{-\lambda}\lambda^k/k!$ for $k = 0, 1, \ldots$. This distribution has mean λ as was already shown in Section 4.2.1. What is the variance of X?

A remarkable property of the Poisson distribution is that its variance has the same value as its mean. That is,

$$\text{var}(X) = \lambda.$$

The proof is simple. To evaluate $E(X^2)$, use the identity $k^2 = k(k-1) + k$. This gives

$$\begin{aligned}
E(X^2) &= \sum_{k=0}^{\infty} k^2 P(X = k) \\
&= \sum_{k=1}^{\infty} k(k-1)P(X = k) + \sum_{k=1}^{\infty} kP(X = k) \\
&= \sum_{k=1}^{\infty} k(k-1)e^{-\lambda}\frac{\lambda^k}{k!} + E(X) = \lambda^2 \sum_{k=2}^{\infty} e^{-\lambda}\frac{\lambda^{k-2}}{(k-2)!} + \lambda.
\end{aligned}$$

Since $\sum_{k=2}^{\infty} e^{-\lambda}\lambda^{k-2}/(k-2)! = \sum_{n=0}^{\infty} e^{-\lambda}\lambda^n/n! = 1$, we obtain $E(X^2) = \lambda^2 + \lambda$. Next, by $\text{var}(X) = E(X^2) - (E(X))^2$, the desired result follows.

Problem 9.6 Three friends go to the cinema together every week. Each week, in order to decide which friend will pay for the other two, they all toss a fair coin into the air simultaneously. They continue to toss coins until one of the three gets a different outcome from the other two. What are the expected value and the standard deviation of the number of tosses necessary?

An instructive illustration of the substitution rule is provided by the next example. This example deals with the famous newsboy problem.

Example 9.3 Every morning, rain or shine, young Billy Gates can be found at the entrance to the metro, hawking copies of *The Morningstar*. Demand for newspapers varies from day to day, but Billy's regular early morning haul yields him 200 copies. He purchases these copies for \$1 per paper, and sells them for \$1.50 apiece. Billy goes home at the end of the morning, or earlier if he sells out. He can return unsold papers to the distributor for \$0.50 apiece. From experience, Billy knows that demand for papers on any given morning is uniformly distributed between 150 and 250, where each of the possible values $150, \ldots, 250$ is equally likely. What are the expected value and the standard deviation of Billy's net earnings on any given morning?

Denote by the random variable X the number of copies Billy would have sold on a given morning if he had ample supply. The actual number of copies sold by Billy is X if $X \leq 200$ and 200 otherwise. The probability mass function of X is given by $P(X = k) = \frac{1}{101}$ for $k = 150, \ldots, 250$. Billy's net earnings on any given morning is a random variable $g(X)$, where the function $g(x)$ is given by

$$g(x) = \begin{cases} -200 + 1.5x + 0.5(200 - x), & x \leq 200 \\ -200 + 1.5 \times 200, & x > 200. \end{cases}$$

Applying the substitution rule, we find

$$\begin{aligned} E[g(X)] &= \sum_{k=150}^{250} g(k)P(X = k) \\ &= \frac{1}{101} \sum_{k=150}^{200} (-100 + k) + \frac{1}{101} \sum_{k=201}^{250} 100 \\ &= \frac{3{,}825}{101} + \frac{5{,}000}{101} = 87.3762. \end{aligned}$$

To find the standard deviation of $g(X)$, we apply the formula $\text{var}(Z) = E(Z^2) - (E(Z))^2$ with $Z = g(X)$. This gives

$$\text{var}[g(X)] = E[(g(X))^2] - (E[g(X)])^2.$$

Letting $h(x) = (g(x))^2$, then $h(x) = (-100 + x)^2$ for $x \leq 200$ and $h(x) = 100^2$ for $x > 200$. By applying the substitution rule again,

$$\begin{aligned} E[h(X)] &= \sum_{k=150}^{250} h(k)P(X = k) \\ &= \frac{1}{101} \sum_{k=150}^{200} (-100 + k)^2 + \frac{1}{101} \sum_{k=201}^{250} 100^2 \\ &= \frac{297{,}925}{101} + \frac{500{,}000}{101} = 7900.2475. \end{aligned}$$

Hence, the variance of Billy's net earnings on any given morning is

$$\text{var}[g(X)] = 7900.2475 - (87.3762)^2 = 265.64.$$

Concluding, Billy's net earnings on any given morning has an expected value of 87.38 dollars and a standard deviation of $\sqrt{265.64} = 16.30$ dollars.

Problem 9.7 At the beginning of every month, a pharmacist orders an amount of a certain costly medicine that comes in strips of individually packed tablets. The wholesale price per strip is \$100, and the retail price per strip is \$400. The medicine has a limited shelf life. Strips not purchased by month's end will have reached their expiration date and must be discarded. When it so happens that demand for the item exceeds the pharmacist's supply, he may place an emergency order for \$350 per strip. The monthly demand for this medicine takes on the possible values 3, 4, 5, 6, 7, 8, 9, and 10 with respective probabilities 0.3, 0.1, 0.2, 0.2, 0.05, 0.05, 0.05, and 0.05. The pharmacist decides to order eight strips at the start of each month. What are the expected value and the standard deviation of the net profit made by the pharmacist on this medicine in any given month?

Problem 9.8 The University of Gotham City renegotiates its maintenance contract with a particular copy machine distributor on a yearly basis. For the coming year, the distributor has come up with the following offer. For a prepaid cost of \$50 per repair call, the university can opt for a fixed number of calls. For each visit beyond that fixed number, the university will pay \$100. If the actual number of calls made by a repairman remains below the fixed number, no money will be refunded. Based on previous experience, the university estimates that the number of repairs that will be necessary in the coming year will have a Poisson distribution with an expected value of 150. The university signs a contract with a fixed number of 170 repair calls. What are the expected value and the standard deviation of the total maintenance costs for the copy machines in the coming year?

9.4 Independence of random variables

In Chapter 8, we dealt with the concept of independent events. It makes intuitive sense to say that random variables are independent when the underlying events are independent. Let X and Y be two random variables that are defined on the same sample space with probability measure P.

Definition 9.2 *The random variables X and Y are said to be independent if*

$$P(X \leq x, Y \leq y) = P(X \leq x)P(Y \leq y)$$

for any two real numbers x and y, where $P(X \leq x, Y \leq y)$ represents the probability of occurrence of both event $\{X \leq x\}$ and event $\{Y \leq y\}$.[1]

In words, the random variables X and Y are independent if the event of the random variable X taking on a value smaller than or equal to x and the event

[1] In general, the n random variables $X_1, \ldots, X_n$ are said to be independent if $P(X_1 \leq x_1, \ldots, X_n \leq x_n) = P(X_1 \leq x_1) \cdots P(X_n \leq x_n)$ for each n-tuple of real numbers $x_1, \ldots, x_n$. An infinite collection of random variables is said to be independent if every finite subcollection of them is independent.

of the random variable Y taking on a value smaller than or equal to y are independent for all real numbers x, y. Using the axioms of probability theory, it can be shown that Definition 9.2 is equivalent to

$$P(X \in A, Y \in B) = P(X \in A)P(Y \in B)$$

for any two sets A and B of real numbers. The technical proof is omitted. It is not difficult to verify the following two rules from the alternative definition of independence.

Rule 9.3 *If X and Y are independent random variables, then the random variables $f(X)$ and $g(Y)$ are independent for any two functions f and g.*

In case X and Y are discrete random variables, another representation of independence can be given.

Rule 9.4 *Discrete random variables X and Y are independent if and only if*

$$P(X = x, Y = y) = P(X = x)P(Y = y) \quad \text{for} \quad \text{all } x, y.$$

A very useful rule applies to the calculation of the expected value of the product of two independent random variables.

Rule 9.5 *If the random variables X and Y are independent, then*

$$E(XY) = E(X)E(Y),$$

assuming that $E(X)$ and $E(Y)$ exist.

We prove this important result for the case of discrete random variables X and Y. Let I and J denote the sets of possible values of the random variables X and Y. Define the random variable Z by $Z = XY$, then

$$\begin{aligned}
E(Z) &= \sum_z zP(Z = z) = \sum_z z \sum_{\substack{x \in I, y \in J: \\ xy = z}} P(X = x, Y = y) \\
&= \sum_z \sum_{\substack{x \in I, y \in J: \\ xy = z}} xyP(X = x, Y = y) \\
&= \sum_{x \in I, y \in J} xyP(X = x, Y = y) = \sum_{x \in I, y \in J} xyP(X = x)P(Y = y) \\
&= \sum_{x \in I} xP(X = x) \sum_{y \in J} yP(Y = y) = E(X)E(Y).
\end{aligned}$$

The converse of the above result is not truc. It is possible that $E(XY) = E(X)E(Y)$, while X and Y are not independent. A simple example is as follows. Suppose two fair dice are tossed. Denote by the random variable V_1 the number appearing on the first die and by the random variable V_2 the number appearing on the second die. Let $X = V_1 + V_2$ and $Y = V_1 - V_2$. It is readily seen that the random variables X and Y are not independent. We leave it to the reader to verify that $E(X) = 7$, $E(Y) = 0$, and $E(XY) = E(V_1^2 - V_2^2) = 0$ and so $E(XY) = E(X)E(Y)$.

Problem 9.9 Two fair dice are tossed. Let the random variable X denote the sum of the two numbers shown by the dice and let Y be the largest of these two numbers. Are the random variables X and Y independent? What are the values of $E(XY)$ and $E(X)E(Y)$?

Convolution formula

Suppose X and Y are two discrete random variables each having the set of nonnegative integers as the range of possible values. A useful rule is

Rule 9.6 *If the nonnegative random variables X and Y are independent, then*

$$P(X + Y = k) = \sum_{j=0}^{k} P(X = j)P(Y = k - j) \quad \text{for} \quad k = 0, 1, \ldots.$$

This rule is known as the *convolution rule*. The proof is as follows. Fix k. Let A be the event that $X + Y = k$ and let B_j be the event that $X = j$ for $j = 0, 1, \ldots$. The events $AB_0, AB_1, \ldots$ are mutually exclusive and so, by Axiom 7.3 in Chapter 7,

$$P(A) = \sum_{j=0}^{\infty} P(AB_j).$$

Obviously,

$$\begin{aligned} P(AB_j) &= P(X + Y = k, X = j) = P(X = j, Y = k - j) \\ &= P(X = j)P(Y = k - j), \end{aligned}$$

where the last equality uses the independence of X and Y. Thus,

$$P(X + Y = k) = \sum_{j=0}^{\infty} P(X = j)P(Y = k - j).$$

Since $P(Y = k - j) = 0$ for $j > k$, the convolution formula next follows.

Example 9.4 Suppose the random variables X and Y are independent and have Poisson distributions with respective means λ and μ. What is the probability distribution of $X + Y$? To answer this question, we apply the convolution formula.

This gives

$$P(X+Y=k)=\sum_{j=0}^{k} e^{-\lambda}\frac{\lambda^j}{j!}e^{-\mu}\frac{\mu^{k-j}}{(k-j)!}$$

$$=\frac{e^{-(\lambda+\mu)}}{k!}\sum_{j=0}^{k}\binom{k}{j}\lambda^j\mu^{k-j},$$

where the second equality uses the fact that $\binom{k}{j}=\frac{k!}{j!(k-j)!}$. Next, by Newton's binomial $(a+b)^k=\sum_{j=0}^{k}\binom{k}{j}a^j b^{k-j}$, we find

$$P(X+Y=k)=e^{-(\lambda+\mu)}\frac{(\lambda+\mu)^k}{k!}\quad \text{for}\quad k=0,1,\ldots.$$

Hence, $X+Y$ is Poisson distributed with mean $\lambda+\mu$.

We conclude this chapter with an example in which all the above rules for the expectation operator pass in review.

Example 9.5 A drunkard is standing in the middle of a very large town square. He begins to walk. Each step is a unit distance in one of the four directions East, West, North, and South. All four possible directions are equally probable. The direction for each step is chosen independently of the direction of the others. The drunkard takes a total of n steps. What is the expected value of the quadratic distance of the drunkard to his starting point after n steps?

Denote the drunkard's starting point by $(0,0)$. The drunkard is in the point $(X_1+\cdots+X_n, Y_1+\cdots+Y_n)$ after n steps, where the random variables X_i and Y_i denote the changes in the x-coordinate and the y-coordinate of the position of the drunkard caused by his ith step. Some reflections show that

$$P(X_i=1)=P(Y_i=1)=\frac{1}{4},\ P(X_i=-1)=P(Y_i=-1)=\frac{1}{4},$$

$$P(X_i=0)=P(Y_i=0)=\frac{1}{2}.$$

The random variables $X_1,\ldots,X_n$ are independent, as are random variables $Y_1,\ldots,Y_n$. The quadratic distance of the drunkard to his starting point after n steps is distributed as the random variable $(X_1+\cdots+X_n)^2+(Y_1+\cdots+Y_n)^2$. By the linearity of the expectation operator,

$$E[(X_1+\cdots+X_n)^2+(Y_1+\cdots+Y_n)^2]$$

$$=E[(X_1+\cdots+X_n)^2]+E[(Y_1+\cdots+Y_n)^2].$$

Using the algebraic formula,

$$(X_1 + \cdots + X_n)^2 = \sum_{i=1}^{n} X_i^2 + 2 \sum_{i=1}^{n-1} \sum_{j=i+1}^{n} X_i X_j$$

it follows that

$$E[(X_1 + \cdots + X_n)^2] = \sum_{i=1}^{n} E(X_i^2) + 2 \sum_{i=1}^{n-1} \sum_{j=i+1}^{n} E(X_i X_j)$$

$$= \sum_{i=1}^{n} E(X_i^2) + 2 \sum_{i=1}^{n-1} \sum_{j=i+1}^{n} E(X_i) E(X_j),$$

where the last equality uses the fact that $E(X_i X_j) = E(X_i)E(X_j)$ by the independence of X_i and X_j for $i \neq j$. For each i,

$$E(X_i) = 1 \times \frac{1}{4} + (-1) \times \frac{1}{4} = 0$$

$$\text{and} \qquad E(X_i^2) = 1^2 \times \frac{1}{4} + (-1)^2 \times \frac{1}{4} = \frac{1}{2}.$$

This gives

$$E[(X_1 + \cdots + X_n)^2] = \frac{1}{2} n.$$

In the same way, $E[(Y_1 + \cdots + Y_n)^2] = \frac{1}{2} n$. Hence, we find the interesting result that the expected value of the *quadratic* distance of the drunkard to his starting point after n steps is equal to n, irrespective of the value of n. It was already pointed out in Section 2.4 of Chapter 2 that it is false to conclude from this result that the distance of the drunkard to his starting point after n steps has $\sqrt{n}$ as expected value. In Chapter 12, it will be shown that the expected value of this distance is approximately equal $0.886\sqrt{n}$ for n sufficiently large.

Problem 9.10 Use the definition of variance to explain why the expected value of the distance of the drunkard to his starting point after n steps cannot be equal to $\sqrt{n}$.

10

Continuous random variables

In many practical applications of probability, physical situations are better described by random variables that can take on a *continuum* of possible values rather than a *discrete* number of values. Examples are the decay time of a radioactive particle, the time until the occurrence of the next earthquake in a certain region, the lifetime of a battery, the annual rainfall in London, and so on. These examples make clear what the fundamental difference is between discrete random variables and continuous random variables. Whereas a discrete random variable associates *positive* probabilities to its individual values, any individual value has probability *zero* for a continuous random variable. It is only meaningful to speak of the probability of a continuous random variable taking on a value in some interval. Taking the lifetime of a battery as an example, it will be intuitively clear that the probability of this lifetime taking on a specific value becomes zero when a finer and finer unit of time is used. If you can measure the heights of people with infinite precision, the height of a randomly chosen person is a continuous random variable. In reality, heights cannot be measured with infinite precision, but the mathematical analysis of the distribution of heights of people is greatly simplified when using a mathematical model in which the height of a randomly chosen person is modeled as a continuous random variable. Integral calculus is required to formulate the continuous analog of a probability mass function. The purpose of this chapter is to familiarize the reader with the concept of probability density function of a continuous random variable. This is always a difficult concept for the beginning student. However, integral calculus enables us to give an enlightening interpretation of a probability density. Also, this chapter summarizes the most important probability densities used in practice. Finally, the inverse-transformation method for generating a random observation from a continuous random variable will be discussed.

10.1 Concept of probability density

Let X be a random variable that is defined on a sample space with probability measure P. It is assumed that the set of possible values of X is a continuous interval (e.g., the set of all real numbers).

Definition 10.1 *The random variable X is said to be (absolutely) continuously distributed if a function $f(x)$ exists such that*

$$P(X \leq a) = \int_{-\infty}^{a} f(x)\,dx \quad \textit{for} \quad \textit{each real number } a,$$

where the function $f(x)$ satisfies

$$f(x) \geq 0 \quad \textit{for} \quad \textit{all } x \qquad \textit{and} \qquad \int_{-\infty}^{\infty} f(x)\,dx = 1.$$

The notation $P(X \leq a)$ stands for the probability that is assigned by the probability measure P to the set of all outcomes ω for which $X(\omega) \leq a$. The function $P(X \leq x)$ is called the *(cumulative) probability distribution function* of the random variable X, and the function $f(x)$ is called the *probability density function* of X.

Beginning students often misinterpret the nonnegative number $f(a)$ as a probability, namely as the probability $P(X = a)$. This interpretation is wrong. Nevertheless, it is possible to give an intuitive interpretation of the nonnegative number $f(a)$ in terms of probabilities. Before doing this, we present an example of a continuous random variable with a probability density function.

Example 10.1 Suppose that the lifetime X of a battery has the cumulative probability distribution function

$$P(X \leq x) = \begin{cases} 0 & \text{for} \quad x < 0, \\ \frac{1}{4}x^2 & \text{for} \quad 0 \leq x \leq 2, \\ 1 & \text{for} \quad x > 2. \end{cases}$$

The probability distribution function $P(X \leq x)$ is continuous and is differentiable at each point x except for the two points $x = 0$ and $x = 2$. The derivative is continuous at each point at which it exists. We can now conclude from the Fundamental Theorem of integral calculus that the random variable X has a probability density function. This probability density function is obtained by differentiation of the probability distribution function and is given by

$$f(x) = \begin{cases} \frac{1}{2}x & \text{for} \quad 0 < x < 2, \\ 0 & \text{otherwise.} \end{cases}$$

In each of the finite number of points x at which $P(X \leq x)$ has no derivative, it does not matter what value we give $f(x)$. These values do not affect $\int_{-\infty}^{a} f(x)\,dx$. Usually, we give $f(x)$ the value 0 at any of these exceptional points.

10.1.1 Interpretation of the probability density

The use of the word "density" originated with the analogy to the distribution of matter in space. In physics, any finite volume, no matter how small, has a positive mass, but there is no mass at a single point. A similar description applies to continuous random variables. To make this more precise, we first express $P(a < X \leq b)$ in terms of the density $f(x)$ for any constants a and b with $a < b$. Noting that the event $\{X \leq b\}$ is the union of the two disjoint events $\{a < X \leq b\}$ and $\{X \leq a\}$, it follows that $P(X \leq b) = P(a < X \leq b) + P(X \leq a)$. Hence,

$$\begin{aligned} P(a < X \leq b) &= P(X \leq b) - P(X \leq a) \\ &= \int_{-\infty}^{b} f(x)\,dx - \int_{-\infty}^{a} f(x)\,dx \quad \text{for} \quad a < b \end{aligned}$$

and so

$$P(a < X \leq b) = \int_{a}^{b} f(x)\,dx \quad \text{for} \quad a < b.$$

In other words, the area under the graph of $f(x)$ between the points a and b gives the probability $P(a < X \leq b)$. Next, we find that

$$\begin{aligned} P(X = a) &= \lim_{n\to\infty} P(a - \frac{1}{n} < X \leq a) \\ &= \lim_{n\to\infty} \int_{a-\frac{1}{n}}^{a} f(x)\,dx = \int_{a}^{a} f(x)dx, \end{aligned}$$

using the continuity property of the probability measure P stating that $\lim_{n\to\infty} P(A_n) = P(\lim_{n\to\infty} A_n)$ for any nonincreasing sequence of events A_n (see Remark 7.1). Hence, we arrive at the conclusion

$$P(X = a) = 0 \quad \text{for} \quad \text{each real number } a.$$

This formally proves that, for a continuous random variable X, it makes no sense to speak of the probability that the random variable X will take on a *specific* value. This probability is always zero. It only makes sense to speak of the probability that the continuous random variable X will take on a value in some interval. Incidentally, since $P(X = c) = 0$ for any number c, the probability that X takes on a value in an interval with endpoints a and b is not influenced by whether or not the endpoints are included. In other words, for any two real numbers a and b with $a < b$, we have

$$P(a \leq X \leq b) = P(a < X \leq b) = P(a \leq X < b) = P(a < X < b).$$

The fact that the area under the graph of $f(x)$ can be interpreted as a probability leads to an intuitive interpretation of $f(a)$. Let a be a given continuity point of $f(x)$. Consider now a small interval of length Δa around the point a, say $[a - \frac{1}{2}\Delta a, a + \frac{1}{2}\Delta a]$. Since

$$P\left(a - \frac{1}{2}\Delta a \leq X \leq a + \frac{1}{2}\Delta a\right) = \int_{a-\frac{1}{2}\Delta a}^{a+\frac{1}{2}\Delta a} f(x)\,dx$$

and

$$\int_{a-\frac{1}{2}\Delta a}^{a+\frac{1}{2}\Delta a} f(x)\,dx \approx f(a)\Delta a \quad \text{for} \quad \Delta a \text{ small,}$$

we obtain that

$$P\left(a - \frac{1}{2}\Delta a \leq X \leq a + \frac{1}{2}\Delta a\right) \approx f(a)\Delta a \quad \text{for} \quad \Delta a \text{ small.}$$

In other words, the probability of random variable X taking on a value in a *small* interval around point a is approximately equal to $f(a)\Delta a$ when Δa is the length of the interval. You see that the number $f(a)$ itself is *not* a probability, but it is a relative measure for the likelihood that random variable X will take on a value in the immediate neighborhood of point a. Stated differently, the probability density function $f(x)$ expresses how densely the probability mass of random variable X is smeared out in the neighborhood of point x. Hence, the name of density function. The probability density function provides the most useful description of a continuous random variable. The graph of the density function provides a good picture of the likelihood of the possible values of the random variable.

10.1.2 Verification of a probability density

In general, how can we verify whether a random variable X has a probability density? In concrete situations, we first determine the cumulative distribution function $F(a) = P(X \leq a)$ and next we verify whether $F(a)$ can be written in the form of $F(a) = \int_{-\infty}^{a} f(x)\,dx$. A sufficient condition is that $F(x)$ is continuous at every point x and is differentiable except for a finite number of points x. The following three examples are given in illustration of this point.

Example 10.2 Let the random variable be given by $X = -\frac{1}{\lambda}\ln(U)$, where U is a random number between 0 and 1 and λ is a given positive number. What is the probability density function of X?

To answer this question, note first that X is a positive random variable. Hence, $P(X \leq x) = 0$ for $x \leq 0$. For any $x > 0$,

$$P(X \leq x) = P(-\frac{1}{\lambda}\ln(U) \leq x) = P(\ln(U) \geq -\lambda x)$$
$$= P(U \geq e^{-\lambda x}) = 1 - P(U \leq e^{-\lambda x}),$$

where the last equality uses the fact that $P(U < u) = P(U \leq u)$ for the continuous random variable U. Since $P(U \leq u) = u$ for $0 < u < 1$, it follows that

$$P(X \leq x) = 1 - e^{-\lambda x}, \qquad x > 0.$$

Obviously, $P(X \leq x) = 0$ for $x \leq 0$. Noting that the expression for $P(X \leq x)$ is continuous at every point x and is differentiable except at $x = 0$, we obtain by differentiation that X has a probability density function $f(x)$ with $f(x) = \lambda e^{-\lambda x}$ for $x > 0$ and $f(x) = 0$ for $x \leq 0$.

Example 10.3 A point is picked at random in the inside of a circular disk with radius r. Let the random variable X denote the distance from the center of the disk to this point. Does the random variable X have a probability density function and, if so, what is its form?

To answer this question, we first define a sample space with an appropriate probability measure P for the random experiment. The sample space is taken as the set of all points (x, y) in the two-dimensional plane with $x^2 + y^2 \leq r^2$. Since the point inside the circular disk is chosen at random, we assign to each well-defined subset A of the sample space the probability

$$P(A) = \frac{\text{area of region } A}{\pi r^2}.$$

The cumulative probability distribution function $P(X \leq x)$ is easily calculated. The event $X \leq a$ occurs if and only if the randomly picked point falls in the disk of radius a with area πa^2. Therefore

$$P(X \leq a) = \frac{\pi a^2}{\pi r^2} = \frac{a^2}{r^2} \quad \text{for} \quad 0 \leq a \leq r.$$

Obviously, $P(X \leq a) = 0$ for $a < 0$ and $P(X \leq a) = 1$ for $a > r$. Since the expression for $P(X \leq x)$ is continuous at every point x and is differentiable except at the points $x = 0$ and $x = a$, it follows that X has a probability density function which is given by

$$f(x) = \begin{cases} \frac{2x}{r^2} & \text{for} \quad 0 < x < r, \\ 0 & \text{otherwise.} \end{cases}$$

Example 10.4 The numbers U_1 and U_2 are chosen at random from the interval (0, 1), independently of each other. Let the random variables V and W be defined by $V = \min(U_1, U_2)$ and $W = \max(U_1, U_2)$. What are the probability density functions of the random variables V and W?

Let's first consider the random variable W. A key observation is that the largest of two numbers a and b is smaller than or equal to c only if both a and b are smaller than or equal to c. Thus,

$$P(W \leq w) = P(U_1 \leq w, U_2 \leq w) =$$
$$= P(U_1 \leq w)P(U_2 \leq w) = w^2, \qquad 0 \leq w \leq 1,$$

where the second equality uses the independence of U_1 and U_2. Differentiating $P(W \leq w)$ gives that W has a probability density function $g(w)$ with $g(w) = 2w$ for $0 < w < 1$ and $g(w) = 0$ otherwise. The probability density of $V = \min(U_1, U_2)$ is obtained by noting that the smallest of two numbers a and b is larger than c only if both a and b are larger than c. Thus

$$P(V \leq v) = 1 - P(V > v) = 1 - P(U_1 > v, U_2 > v)$$
$$= 1 - P(U_1 > v)P(U_2 > v) = 1 - (1 - v)^2, \qquad 0 \leq v \leq 1.$$

Differentiating $P(V \leq v)$ gives that V has a probability density function $h(v)$ with $h(v) = 2(1 - v)$ for $0 < v < 1$ and $h(v) = 0$ otherwise.

Problem 10.1 The number X is chosen at random between 0 and 1. Determine the probability density function of each of the random variables $V = X/(1 - X)$ and $W = X(1 - X)$.

Problem 10.2 Suppose you decide to take a ride on the ferris wheel at an amusement park. The ferris wheel has a diameter of 30 meters. After several turns, the ferris wheel suddenly stops due to a power outage. What random variable determines your height above the ground when the ferris wheel stops? What is the probability that this height is not more than 22.5 meters? And the probability of no more than 7.5 meters? What is the probability density function of the random variable governing the height above the ground?

10.1.3 Expected value

The expected value of a continuous random variable X with probability density function $f(x)$ is defined by

$$E(X) = \int_{-\infty}^{\infty} x f(x)\, dx$$

provided that the integral $\int_{-\infty}^{\infty} |x| f(x)\, dx$ is finite (the latter integral is always well-defined by the nonnegativity of the integrand). It is then said that $E(X)$ exists. In the case that X is a nonnegative random variable, the integral $\int_0^{\infty} x f(x)\, dx$ is always defined when allowing ∞ as possible value. In this case, it is convenient to say that $E(X) = \int_0^{\infty} x f(x)\, dx$ always exists. The definition of expected value in the continuous case parallels the definition $E(X) = \sum x_i p(x_i)$ for a discrete random variable X with $x_1, x_2, \ldots$ as possible values and $p(x_i) = P(X = x_i)$. For dx small, the quantity $f(x)\, dx$ in a discrete approximation of the continuous case corresponds with $p(x)$ in the discrete case. The summation becomes an integral when dx approaches zero. Results for discrete random variables are typically expressed as sums. The corresponding results for continuous random variables are expressed as integrals.

As an illustration, consider the random variable X from Example 10.3. The expected value of the distance X equals

$$E(X) = \int_0^r x \frac{2x}{r^2}\, dx = \frac{2}{3} r.$$

We could also determine the result that the average distance of a randomly selected point from the origin is equal to $\frac{2}{3}r$ empirically with the help of computer simulation and without making use of the mathematical constructs of probability density and expected value of a continuous random variable. We would have obtained the same answer. This confirms that the chosen definitions of probability density and expected value are adequate descriptions of reality.

Problem 10.3 A point is chosen at random inside the unit square $\{(x, y) | 0 \leq x, y \leq 1\}$. What is the expected value of the distance from this point to the origin?

Problem 10.4 Let X be a nonnegative continuous random variable with density function $f(x)$. Use an interchange of the order of integration to verify that $E(X) = \int_0^{\infty} P(X > u)\, du$.

10.1.4 Substitution rule

The substitution rule 9.2 for the discrete case has an obvious analog for the continuous case. Let X be a continuous random variable with density $f(x)$. For any given function $g(x)$, the expected value of the random variable $g(X)$ can be calculated from

$$E[g(X)] = \int_{-\infty}^{\infty} g(x)f(x)\,dx$$

provided that the integral exists. In particular, letting $\mu = E(X)$, the variance $\text{var}(X) = E[(X-\mu)^2]$ of the random variable X is given by

$$\text{var}(X) = \int_{-\infty}^{\infty} (x-\mu)^2 f(x)\,dx.$$

The variance of X is usually calculated by using the formula $\text{var}(X) = E(X^2) - \mu^2$. As an illustration, we calculate the variance of the random variable X from Example 10.3:

$$\text{var}(X) = \int_0^r x^2 \frac{2x}{r^2}\,dx - \left(\frac{2}{3}r\right)^2 = \frac{2r^2}{4} - \frac{4}{9}r^2 = \frac{1}{18}r^2.$$

The standard deviation of the distance from the randomly selected point inside the circle to the origin is $\sigma(X) = \sqrt{\text{var}(X)} = 0.2357r$.

Problem 10.5 Let X be a continuous random variable whose probability mass is concentrated on the interval (a, b). You may choose any point in the interval (a, b). After you have made your choice, a point in (a, b) is selected by sampling from the distribution of X. Your loss is equal to the squared distance between the sampled point and the point you have chosen. How do you choose your point such that the expected loss is minimized? What is the minimal value of the expected loss?

Problem 10.6 Consider Problem 10.2 again. Calculate the expected value and standard deviation of the height above the ground when the ferris wheel stops.

Problem 10.7 In an inventory system, a replenishment order is placed when the stock on hand of a certain product drops to the level s, where the reorder point s is a given positive number. The total demand for the product during the lead time of the replenishment order has the probability density $f(x) = \lambda e^{-\lambda x}$ for $x > 0$. What are the expected value and standard deviation of the shortage (if any) when the replenishment order arrives?

Problem 10.8 Suppose that the continuous random variable X has the probability density function $f(x) = (\alpha/\beta)(\beta/x)^{\alpha+1}$ for $x > \beta$ and $f(x) = 0$ for

$x \leq \beta$ for given values of the parameters $\alpha > 0$ and $\beta > 0$. This density is called the *Pareto* density, which provides a useful probability model for income distributions among others.

(a) Calculate the expected value and standard deviation of X.

(b) Assume that the annual income of employed measured in thousands of dollars in a given country follows a Pareto distribution with $\alpha = 2.25$ and $\beta = 2.5$. What percentage of the working population has an annual income of between 25 and 40 thousand dollars?

(c) Why do you think the Pareto distribution is a good model for income distributions? *Hint*: use the probabilistic interpretation of the density function $f(x)$.

Problem 10.9 A stick of unit length is broken at random into two pieces. Let the random variable X represent the length of the shorter piece.

(a) Use the substitution rule to calculate $E(X)$, $E[X/(1-X)]$, and $E[(1-X)/X]$. Also, calculate $\text{var}(X)$ and $\text{var}[X/(1-X)]$.

(b) Determine the probability density function of each of the random variables X, $X/(1-X)$, and $(1-X)/X$.

10.2 Important probability densities

Any nonnegative function $f(x)$ whose integral over the interval $(-\infty, \infty)$ equals 1 can be regarded as a probability density function of a random variable. In real-world applications, however, special mathematical forms naturally show up. In this section, we introduce several families of continuous random variables that frequently appear in practical applications. The probability densities of the members of each family all have the same mathematical form but differ only in one or more parameters. Uses of the densities in practical applications are indicated. Also, the expected values and the variances of the densities are listed without proof. A convenient method to obtain the expected values and the variances of special probability densities is the moment-generating function method to be discussed in Chapter 14.

10.2.1 Uniform density

A continuous random variable X is said to have a *uniform* density over the interval (a, b) if its probability density function is given by

$$f(x) = \begin{cases} \frac{1}{b-a} & \text{for } a < x < b \\ 0 & \text{otherwise.} \end{cases}$$

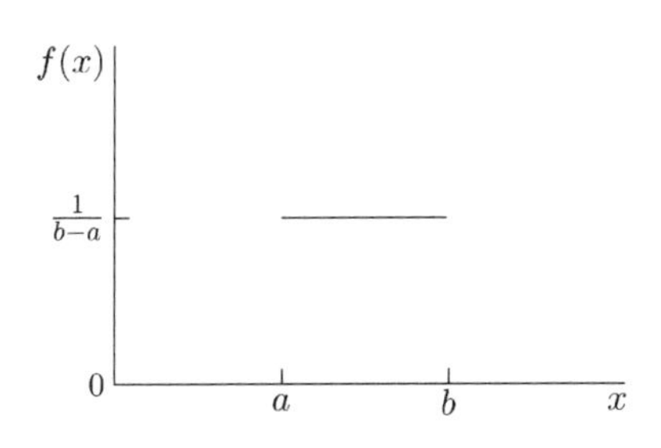

Figure 10.1 Uniform density.

This density has two parameters a and b with $b > a$. Figure 10.1 gives the graph of the uniform density function. The uniform distribution provides a probability model for selecting a point at random from the interval (a, b). It is also used as a model for a quantity that is known to vary randomly between a and b but about which little else is known. Since $f(x) = 0$ outside the interval (a, b), the random variable X must assume a value in (a, b). Also, since $f(x)$ is constant over the interval (a, b), the random variable X is just as likely to be near any value in (a, b) as any other value. This property is also expressed by

$$P\left(c - \frac{1}{2}\Delta \leq X \leq c + \frac{1}{2}\Delta\right) = \int_{c-\frac{1}{2}\Delta}^{c+\frac{1}{2}\Delta} \frac{1}{b-a}\, dx = \frac{\Delta}{b-a},$$

regardless of c provided that the points $c - \frac{1}{2}\Delta$ and $c + \frac{1}{2}\Delta$ belong to the interval (a, b). The expected value and the variance of the random variable X are given by

$$E(X) = \frac{1}{2}(a+b) \qquad \text{and} \qquad \operatorname{var}(X) = \frac{1}{12}(b-a)^2.$$

10.2.2 Triangular density

A continuous random variable X is said to have a *triangular* density over the interval (a, b) if its probability density function is given by

$$f(x) = \begin{cases} h\frac{x-a}{m-a} & \text{for} \quad a < x \leq m \\ h\frac{b-x}{b-m} & \text{for} \quad m \leq x < b \\ 0 & \text{otherwise.} \end{cases}$$

This density has three parameters a, b, and m with $a < m < b$. The constant $h > 0$ is determined by $\int_a^b f(x)\, dx = 1$, and so

$$h = \frac{2}{b-a}.$$

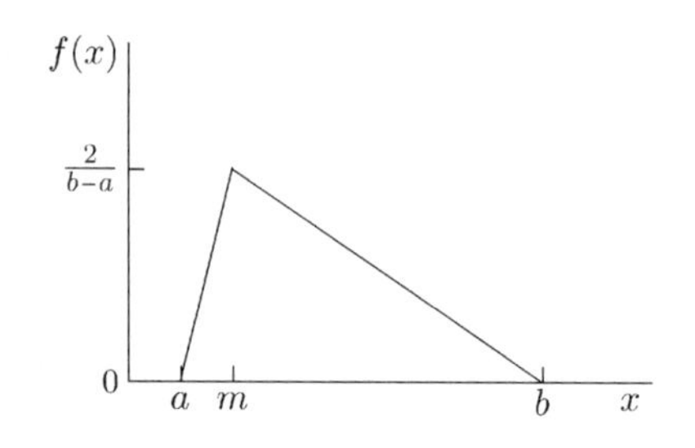

Figure 10.2 Triangular density.

Figure 10.2 gives the graph of the triangular density function. The density function increases linearly on the interval $[a, m]$ and decreases linearly on the interval $[m, b]$. The triangular distribution is often used as probability model when little information is available about the quantity of interest but one knows its lowest possible value a, its most likely value m, and its highest possible value b. The expected value and the variance of the random variable X are given by

$$E(X) = \frac{1}{3}(a + b + m), \qquad \mathrm{var}(X) = \frac{1}{18}(a^2 + b^2 + m^2 - ab - am - bm).$$

10.2.3 Exponential density

The continuous random variable X is said to have an *exponential* density with parameter $\lambda > 0$ if its probability density function is of the form

$$f(x) = \begin{cases} \lambda e^{-\lambda x} & \text{for} \quad x > 0 \\ 0 & \text{otherwise.} \end{cases}$$

The parameter λ is a scale parameter. An exponentially distributed random variable X takes on only positive values. Figure 10.3 displays the exponential density function with $\lambda = 1$. The exponential distribution is often used as probability model for the time until a *rare* event occurs. Examples are the time elapsed until the next earthquake in a certain region and the decay time of

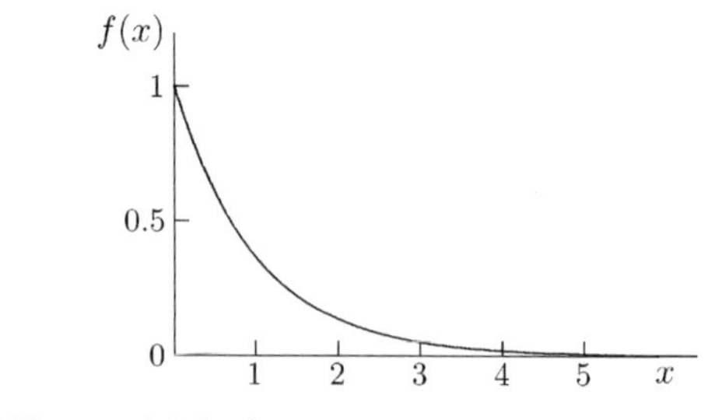

Figure 10.3 Exponential density ($\lambda = 1$).

a radioactive particle. Also, the exponential distribution is frequently used to model times between independent events such as arrivals at a service facility. The exponential distribution is intimately related to the Poisson arrival process that was discussed in Section 4.2.4. The expected value and the variance of the random variable X are given by

$$E(X) = \frac{1}{\lambda} \quad \text{and} \quad \sigma^2(X) = \frac{1}{\lambda^2}.$$

The probability $P(X > t) = \int_t^\infty \lambda e^{-\lambda x} dx$ is easily calculated as $e^{-\lambda t}$. A probability of the form $\alpha e^{-\beta t}$ for constants $\alpha, \beta > 0$ is called an *exponential tail* probability. In many situations, the probability of exceeding some *extreme* level is approximately equal to an exponential tail probability. For example, in queueing systems, the probability of a customer waiting more than a time t is often approximately equal to an exponential tail probability when t is large. Another interesting example concerns the probability that a high tide of h meters or more above sea level will occur in any given year somewhere along the Dutch coastline. This probability is approximately equal to $e^{-2.97h}$ for values of h larger than 1.70 m. This empirical result was used in the design of the Delta works that were built following the 1953 disaster when the sea flooded a number of polders in the Netherlands.

Problem 10.10 Suppose that the lifetime X of a battery has an exponential distribution with parameter λ. The battery is already in use for d time units for a given value of d. Evaluate the probability $P(X > t + d \mid X > d)$ to verify that the remaining lifetime of the battery has the *same* exponential distribution with parameter λ as the original lifetime. This property of the exponential distribution is known as the *memoryless* property ("used is as good as new"). Can you explain this property by looking at the graph in Figure 10.3?

10.2.4 Gamma density

A continuous random variable X is said to have a *gamma* density with parameters $\alpha > 0$ and $\lambda > 0$ if its probability density function is given by

$$f(x) = \begin{cases} c\lambda^\alpha x^{\alpha-1} e^{-\lambda x} & \text{for} \quad x > 0 \\ 0 & \text{otherwise.} \end{cases}$$

The constant c is determined by $\int_0^\infty f(x)dx = 1$. To specify c, we note that in advanced calculus, the so-called *gamma function* is defined by

$$\Gamma(a) = \int_0^\infty e^{-y} y^{a-1} dy \quad \text{for} \quad a > 0.$$

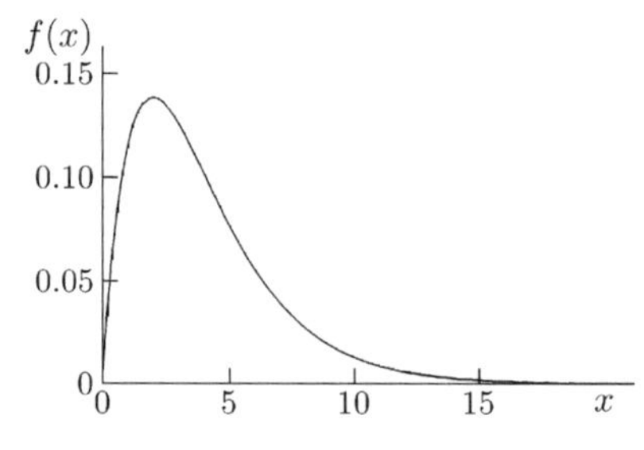

Figure 10.4 Gamma density ($\alpha = 2.5, \lambda = 0.5$).

This famous function has the property that

$$\Gamma(a+1) = a\Gamma(a) \quad \text{for} \quad a > 0.$$

This result is easily verified by partial integration. In particular,

$$\Gamma(a) = (a-1)! \qquad \text{if } a \text{ is a positive integer.}$$

An easy consequence of the definition of $\Gamma(a)$ is that the constant c in the gamma density is given by

$$c = 1/\,\Gamma(\alpha).$$

The parameter α is a shape parameter, and the parameter λ is a scale parameter. A gamma-distributed random variable takes on only positive values. The gamma density with $\alpha = 1$ reduces to the exponential density. Figure 10.4 displays the gamma density with $\alpha = 2.5$ and $\lambda = 0.5$. The graph in Figure 10.4 is representative of the shape of the gamma density if the shape parameter α is larger than 1; otherwise, the shape of the gamma density is similar to that of the exponential density in Figure 10.3. The gamma distribution is a useful model in inventory and queueing applications to model demand sizes and service times. The expected value and the variance of the random variable X are given by

$$E(X) = \frac{\alpha}{\lambda} \qquad \text{and} \qquad \operatorname{var}(X) = \frac{\alpha}{\lambda^2}.$$

10.2.5 Weibull density

A random variable X is said to have a *Weibull* density with parameters $\alpha > 0$ and $\lambda > 0$ if it has a probability density function of the form

$$f(x) = \begin{cases} \alpha\lambda(\lambda x)^{\alpha-1}e^{-(\lambda x)^\alpha} & \text{for} \quad x > 0 \\ 0 & \text{otherwise.} \end{cases}$$

The parameter α is a shape parameter, and the parameter λ is a scale parameter. The Weibull density has a similar shape as the gamma density. The expected value and the variance of the random variable X are given by

$$E(X) = \frac{1}{\lambda}\Gamma(1+\frac{1}{\alpha}), \qquad \text{var}(X) = \frac{1}{\lambda^2}\left[\Gamma(1+\frac{2}{\alpha}) - \left(\Gamma(1+\frac{1}{\alpha})\right)^2\right].$$

The Weibull distribution is a useful probability model for fatigue strengths of materials and is used in reliability models for lifetimes of devices.

Problem 10.11 Use properties of the gamma function to derive $E(X)$ and $E(X^2)$ for a gamma-distributed random variable X. Do the same for a Weibull-distributed random variable X.

10.2.6 Beta density

A random variable X is said to have a *beta* density with parameters $\alpha > 0$ and $\beta > 0$ if its probability density function is of the form

$$f(x) = \begin{cases} \frac{1}{c}x^{\alpha-1}(1-x)^{\beta-1} & \text{for} \quad 0 < x < 1 \\ & \text{otherwise} \end{cases}$$

for an appropriate constant c. The constant c is determined by $\int_0^1 f(x)\,dx = 1$. Using advanced calculus, it can be shown that

$$c = \frac{\Gamma(\alpha)\Gamma(\beta)}{\Gamma(\alpha+\beta)}.$$

Both parameters α and β are shape parameters. The beta distribution, is a flexible distribution, and the graph of the beta density function can assume widely different shapes depending on the values of α and β. An extreme case is the uniform distribution on (0,1) corresponding to $\alpha = \beta = 1$. The graphs of several beta densities are given in Figure 10.5. The expected value and the variance of the random variable X are given by

$$E(X) = \frac{\alpha}{\alpha+\beta} \qquad \text{and} \qquad \text{var}(X) = \frac{\alpha\beta}{(\alpha+\beta)^2(\alpha+\beta+1)}.$$

The beta density is often used to model the distribution of a random proportion. It is common practice in Bayesian statistics to use a beta distribution for the prior distribution of the unknown value of the success probability in a Bernoulli experiment.

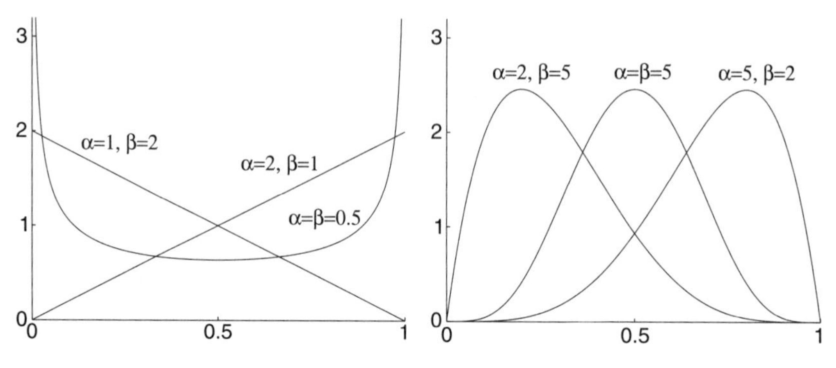

Figure 10.5 Several beta densities.

10.2.7 Normal density

This density was already discussed extensively in Section 5.1. For completeness, we repeat the definition. A continuous random variable X is said to have a *normal* density with parameters μ and $\sigma > 0$ if its probability density function is given by

$$f(x) = \frac{1}{\sigma\sqrt{2\pi}}\, e^{-\frac{1}{2}(x-\mu)^2/\sigma^2} \quad \text{for} \quad -\infty < x < \infty.$$

The parameter σ is a shape parameter, and the parameter μ is a scale parameter. The normal distribution also is referred to frequently as the *Gaussian* distribution. The expected value and the variance of the random variable X are given by

$$E(X) = \mu \qquad \text{and} \qquad \text{var}(X) = \sigma^2.$$

The notation X is $N(\mu,\sigma^2)$ is often used as a shorthand for X is a normally distributed random variable with parameters μ and σ. If $\mu = 0$ and $\sigma = 1$, the random variable X is said to have a *standard normal* distribution. The standard normal density function is $(1/\sqrt{2\pi})e^{-\frac{1}{2}x^2}$. Figure 5.2 in Chapter 5 displays the famous bell-shaped graph of the normal density function. Advanced calculus is required to prove that the area under the graph of the normal density function is indeed 1 (see Problem 10.12). The importance and applications of the normal density were already discussed in Chapter 5. Although a normal random variable theoretically takes on values in the interval $(-\infty,\infty)$, it still may provide a useful model for a variable that takes on only positive values provided that the normal probability mass on the negative axis is negligible.

Problem 10.12 In order to prove that the normal density function integrates to 1 over the interval $(-\infty,\infty)$, evaluate the integral $I = \int_{-\infty}^{\infty} e^{-\frac{1}{2}x^2}\, dx$ for the standard normal density. Verify that $I = \sqrt{2\pi}$ by changing to polar coordinates in the double integral

$$I^2 = \int_{-\infty}^{\infty} \int_{-\infty}^{\infty} e^{-\frac{1}{2}(x^2+y^2)}\, dx\, dy$$

(the polar coordinates r and θ satisfy $x = r\cos(\theta)$ and $y = r\sin(\theta)$ with $dx\, dy = r\, dr\, d\theta$). Also, verify that the change of variable $t = \frac{1}{2}x^2$ in $I = \int_{-\infty}^{\infty} e^{-\frac{1}{2}x^2}\, dx$ leads to $\Gamma(\frac{1}{2}) = \sqrt{\pi}$.

Problem 10.13 Let the random variable X be defined by $X = |W|$, where W is a $N(0,\sigma^2)$ random variable. What is the probability density function of X? Verify that $E(X)$ is equal to $\sigma\sqrt{2}/\sqrt{\pi}$.

10.2.8 Lognormal density

A continuous random variable X is said to have a *lognormal* density with parameters μ and $\sigma > 0$ if its probability density function is given by

$$f(x) = \begin{cases} \frac{1}{\sigma x\sqrt{2\pi}}\, e^{-\frac{1}{2}[\ln(x)-\mu]^2/\sigma^2} & \text{for} \quad x > 0, \\ 0 & \text{otherwise.} \end{cases}$$

A lognormally distributed random variable takes on only positive values. The graph of the lognormal density function with $\mu = 0$ and $\sigma = 1$ is displayed in Figure 10.6. It is not difficult to prove that the random variable X is lognormally distributed with parameters μ and σ if the random variable $\ln(X)$ is $N(\mu, \sigma^2)$ distributed (see Section 10.3). The lognormal distribution with a value of $\sigma \geq 1$ provides a useful probability model for income distributions. The explanation is that its probability density function $f(x)$ is skewed to the left and tends very slowly to zero as x approaches infinity. In other words, most outcomes of this lognormal distribution will be relatively small, but very large outcomes occur

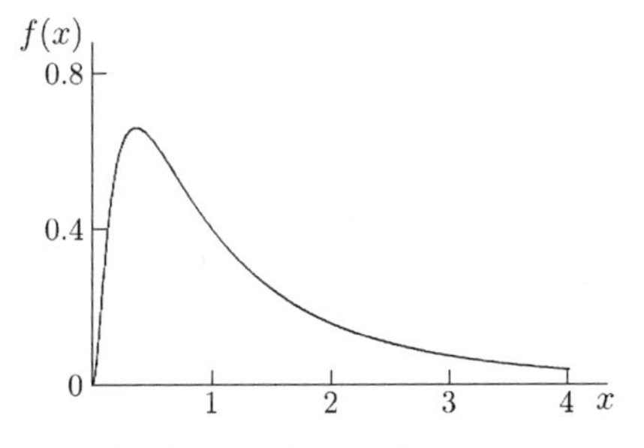

Figure 10.6 Lognormal density ($\mu = 0, \sigma = 1$).

occasionally. The lognormal distribution is also often used to model sizes of insurance claims and future stock prices after a longer period of time. The expected value and the variance of the random variable X are given by

$$E(X) = e^{\mu + \frac{1}{2}\sigma^2} \qquad \text{and} \qquad \text{var}(X) = e^{2\mu + \sigma^2}\left(e^{\sigma^2} - 1\right).$$

10.2.9 Chi-square density

A random variable X is said to have a *chi-square* distribution with *n degrees of freedom* if it can be represented as

$$X = Z_1^2 + Z_2^2 + \cdots + Z_n^2,$$

where $Z_1, Z_2, \ldots, Z_n$ are independent random variables, each having a standard normal distribution. It can be proved that the probability density of X is a gamma density with shape parameter $\alpha = \frac{1}{2}n$ and scale parameter $\lambda = \frac{1}{2}$ (see Chapter 14). The graph of the chi-square density with $n = 5$ is displayed in Figure 10.4. The expected value and the variance of the random variable X are given by

$$E(X) = n \qquad \text{and} \qquad \text{var}(X) = 2n.$$

The chi-square distribution plays an important role in statistics and is best known for its use in the so-called "chi-square tests."

10.2.10 Student-*t* density

A continuous random variable X is said to have a *Student-t* distribution with n degrees of freedom if it can be represented as

$$X = \frac{Z}{\sqrt{U/n}},$$

where Z has a standard normal distribution, U has a chi-square distribution with n degrees of freedom and the random variables Z and U are independent. It can be shown that the probability density function of X is given by

$$f(x) = c\left(1 + \frac{x^2}{n}\right)^{-(n+1)/2} \qquad \text{for} \quad -\infty < x < \infty.$$

The constant c is determined by $\int_{-\infty}^{\infty} f(x)dx = 1$. Using advanced calculus, it can be verified that $c = \frac{1}{\sqrt{\pi n}}\Gamma\left(\frac{1}{2}(n+1)\right) / \Gamma\left(\frac{1}{2}n\right)$. In Figure 10.7, the Student-t density function is displayed for $n = 5$. The density function is very similar to that of the standard normal density but it has a longer tail than the $N(0, 1)$

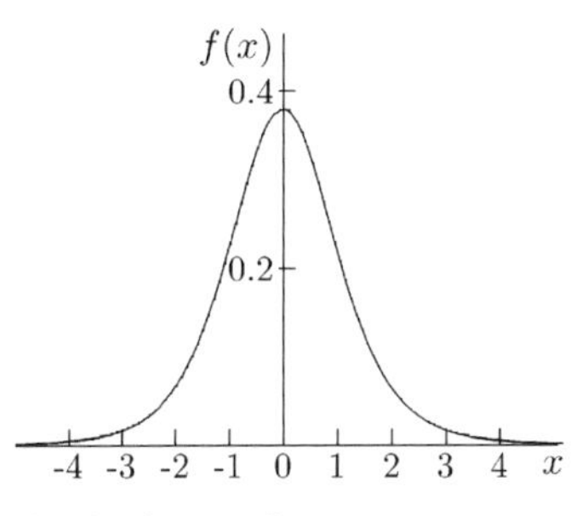

Figure 10.7 Student-t density for $n = 5$.

density. The Student-t distribution is named after William Gosset, who invented this distribution in 1908 and used the pen name "A. Student" in his publication. Gosset worked for the Guiness brewery in Dublin which, at that time, did not allow its employees to publish research papers. The expected value and the variance of the random variable X are given by

$$E(X) = 0 \qquad \text{and} \qquad \sigma^2(X) = \frac{n}{n-2} \quad \text{for } n > 2.$$

The Student-t distribution is used in statistics, primarily when dealing with small samples from a *normal* population. In particular, this distribution is used for constructing an *exact* confidence interval in case the observations are generated from a normal distribution (confidence intervals were discussed in Section 5.7). This goes as follows. Suppose that $Y_1, \ldots, Y_n$ are independent samples from an $N(\mu, \sigma^2)$ distribution with (unknown) expected value μ. The construction of the confidence interval uses the sample mean $\overline{Y}(n)$ and the sample variance $\overline{S}^2(n)$ which are defined by

$$\overline{Y}(n) = \frac{1}{n}\sum_{k=1}^{n} Y_k \qquad \text{and} \qquad \overline{S}^2(n) = \frac{1}{n-1}\sum_{k=1}^{n}\left[Y_k - \overline{Y}(n)\right]^2.$$

It is stated without proof that the random variables $\overline{Y}(n)$ and $\overline{S}^2(n)$ are independent. Moreover, it can be shown that $(\overline{Y}(n) - \mu)/(\sigma/\sqrt{n})$ has a standard normal distribution and $(n-1)\overline{S}^2(n)/\sigma^2$ has a chi-square distribution with $n-1$ degrees of freedom. Thus, the ratio

$$\frac{\overline{Y}(n) - \mu}{\sqrt{\overline{S}^2(n)/n}}$$

has a Student-t distribution with $n-1$ degrees of freedom. This important result holds for any value of n and enables us to give an *exact* $100(1-\alpha)\%$ confidence interval for the unknown expected value μ. This confidence interval is

$$\overline{Y}(n) \pm t_{n-1,1-\frac{1}{2}\alpha}\sqrt{\frac{\overline{S}^2(n)}{n}},$$

where $t_{n-1,1-\frac{1}{2}\alpha}$ is the $(1-\frac{1}{2}\alpha)$th percentile of the Student-t density function with $n-1$ degrees of freedom. That is, the area under the graph of this symmetric density function between the points $-t_{n-1,1-\frac{1}{2}\alpha}$ and $t_{n-1,1-\frac{1}{2}\alpha}$ equals $1-\alpha$. This confidence interval for a sample from a normal population does not require a large n but can be used for any value of n. The statistic $\left(\overline{Y}(n)-\mu\right)/\sqrt{\overline{S}^2(n)/n}$ has the pleasant feature of being *robust*. This means that the statistic is not sensitive for small deviations from the normality assumption.

10.3 Transformation of random variables

In Chapter 2, we saw several methods for simulating random variates from a discrete distribution. Each of these methods used the tool of generating random numbers between 0 and 1. This tool is also indispensable for simulating random variates from a continuous distribution. This will be shown by an example. Let R be a continuous random variable with probability density function $h(r) = r\exp(-\frac{1}{2}r^2)$ for $r > 0$ and $h(r) = 0$ otherwise. This is the Rayleigh density with parameter 1, a much used density in physics. How to generate a random observation of R? To do so, we need the probability distribution function of the positive random variable R. Letting $H(r) = P(R \leq r)$, we have

$$H(r) = \int_0^r xe^{-\frac{1}{2}x^2}\,dx = -\int_0^r de^{-\frac{1}{2}x^2} = -e^{-\frac{1}{2}x^2}|_0^r = 1 - e^{-\frac{1}{2}r^2}.$$

If u is a random number between 0 and 1, then the solution r to the equation

$$H(r) = u$$

is a random observation of R provided that this equation has a unique solution. Since $H(r)$ is strictly increasing on $(0, \infty)$, the equation has a unique solution. Also, the equation can be solved explicitly. The reader may easily verify that the equation $H(r) = u$ has the solution

$$r = \sqrt{-2\ln(1-u)}.$$

It will be clear that the above approach is generally applicable to simulate a random observation of a continuous random variable provided that the probability distribution function of the random variable is strictly increasing and allows for an easily computable inverse function. This approach is known as the *inverse-transformation* method.

Problem 10.14 Verify that a random observation from the Weibull distribution with shape parameter α and scale parameter λ can be simulated by taking $X = \frac{1}{\lambda}[-\ln(1-U)]^{1/\alpha}$, where U is a random number from the interval (0,1). In particular, $X = -\frac{1}{\lambda}\ln(1-U)$ is a random observation from the exponential distribution with parameter λ.

As a by-product of the discussion above, we find that the transformation $\sqrt{-2\ln(1-U)}$ applied to the uniform random variable U on $(0, 1)$ yields a random variable with probability density function $r\exp(-\frac{1}{2}r^2)$ on $(0, \infty)$. This result can be put in a more general framework. Suppose that X is a continuous random variable with probability density function $f(x)$. What is the probability density function of the random variable $Y = v(X)$ for a given function $v(x)$? In general, one best uses first principles, as is done in Example 10.2. However, a simple formula can be given for the density function of $v(X)$ when the function $v(x)$ is monotone. Assuming that the function $v(x)$ is strictly increasing or strictly decreasing on the range of X, the function $v(x)$ has a unique inverse function $a(y)$ (say). That is, for each attainable value y of $Y = v(X)$, the equation $v(x) = y$ has a unique solution $x = a(y)$. It is also assumed that $a(y)$ is continuously differentiable. Under these assumptions we have:

Rule 10.1 *The probability density of the random variable $Y = v(X)$ is given by*

$$f(a(y))|a'(y)|.$$

The proof is simple and instructive. We first give the proof for the case that $v(x)$ is strictly increasing. Then, $v(x) \leq v$ if and only if $x \leq a(v)$. Thus,

$$P(Y \leq y) = P(v(X) \leq y) = P(X \leq a(y)) = F(a(y)),$$

where $F(x)$ denotes the cumulative probability distribution function of X. Differentiating $P(Y \leq y)$ gives

$$\frac{d}{dy}P(Y \leq y) = \frac{d}{da(y)}F(a(y))\frac{da(y)}{dy} = f(a(y))\,a'(y).$$

Since $a(y)$ is strictly increasing, we have $|a'(y)| = a(y) > 0$ and so we obtain the desired result. In the case of a strictly decreasing function $v(x)$, we have $v(x) \leq v$ if and only if $x \geq a(v)$ and so

$$P(Y \leq y) = P(v(X) \leq y) = P(X \geq a(y)) = 1 - F(a(y)).$$

Next, the desired result follows by differentiating $P(Y \leq y)$ and noting that $a'(y) < 0$.

Example 10.5 Consider the transformation $Y = e^X$ for an $N(\mu,\sigma^2)$ distributed random variable X. The inverse of the function $v(x) = e^x$ is given by $a(y) = \ln(y)$. The derivative of $a(y)$ is $1/y$. Applying Rule 10.1 gives that the probability density of Y is given by

$$\frac{1}{\sigma\sqrt{2\pi}} e^{-\frac{1}{2}(\ln(y)-\mu)^2/\sigma^2} \frac{1}{y}.$$

In other words, the random variable Y has a lognormal density with parameters μ and σ.

Problem 10.15 Let X be a continuous random variable with probability density function $f(x)$. Suppose that the probability distribution function $F(x) = P(X \leq x)$ is strictly increasing on the range of X. Define the function $I(u)$ as the inverse function of $F(x)$. That is, for fixed u with $0 < u < 1$, the solution of the equation $u = F(x)$ is given by $x = I(u)$. Verify that

(a) $P(I(U) \leq x) = P(X \leq x)$ for all x, where the continuous random variable U is uniformly distributed on (0, 1).

(b) For any function $g(x)$,

$$E[g(X)] = \int_{-\infty}^{\infty} g(x)f(x)\,dx = \int_0^1 h(u)\,du,$$

where the function $h(u)$ is defined by $h(u) = g(I(u))$ for $0 < u < 1$.

Problem 10.16 Suppose that the random variable X has an exponential density with parameter μ. Let the random variable Y be defined by $Y = X^\beta$ for some $\beta > 0$. Verify that Y has a Weibull density with shape parameter $1/\beta$ and scale parameter μ^β.

11

Jointly distributed random variables

In experiments, one is often interested not only in individual random variables, but also in relationships between two or more random variables. For example, if the experiment is the testing of a new medicine, the researcher might be interested in cholesterol level, blood pressure, and glucose level of a test person. Similarly, a political scientist investigating the behavior of voters might be interested in the income and level of education of a voter. There are many more examples in the physical sciences, medical sciences, and social sciences. In applications, one often wishes to make inferences about one random variable on the basis of observations of other random variables. The purpose of this chapter is to familiarize the student with the notations and the techniques relating to experiments whose outcomes are described by two or more real numbers. The discussion is restricted to the case of pairs of random variables. Extending the notations and techniques to collections of more than two random variables is straightforward.

11.1 Joint probability densities

It is helpful to discuss the joint probability mass function of two discrete random variables before discussing the concept of the joint density of two continuous random variables. In fact, Section 9.4 already dealt with the joint distribution of discrete random variables. If X and Y are two discrete random variables defined on a same sample space with probability measure P, the mass function $p(x, y)$ defined by

$$p(x, y) = P(X = x, Y = y)$$

is called the *joint probability mass function* of X and Y. As noted before, $P(X = x, Y = y)$ is the probability assigned by P to the intersection of the two sets

$A = \{\omega : X(\omega) = x\}$ and $B = \{\omega : Y(\omega) = y\}$, with ω representing an element of the sample space. The joint probability mass function uniquely determines the probability distributions $p_X(x) = P(X = x)$ and $p_Y(y) = P(Y = y)$ by

$$p_X(x) = \sum_y P(X = x, Y = y), \quad p_Y(y) = \sum_x P(X = x, Y = y).$$

These distributions are called the *marginal distributions* of X and Y.

Example 11.1 Two fair dice are rolled. Let the random variable X represent the smallest of the outcomes of the two rolls, and let Y represent the sum of the outcomes of the two rolls. The random variables X and Y are defined on a same sample space. The sample space is the set of all 36 pairs (i, j) for $i, j = 1, \ldots, 6$, where i and j are the outcomes of the first and second dice. A probability of $\frac{1}{36}$ is assigned to each element of the sample space. In Table 11.1, we give the joint probability mass function $p(x, y) = P(X = x, Y = y)$. For example, $P(X = 2, Y = 5)$ is the probability of the intersection of the sets $A = \{(2, 2), (2, 3), (2, 4), (2, 5), (2, 6), (3, 2), (4, 2), (5, 2), (6, 2)\}$ and $B = \{(1, 4), (4, 1), (2, 3), (3, 2)\}$. The set $\{(2, 3), (3, 2)\}$ is the intersection of these two sets and has probability $\frac{2}{36}$.

Problem 11.1 You roll a pair of dice. What is the joint probability mass function of the low and high points rolled?

Problem 11.2 Let X denote the number of hearts and Y the number of diamonds in a bridge hand. What is the joint probability mass function of X and Y?

Table 11.1 *The joint probability mass function $p(x, y)$*

$x \backslash y$	2	3	4	5	6	7	8	9	10	11	12	$p_X(x)$
1	$\frac{1}{36}$	$\frac{2}{36}$	$\frac{2}{36}$	$\frac{2}{36}$	$\frac{2}{36}$	$\frac{2}{36}$	0	0	0	0	0	$\frac{11}{36}$
2	0	0	$\frac{1}{36}$	$\frac{2}{36}$	$\frac{2}{36}$	$\frac{2}{36}$	$\frac{2}{36}$	0	0	0	0	$\frac{9}{36}$
3	0	0	0	0	$\frac{1}{36}$	$\frac{2}{36}$	$\frac{2}{36}$	$\frac{2}{36}$	0	0	0	$\frac{7}{36}$
4	0	0	0	0	0	0	$\frac{1}{36}$	$\frac{2}{36}$	$\frac{2}{36}$	0	0	$\frac{5}{36}$
5	0	0	0	0	0	0	0	0	$\frac{1}{36}$	$\frac{2}{36}$	0	$\frac{3}{36}$
6	0	0	0	0	0	0	0	0	0	0	$\frac{1}{36}$	$\frac{1}{36}$
$p_Y(y)$	$\frac{1}{36}$	$\frac{2}{36}$	$\frac{3}{36}$	$\frac{4}{36}$	$\frac{5}{36}$	$\frac{6}{36}$	$\frac{5}{36}$	$\frac{4}{36}$	$\frac{3}{36}$	$\frac{2}{36}$	$\frac{1}{36}$	sum = 1

The following example provides a good starting point for a discussion of joint probability densities.

Example 11.2 A point is picked at random inside a circular disc with radius r. Let the random variable X denote the length of the line segment between the center of the disc and the randomly picked point, and let the random variable Y denote the angle between this line segment and the horizontal axis (Y is measured in radians and so $0 \leq Y < 2\pi$). What is the joint distribution of X and Y?

The two continuous random variables X and Y are defined on a common sample space. The sample space consists of all points (v, w) in the two-dimensional plane with $v^2 + w^2 \leq 1$, where the point $(0, 0)$ represents the center of the disc. The probability $P(A)$ assigned to each well-defined subset A of the sample space is taken as the area of region A divided by πr^2. The probability of the event of X taking on a value less than or equal to a and Y taking on a value less than or equal to b is denoted by $P(X \leq a, Y \leq b)$. This event occurs only if the randomly picked point falls inside the disc segment with radius a and angle b. The area of this disc segment is $\frac{b}{2\pi}\pi a^2$. Dividing this by πr^2 gives

$$P(X \leq a, Y \leq b) = \frac{b}{2\pi}\frac{a^2}{r^2} \quad \text{for} \quad 0 \leq a \leq r \quad \text{and} \quad 0 \leq b \leq 2\pi.$$

We are now in a position to introduce the concept of joint density. Let X and Y be two random variables that are defined on a same sample space with probability measure P. The *joint cumulative probability distribution function* of X and Y is defined by $P(X \leq x, Y \leq y)$ for all x, y, where $P(X \leq x, Y \leq y)$ is a shorthand for $P(\{\omega : X(\omega) \leq x \text{ and } Y(\omega) \leq y\})$ and the symbol ω represents an element of the sample space.

Definition 11.1 *The continuous random variables X and Y have a joint probability density function $f(x, y)$ if the joint cumulative probability distribution function $P(X \leq a, Y \leq b)$ allows for the representation*

$$P(X \leq a, Y \leq b) = \int_{x=-\infty}^{a}\int_{y=-\infty}^{b} f(x, y)\,dx\,dy, \qquad -\infty < a, b < \infty,$$

where the function $f(x, y)$ satisfies

$$f(x, y) \geq 0 \quad \textit{for} \quad \textit{all } x, y \qquad \textit{and} \qquad \int_{-\infty}^{\infty}\int_{-\infty}^{\infty} f(x, y)\,dx\,dy = 1.$$

The function $f(x, y)$ is called the *joint probability density function* of X and Y. Just as in the one-dimensional case, $f(a, b)$ allows for the interpretation:

$$f(a, b)\, \Delta a\, \Delta b$$

$$\approx P\left(a - \frac{1}{2}\Delta a \leq X \leq a + \frac{1}{2}\Delta a, b - \frac{1}{2}\Delta b \leq Y \leq b + \frac{1}{2}\Delta b\right)$$

for small positive values of Δa and Δb provided that $f(x, y)$ is continuous in the point (a, b). In other words, the probability that the random point (X, Y) falls into a small rectangle with sides of lengths Δa, Δb around the point (a, b) is approximately given by $f(a, b)\, \Delta a\, \Delta b$.

To obtain the joint probability density function $f(x, y)$ of the random variables X and Y in Example 11.2, we take the partial derivatives of $P(X \leq x, Y \leq y)$ with respect to x and y. It then follows from

$$f(x, y) = \frac{\partial^2}{\partial x \partial y} P(X \leq x, Y \leq y)$$

that

$$f(x, y) = \begin{cases} \frac{1}{2\pi} \frac{2x}{r^2} & \text{for} \quad 0 < x < r \quad \text{and} \quad 0 < y < 2\pi, \\ 0 & \text{otherwise.} \end{cases}$$

In general the joint probability density function is found by determining first the cumulative joint probability distribution function and taking next the partial derivatives. However, sometimes it is easier to find the joint probability density function by using its probabilistic interpretation. This is illustrated with the next example.

Example 11.3 The pointer of a spinner of radius r is spun three times. The three spins are performed independently of each other. With each spin, the pointer stops at an unpredictable point on the circle. The random variable L_i corresponds to the length of the arc from the top of the circle to the point where the pointer stops on the ith spin. The length of the arc is measured clockwise. Let $X = \min(L_1, L_2, L_3)$ and $Y = \max(L_1, L_2, L_3)$. What is the joint probability density function of the two continuous random variables X and Y?

We can derive the joint probability density function $f(x, y)$ by using the interpretation that the probability $P(x < X \leq x + \Delta x, y < Y \leq y + \Delta y)$ is approximately equal to $f(x, y)\Delta x \Delta y$ for Δx and Δy small. The event $\{x < X \leq x + \Delta x, y < Y \leq y + \Delta y\}$ occurs only if one of the L_i takes on a value between x and $x + \Delta x$, one of the L_i a value between y and $y + \Delta y$, and the remaining L_i a value between x and y, where $0 < x < y$. There are $3 \times 2 \times 1 = 6$

ways at which L_1, L_2, L_3 can be arranged and the probability that for fixed i the random variable L_i takes on a value between a and b equals $(b-a)/(2\pi r)$ for $0 < a < b < 2\pi r$ (explain!). Thus, by the independence of L_1, L_2, and L_3,

$$P(x < X \leq x + \Delta x, y < Y \leq y + \Delta y)$$
$$= 6\frac{(x + \Delta x - x)}{2\pi r}\frac{(y + \Delta y - y)}{2\pi r}\frac{(y - x)}{2\pi r}.$$

Hence the joint probability density function of X and Y is given by

$$f(x, y) = \begin{cases} \frac{6(y-x)}{(2\pi r)^3} & \text{for } 0 < x < y < 2\pi r \\ 0 & \text{otherwise.} \end{cases}$$

In general, if the random variables X and Y have a joint probability density function $f(x, y)$,

$$P((X, Y) \in C) = \iint_C f(x, y)\, dx\, dy$$

for any set C of pairs of real numbers. In calculating a double integral over a nonnegative integrand, it does not matter whether we integrate over x first or over y first. This is a basic fact from calculus. The double integral can be written as a repeated one-dimensional integral. To illustrate this, we derive the useful result that the sum $Z = X + Y$ has the probability density

$$f_Z(z) = \int_{-\infty}^{\infty} f(u, z - u)\, du.$$

To do so, note that

$$P(Z \leq z) = \iint_{\substack{(x,y):\\ x+y\leq z}} f(x, y)\, dx\, dy = \int_{x=-\infty}^{\infty} \int_{y=-\infty}^{z-x} f(x, y)\, dx\, dy$$
$$= \int_{v=-\infty}^{z} \int_{u=-\infty}^{\infty} f(u, v - u)\, du\, dv,$$

using the change of variables $u = x$ and $v = x + y$. Next, differentiation of $P(Z \leq z)$ with respect to z yields the convolution formula for $f_Z(z)$.

Another useful result is the following. Suppose that a point (X, Y) is picked at random inside a bounded region R in the two-dimensional plane. Then, the joint probability density function $f(x, y)$ of X and Y is given by

$$f(x, y) = \begin{cases} 1/(\text{area of region } R) & \text{for } (x, y) \in R \\ 0 & \text{otherwise.} \end{cases}$$

The proof is simple. For any subset C of R, it holds that $P((X, Y) \in C) =$ (area of C)/(area of R). Moreover, area of $C = \iint_C dx\, dy$. This verifies the representation $P((X, Y) \in C) = \iint_C f(x, y)\, dx\, dy$ with $f(x, y) = 1/$(area of R) for $(x, y) \in R$.

Problem 11.3 A point (X, Y) is picked at random inside the triangle consisting of the points (x, y) in the plane with $x, y \geq 0$ and $x + y \leq 1$. What is the joint probability density of the point (X, Y)? Determine the probability density of each of the random variables $X + Y$ and $\max(X, Y)$.

Problem 11.4 Let X and Y be two random variables with a joint probability density

$$f(x, y) = \begin{cases} \frac{1}{(x+y)^3} & \text{for} \quad x, y > c \\ 0 & \text{otherwise,} \end{cases}$$

for an appropriate constant c. Verify that $c = \frac{1}{4}$ and calculate the probability $P(X > a, Y > b)$ for $a, b > c$.

Problem 11.5 Independently of each other, two numbers X and Y are chosen at random in the interval (0, 1). Let $Z = X/Y$ be the ratio of these two random numbers.

(a) Use the joint density of X and Y to verify that $P(Z \leq z)$ equals $\frac{1}{2}z$ for $0 < z < 1$ and equals $1 - 1/(2z)$ for $z \geq 1$.
(b) What is the probability that the first significant (nonzero) digit of Z equals 1? What about the digits $2, \ldots, 9$?
(c) What is the answer to Question (b) for the random variable $V = XY$?
(d) What is the probability density function of the random variable $(X/Y)U$ when U is a random number from (0, 1) that is independent of X and Y?

Problem 11.6 Independently of each other, two points are chosen at random in the interval (0, 1). Show that the smallest of the three resulting intervals is larger than a with probability $(1 - 3a)^2$ for $0 < a < 1/3$. *Hint*: verify first that the joint density function of X and Y is $f(x, y) = 2$ for $0 < x < y < 1$, where X is the smallest of the two random numbers chosen in (0, 1) and Y is the largest of these two numbers.

11.2 Marginal probability densities

If the two random variables X and Y have a joint probability density function $f(x, y)$, then each of the random variables X and Y has a probability density itself. Using the fact that $\lim_{n\to\infty} P(A_n) = P(\lim_{n\to\infty} A_n)$ for any nondecreasing sequence of events A_n, it follows that

$$P(X \leq a) = \lim_{b\to\infty} P(X \leq a, Y \leq b)$$
$$= \int_{-\infty}^{a} \left[\int_{-\infty}^{\infty} f(x, y)\, dy \right] dx.$$

This representation shows that X has probability density function

$$f_X(x) = \int_{-\infty}^{\infty} f(x, y)\, dy, \qquad -\infty < x < \infty.$$

In the same way, the random variable Y has probability density function

$$f_Y(y) = \int_{-\infty}^{\infty} f(x, y)\, dx, \qquad -\infty < y < \infty.$$

The probability density functions $f_X(x)$ and $f_Y(y)$ are called the *marginal probability density functions* of X and Y.

Example 11.4 A point (X, Y) is chosen at random inside the unit circle. What is the marginal density of X (the marginal density of Y is the same as that of X)?

Denote by $C = \{(x, y) | x^2 + y^2 \leq 1\}$ the unit circle. The joint probability density function $f(x, y)$ of X and Y is given by $f(x, y) = 1/(\text{area of } C)$ for $(x, y) \in C$. Hence,

$$f(x, y) = \begin{cases} \frac{1}{\pi} & \text{for } (x, y) \in C \\ 0 & \text{otherwise.} \end{cases}$$

Using the fact that $f(x, y)$ is equal to zero for those y satisfying $y^2 > 1 - x^2$, if follows that

$$f_X(x) = \int_{-\infty}^{\infty} f(x, y)\, dy = \int_{-\sqrt{1-x^2}}^{\sqrt{1-x^2}} \frac{1}{\pi}\, dy$$

and so

$$f_X(x) = \begin{cases} \frac{2}{\pi}\sqrt{1 - x^2} & \text{for } -1 < x < 1 \\ 0 & \text{otherwise.} \end{cases}$$

Can you explain why the marginal density of X is not the uniform density on $(-1, 1)$? *Hint*: interpret $P(x < X \leq x + \Delta x)$ as the area of a vertical strip in the unit circle.

Problem 11.7 A point (X, Y) is chosen at random in the equilateral triangle having $(0, 0)$, $(1, 0)$, and $(\frac{1}{2}, \frac{1}{2}\sqrt{3})$ as corner points. Determine the marginal densities of X and Y. Before determining the function $f_X(x)$, can you explain why $f_X(x)$ must be largest at $x = \frac{1}{2}$?

A general condition for the independence of the jointly distributed random variables X and Y is stated in Definition 9.2. In terms of the marginal densities, the continuous analog of Rule 9.3 for the discrete case is:

Rule 11.1 *The jointly distributed random variables X and Y are independent if and only if*

$$f(x, y) = f_X(x) f_Y(y) \quad \textit{for} \quad \textit{all } x, y.$$

Let us illustrate this with the random variables X and Y from Example 11.2. Then, we obtain from

$$f_X(x) = \int_0^{2\pi} \frac{x}{\pi r^2}\, dy$$

that

$$f_X(x) = \begin{cases} \frac{2x}{r^2} & \text{for} \quad 0 < x < r, \\ 0 & \text{otherwise,} \end{cases}$$

In the same way, we obtain from $f_Y(y) = \int_0^r x/(\pi r^2)\, dx$ that

$$f_Y(y) = \begin{cases} \frac{1}{2\pi} & \text{for} \quad 0 < y < 2\pi, \\ 0 & \text{otherwise.} \end{cases}$$

The calculations lead to the intuitively obvious result that the angle Y has a uniform distribution on $(0, 2\pi)$. A somewhat more surprising result is that the distance X and the angle Y are independent random variables, though there is dependence between the components of the randomly picked point. The independence of X and Y follows from the observation that $f(x, y) = f_X(x) f_Y(y)$ for all x, y.

11.2.1 Substitution rule

The expected value of a given function of jointly distributed random variables X and Y can be calculated by the two-dimensional substitution rule. In the continuous case, we have

Rule 11.2 *If the random variables X and Y have a joint probability density function $f(x, y)$, then*

$$E\left[g(X, Y)\right] = \int_{-\infty}^{\infty}\int_{-\infty}^{\infty} g(x, y) f(x, y)\, dx\, dy$$

for any function $g(x, y)$ provided that the above integral is well defined.

An easy consequence of Rule 11.2 is that

$$E(aX + bY) = aE(X) + bE(Y)$$

for any constants a, b provided that $E(X)$ and $E(Y)$ exist. To see this, note that

$$\begin{aligned}
&\int_{-\infty}^{\infty}\int_{-\infty}^{\infty} (ax + by) f(x, y)\, dx\, dy \\
&= \int_{-\infty}^{\infty}\int_{-\infty}^{\infty} axf(x, y)\, dx\, dy + \int_{-\infty}^{\infty}\int_{-\infty}^{\infty} byf(x, y)\, dx\, dy \\
&= \int_{x=-\infty}^{\infty} ax\, dx \int_{y=-\infty}^{\infty} f(x, y)\, dy + \int_{y=-\infty}^{\infty} by\, dy \int_{x=-\infty}^{\infty} f(x, y)\, dx \\
&= a\int_{-\infty}^{\infty} xf_X(x)\, dx + b\int_{-\infty}^{\infty} yf_Y(y)\, dy,
\end{aligned}$$

which proves the desired result.

An illustration of the substitution rule is provided by Problem 2.23: what is the expected value of the distance between two points that are chosen at random in the interval (0, 1)? To answer this question, let X and Y be two independent random variables that are uniformly distributed on (0, 1). The joint density function of X and Y is given by $f(x, y) = 1$ for all $0 < x, y < 1$. The substitution rule gives

$$\begin{aligned}
E(|X - Y|) &= \int_0^1\int_0^1 |x - y|\, dx\, dy \\
&= \int_0^1 dx\left[\int_0^x (x - y)\, dy + \int_x^1 (y - x)\, dy)\right] \\
&= \int_0^1 \left[\frac{1}{2}x^2 + \frac{1}{2} - \frac{1}{2}x^2 - x(1 - x)\right] dx = \frac{1}{3}.
\end{aligned}$$

Hence, the answer to the question is $\frac{1}{3}$.

As another illustration of Rule 11.2, consider Example 11.2 again. In this example, a point is picked at random inside a circular disk with radius r and the point (0, 0) as center. What is the expected value of the rectangular distance from the randomly picked point to the center of the disk? This rectangular distance is given by $|X\cos(Y)| + |X\sin(Y)|$ (the rectangular distance from point (a, b) to $(0, 0)$ is defined by $|a| + |b|$). For the function $g(x, y) = |x\cos(y)| + |x\sin(y)|$, we find

$$
\begin{aligned}
E[g(X, Y)] & \\
&= \int_0^r \int_0^{2\pi} \{x|\cos(y)| + x|\sin(y)|\} \frac{x}{\pi r^2}\, dx\, dy \\
&= \frac{1}{\pi r^2} \int_0^{2\pi} |\cos(y)|\, dy \int_0^r x^2\, dx + \frac{1}{\pi r^2} \int_0^{2\pi} |\sin(y)|\, dy \int_0^r x^2\, dx \\
&= \frac{r^3}{3\pi r^2} \left[\int_0^{2\pi} |\cos(y)|\, dy + \int_0^{2\pi} |\sin(y)|\, dy \right] = \frac{8r}{3\pi}.
\end{aligned}
$$

11.3 Transformation of random variables

In statistical applications, one sometimes needs the joint density of two random variables V and W that are defined as functions of two other random variables X and Y having a joint density $f(x, y)$. Suppose that the random variables V and W are defined by $V = g(X, Y)$ and $W = h(X, Y)$ for given functions g and h. What is the joint probability density function of V and W? An answer to this question will be given under the assumption that the transformation is one-to-one. That is, it is assumed that the equations $v = g(x, y)$ and $w = h(x, y)$ can be solved uniquely to yield functions $x = a(v, w)$ and $y = b(v, w)$. Also assume that the partial derivatives of the functions $a(v, w)$ and $b(v, w)$ with respect to v and w are continuous in (v, w). Then the following transformation rule holds:

Rule 11.3 *The joint probability density function of V and W is given by*

$$f(a(v, w), b(v, w))\, |J(v, w)|,$$

where the Jacobian $J(v, w)$ is given by the determinant

$$
\begin{vmatrix}
\dfrac{\partial a(v, w)}{\partial v} & \dfrac{\partial a(v, w)}{\partial w} \\[2ex]
\dfrac{\partial b(v, w)}{\partial v} & \dfrac{\partial b(v, w)}{\partial w}
\end{vmatrix}
= \frac{\partial a(v, w)}{\partial v} \frac{\partial b(v, w)}{\partial w} - \frac{\partial a(v, w)}{\partial w} \frac{\partial b(v, w)}{\partial v}.
$$

The proof of this rule is omitted. This transformation rule looks intimidating, but is easy to use in many applications. An interesting application is the following.

11.3.1 Simulating from a normal distribution

A natural transformation of two independent standard normal random variables leads to a practically useful method for simulating random observations from the standard normal distribution. Suppose that X and Y are independent random variables each having the standard normal distribution. Using Rule 11.1, the joint probability density function of X and Y is given by

$$f(x, y) = \frac{1}{2\pi} e^{-\frac{1}{2}(x^2+y^2)}.$$

The random vector (X, Y) can be considered as a point in the two-dimensional plane. Let the random variable V be the distance from the point $(0, 0)$ to the point (X, Y) and let W be the angle that the line through the points $(0, 0)$ and (X, Y) makes with the horizontal axis. The random variables V and W are functions of X and Y (the function $g(x, y) = \sqrt{x^2 + y^2}$ and $h(x, y) = \arctan(y/x)$). The inverse functions $a(v, w)$ and $b(v, w)$ are very simple. By basic geometry, $x = v\cos(w)$ and $y = v\sin(w)$. We thus obtain the Jacobian

$$\begin{vmatrix} \cos(w) & -v\sin(w) \\ \sin(w) & v\cos(w) \end{vmatrix} = v\cos^2(w) + v\sin^2(w) = v,$$

using the celebrated identity $\cos^2(w) + \sin^2(w) = 1$. Hence, the joint probability density function of V and W is given by

$$f_{VW}(v, w) = \frac{v}{2\pi} e^{-\frac{1}{2}\left(v^2\cos^2(w)+v^2\sin^2(w)\right)} = \frac{v}{2\pi} e^{-\frac{1}{2}v^2},$$

for $0 < v < \infty$ and $0 < w < 2\pi$. The marginal densities of V and W are given by

$$f_V(v) = \frac{1}{2\pi}\int_0^{2\pi} ve^{-\frac{1}{2}v^2}\,dw = ve^{-\frac{1}{2}v^2}, \qquad 0 < v < \infty$$

and

$$f_W(w) = \frac{1}{2\pi}\int_0^{\infty} ve^{-\frac{1}{2}v^2}\,dv = \frac{1}{2\pi}, \qquad 0 < w < 2\pi.$$

Since $f_{VW}(v, w) = f_V(v)f_W(w)$, we have the remarkable finding that V and W are independent random variables. The random variable V has the probability density function $ve^{-\frac{1}{2}v^2}$ for $v > 0$ and W is uniformly distributed on $(0, 2\pi)$. This result is extremely useful for simulation purposes. Using the inverse-transformation method from Section 10.3, it is a simple matter to simulate random observations from the probability distributions of V and W. If we let U_1 and U_2 denote two independent random numbers from the interval (0,1), it follows from results in Section 10.3 that random observations of V and W are given by

$$V = \sqrt{-2\ln(1 - U_1)} \qquad \text{and} \qquad W = 2\pi U_2.$$

Next, one obtains two random observations X and Y from the standard normal distribution by taking

$$X = V\cos(W) \qquad \text{and} \qquad Y = V\sin(W).$$

Theoretically, X and Y are independent of each other. However, if a pseudo-random generator is used to generate U_1 and U_2, one uses only one of two variates X and Y. It surprisingly appears that the points (X, Y) lie on a spiral in the plane when a multiplicative generator is used for the pseudo-random numbers. The explanation of this subtle dependency lies in the fact that pseudo-random numbers are not truly random. The method described above for generating normal variates is known as the Box-Muller method.

Problem 11.8 The random variables X and Y are independent and both have an exponential density with parameter λ. Verify that the random variables $V = X + Y$ and $W = X/(X + Y)$ are independent and determine their densities.

Problem 11.9 Let (X, Y) be a point chosen at random inside the unit circle. Define V and W by $V = X\sqrt{-2\ln(R^2)/R^2}$ and $W = Y\sqrt{-2\ln(R^2)/R^2}$, where $R^2 = X^2 + Y^2$. Verify that the random variables V and W are independent and $N(0, 1)$ distributed. This method for generating normal variates is known as Marsaglia's polar method.

11.4 Covariance and correlation coefficient

Let the random variables X and Y be defined on a same sample space with probability measure P. A basic rule in probability is that the expected value of the sum $X + Y$ equals the sum of the expected values of X and Y. Does a similar rule hold for the variance of the sum $X + Y$? To answer this question, we apply the definition of variance of a random variable:

$$\begin{aligned}
&\text{var}(X + Y) \\
&= E\left[\left(X + Y - E(X + Y)\right)^2\right] \\
&= E\left[\left(X - E(X) + Y - E(Y)\right)^2\right] \\
&= E\left[\left(X - E(X)\right)^2 + 2\left(X - E(X)\right)\left(Y - E(Y)\right) + \left(Y - E(Y)\right)^2\right] \\
&= \text{var}(X) + 2E[(X - E(X))(Y - E(Y))] + \text{var}(Y).
\end{aligned}$$

This leads to the following general definition.

Definition 11.2 *The covariance* $\text{cov}(X, Y)$ *of two random variables* X *and* Y *is defined by*

$$\text{cov}(X, Y) = E[(X - E(X))(Y - E(Y))]$$

whenever the expectations exist.

The formula for $\text{cov}(X, Y)$ can be written in the equivalent form

$$\text{cov}(X, Y) = E(XY) - E(X)E(Y)$$

by expanding $(X - E(X))(Y - E(Y))$ into $XY - XE(Y) - YE(X) + E(X) E(Y)$ and noting that the expectation is a linear operator. Using the fact that $E(XY) = E(X)E(Y)$ for independent random variables, the alternative formula for $\text{cov}(X, Y)$ has as direct consequence:

Rule 11.4 *If* X *and* Y *are independent random variables, then*

$$\text{cov}(X, Y) = 0.$$

However, the converse of this result is not always true. A simple example of two dependent random variables X and Y having covariance zero is given in Section 9.4. Another counterexample is provided by the random variables $X = Z$ and $Y = Z^2$, where Z has the standard normal distribution. Despite this, $\text{cov}(X, Y)$ is often used as a measure of the dependence of X and Y. The covariance appears over and over in practical applications; see the discussion in Section 5.2. Using the definition of covariance, we find the general rule:

Rule 11.5 *The variance of the sum of two random variables* X *and* Y *is given by*

$$\text{var}(X + Y) = \text{var}(X) + 2\text{cov}(X, Y) + \text{var}(Y).$$

If the random variables X *and* Y *are independent, then*

$$\text{var}(X + Y) = \text{var}(X) + \text{var}(Y).$$

The units of $\text{cov}(X, Y)$ are not the same as the units of $E(X)$ and $E(Y)$. Therefore, it is often more convenient to use the *correlation coefficient* of X and Y which is defined by

$$\rho(X, Y) = \frac{\text{cov}(X, Y)}{\sqrt{\text{var}(X)}\sqrt{\text{var}(Y)}},$$

provided that $\text{var}(X) > 0$ and $\text{var}(Y) > 0$. The correlation coefficient is a dimensionless quantity with the property that

$$-1 \leq \rho(X, Y) \leq 1.$$

The reader is asked to prove this property in Problem 11.10. The random variables X and Y are said to be *uncorrelated* if $\rho(X, Y) = 0$. Independent random variables are always uncorrelated, but the converse is not always true. If $\rho(X, Y) = \pm 1$, then Y is fully determined by X. In this case it can be shown that $Y = aX + b$ for constants a and b with $a \neq 0$.

The problem section of Chapter 5 contains several exercises on the covariance and correlation coefficient. Here are some more exercises.

Problem 11.10 Verify that

$$\text{var}(aX + b) = a^2 \text{var}(X) \qquad \text{and} \qquad \text{cov}(aX, bY) = ab\text{cov}(X, Y)$$

for any constants a, b. Next evaluate the variance of the random variable $Z = Y/\sqrt{\text{var}(Y)} - \rho(X, Y)X/\sqrt{\text{var}(X)}$ to prove that $-1 \leq \rho(X, Y) \leq 1$.

Problem 11.11 The amounts of rainfall in Amsterdam during each of the months January, February, ..., December are independent random variables with expected values of 62.1, 43.4, 58.9, 41.0, 48.3, 67.5, 65.8, 61.4, 82.1, 85.1, 89.0, and 74.9 mm and with standard deviations of 33.9, 27.8, 31.1, 24.1, 29.3, 33.8, 36.8, 32.1, 46.6, 42.4, 40.0, and 36.2 mm. What are the expected value and the standard deviation of the annual rainfall in Amsterdam?

Problem 11.12 Let the random variables $X_1, \ldots, X_n$ be defined on a common probability space. Prove that

$$\text{var}(X_1 + \cdots + X_n) = \sum_{i=1}^{n} \text{var}(X_i) + 2 \sum_{i=1}^{n} \sum_{j=i+1}^{n} \text{cov}(X_i, X_j).$$

Problem 11.13 You roll a pair of dice. What is the correlation coefficient of the high and low points rolled?

Problem 11.14 What is the correlation coefficient of the Cartesian coordinates of a point picked at random in the unit circle?

11.4.1 Linear predictor

Suppose that X and Y are two dependent random variables. In statistical applications, it is often the case that we can observe the random variable X but we want to know the dependent random variable Y. A basic question in statistics is: what is the best *linear* predictor of Y with respect to X? That is, for which linear function $y = \alpha + \beta x$ is

$$E\left[(Y - \alpha - \beta X)^2\right]$$

minimal? The answer to this question is

$$y = \mu_Y + \rho_{XY} \frac{\sigma_Y}{\sigma_X}(x - \mu_X),$$

where $\mu_X = E(X)$, $\mu_Y = E(Y)$, $\sigma_X = \sqrt{\text{var}(X)}$, $\sigma_Y = \sqrt{\text{var}(Y)}$, and $\rho_{XY} = \rho(X, Y)$. The derivation is simple. Rewriting $y = \alpha + \beta x$ as $y = \mu_Y + \beta(x - \mu_X) - (\mu_Y - \alpha - \beta\mu_X)$, it follows after some algebra (verify)

$$\begin{aligned} &E\left[(Y - \alpha - \beta X)^2\right] \\ &= E\left[\left(Y - \mu_Y - \beta(X - \mu_X) + (\mu_Y - \alpha - \beta\mu_X)\right)^2\right] \\ &= \sigma_Y^2 + \beta^2\sigma_X^2 - 2\beta\rho_{XY}\sigma_X\sigma_Y + (\mu_Y - \alpha - \beta\mu_X)^2. \end{aligned}$$

In order to minimize this quadratic function in α and β, we put the partial derivatives of the function with respect to α en β equal to zero. This leads after some simple algebra to

$$\beta = \frac{\rho_{XY}\sigma_Y}{\sigma_X} \qquad \text{and} \qquad \alpha = \mu_Y - \frac{\rho_{XY}\sigma_Y}{\sigma_X}\mu_X.$$

For these values of α and β, we have the minimal value

$$E\left[(Y - \alpha - \beta X)^2\right] = \sigma_Y^2(1 - \rho_{XY}^2).$$

This minimum is sometimes called the residual variance of Y.

The phenomenon of *regression to the mean* can be explained with the help of the best linear predictor. Think of X as the height of a 25-year-old father and think of Y as the height his newborn son will have at the age of 25 years. It is reasonable to assume that $\mu_X = \mu_Y = \mu$, $\sigma_X = \sigma_Y = \sigma$, and $\rho = \rho(X, Y)$ is positive. The best linear predictor $\hat{Y}$ of Y then satisfies $\hat{Y} - \mu = \rho(X - \mu)$ with $0 < \rho < 1$. If the height of the father scores above the mean, the best linear prediction is that the height of the son will score closer to the mean. Very tall fathers tend to have somewhat shorter sons and very short fathers somewhat taller ones! Regression to the mean shows up in a wide variety of places: it helps explain why great movies have often disappointing sequels, and disastrous presidents have often better successors.

12
Multivariate normal distribution

Do the one-dimensional normal distribution and the one-dimensional central limit theorem allow for a generalization to dimension two or higher? The answer is yes. Just as the one-dimensional normal density is completely determined by its expected value and variance, the bivariate normal density is completely specified by the expected values and the variances of its marginal densities and by its correlation coefficient. The bivariate normal distribution can be extended to the multivariate normal distribution in higher dimensions. The multivariate normal distribution arises when when you take the sum of a large number of independent random vectors. To get this distribution, all you have to do is to compute a vector of expected values and a matrix of covariances. The multivariate central limit theorem explains why so many natural phenomena have the multivariate normal distribution. A nice feature of the multivariate normal distribution is its mathematical tractability. The fact that any linear combination of multivariate normal random variables has a univariate normal distribution makes the multivariate normal distribution very convenient for financial portfolio analysis, among others.

12.1 Bivariate normal distribution

A random vector (X, Y) is said to have a *standard bivariate normal distribution* if it has a joint probability density function of the form

$$f(x, y) = \frac{1}{2\pi\sqrt{1-\rho^2}} e^{-\frac{1}{2}(x^2-2\rho xy+y^2)/(1-\rho^2)}, \qquad -\infty < x, y < \infty,$$

where ρ is a constant with $-1 < \rho < 1$. The bivariate normal distribution appears in many applied probability problems. Before showing that ρ can be interpreted as the correlation coefficient of X and Y, we derive the marginal densities of X and Y. Therefore, we first decompose the joint density function $f(x, y)$ as:

$$f(x, y) = \frac{1}{\sqrt{2\pi}} e^{-\frac{1}{2}x^2} \frac{1}{\tau\sqrt{2\pi}} e^{-\frac{1}{2}(y-\mu_x)^2/\tau^2},$$

where $\tau = \sqrt{1-\rho^2}$ and $\mu_x = \rho x$. Next observe that, for *fixed* x,

$$g(y) = \frac{1}{\tau\sqrt{2\pi}} e^{-\frac{1}{2}(y-\mu_x)^2/\tau^2}$$

is an $N(\mu_x, \tau^2)$ probability density function. This implies that $\int_{-\infty}^{\infty} g(y)\, dy = 1$ and so

$$f_X(x) = \int_{-\infty}^{\infty} f(x, y)\, dy = \frac{1}{\sqrt{2\pi}} e^{-\frac{1}{2}x^2}, \qquad -\infty < x < \infty.$$

In other words, the marginal density $f_X(x)$ of X is the standard normal density. Also, for reasons of symmetry, the marginal density $f_Y(y)$ of Y is the standard normal density. Next, we prove that ρ is the correlation coefficient $\rho(X, Y)$ of X and Y. Since $\text{var}(X) = \text{var}(Y) = 1$, it suffices to verify that $\text{cov}(X, Y) = \rho$. To do so, we use again the above decomposition of $f(x, y)$ together with the fact that $E(X) = E(Y) = 0$. This gives

$$\begin{aligned}
\text{cov}(X, Y) &= \int_{-\infty}^{\infty}\int_{-\infty}^{\infty} xyf(x, y)\, dx\, dy \\
&= \int_{x=-\infty}^{\infty} x\frac{1}{\sqrt{2\pi}} e^{-\frac{1}{2}x^2}\, dx \int_{y=-\infty}^{\infty} y\frac{1}{\tau\sqrt{2\pi}} e^{-\frac{1}{2}(y-\mu_x)^2/\tau^2}\, dy \\
&= \int_{-\infty}^{\infty} \rho x^2 \frac{1}{\sqrt{2\pi}} e^{-\frac{1}{2}x^2}\, dx = \rho,
\end{aligned}$$

where the third equality uses the fact that the expected value of an $N(\mu_x, \tau^2)$ random variable is $\mu_x = \rho x$ and the last equality uses the fact that $E(Z^2) = \sigma^2(Z) = 1$ for a standard normal random variable Z.

A random vector (X, Y) is said to be *bivariate normal* distributed with parameters $(\mu_1, \mu_2, \sigma_1^2, \sigma_2^2, \rho)$ if the standardized random vector

$$\left(\frac{X-\mu_1}{\sigma_1}, \frac{Y-\mu_2}{\sigma_2}\right)$$

has the standard bivariate normal distribution with parameter ρ. In this case, the joint density $f(x, y)$ of the random variables X and Y is given by

$$f(x, y) = \frac{1}{2\pi\sigma_1\sigma_2\sqrt{1-\rho^2}} e^{-\frac{1}{2}\left[\left(\frac{x-\mu_1}{\sigma_1}\right)^2 - 2\rho\left(\frac{x-\mu_1}{\sigma_1}\right)\left(\frac{y-\mu_2}{\sigma_2}\right) + \left(\frac{y-\mu_2}{\sigma_2}\right)^2\right]/(1-\rho^2)}.$$

The marginal densities $f_X(x)$ and $f_Y(y)$ of X and Y are the $N(\mu_1, \sigma_1^2)$ density and the $N(\mu_2, \sigma_2^2)$ density. The parameter ρ is the correlation coefficient of X and Y. The proof of this result is simple. As shown above, the covariance of $(X - \mu_1)/\sigma_1$ and $(Y - \mu_2)/\sigma_2$ is ρ. Moreover,

$$\text{cov}\left(\frac{X - \mu_1}{\sigma_1}, \frac{Y - \mu_2}{\sigma_2}\right) = \frac{1}{\sigma_1\sigma_2}\text{cov}(X - \mu_1, Y - \mu_2) = \frac{1}{\sigma_1\sigma_2}\text{cov}(X, Y),$$

showing that $\rho(X, Y) = \rho$.

In general, uncorrelatedness is a necessary but not sufficient condition for independence of two random variables. However, for a bivariate normal distribution, uncorrelatedness is a necessary and sufficient condition for independence. This important result is a direct consequence of Rule 11.1, since $f(x, y) = f_X(x)f_Y(y)$ if and only if $\rho = 0$. As already pointed out, the bivariate normal distribution has the important property that its marginal distributions are one-dimensional normal distributions. More generally, it can be shown that the random variables X and Y have a bivariate normal distribution if and only if $aX + bY$ is normally distributed for any constants a and b not both equal to zero. The reader is asked to prove the "only if" part of this result in Problem 12.1. To conclude that (X, Y) has a bivariate normal distribution, it is not sufficient that X and Y are normally distributed, but normality of $aX + bY$ should be required for all constants a and b not both equal to zero.

Problem 12.1 Prove that $aX + bY$ is normally distributed for any constants a and b not both equal to zero if (X, Y) has a bivariate normal distribution.

Problem 12.2 The rates of return on two stocks A and B have a bivariate normal distribution with parameters $\mu_1 = 0.08$, $\mu_2 = 0.12$, $\sigma_1 = 0.05$, $\sigma_2 = 0.15$, and $\rho = -0.50$. What is the probability that the average of the returns on stocks A and B will be larger than 0.11?

Problem 12.3 Suppose that the probability density function $f(x, y)$ of the random variables X and Y is given by the bivariate standard normal density with parameter ρ. Verify that the probability density function $f_Z(z)$ of the ratio $Z = X/Y$ is given by the so-called *Cauchy* density

$$\int_{-\infty}^{\infty} |y| f(zy, y)\, dy = \frac{(1/\pi)\sqrt{1 - \rho^2}}{(z - \rho)^2 + 1 - \rho^2}, \qquad -\infty < z < \infty.$$

12.1.1 The drunkard's walk

The drunkard's walk is one of the most useful probability models in the physical sciences. Let us formulate this model in terms of a particle moving on the two-dimensional plane. The particle starts at the origin (0, 0). In each step, the particle travels a unit distance in a randomly chosen direction between 0 and 2π. The direction of each successive step is determined independently of the others. What is the joint probability density function of the (x, y)-coordinates of the position of the particle after n steps?

Let the random variable Θ denote the direction taken by the particle in any step. In each step, the x-coordinate of the position of the particle changes with an amount that is distributed as $\cos(\Theta)$ and the y-coordinate with an amount that is distributed as $\sin(\Theta)$. The continuous random variable Θ has a uniform distribution on $(0, 2\pi)$. Let X_k and Y_k be the changes of the x-coordinate and the y-coordinate of the position of the particle in the kth step. Then the position of the particle after n steps can be represented by the random vector (S_{n1}, S_{n2}), where

$$S_{n1} = X_1 + \cdots + X_n \qquad \text{and} \qquad S_{n2} = Y_1 + \cdots + Y_n.$$

For each n the random vectors $(X_1, Y_1), \ldots, (X_n, Y_n)$ are independent and have the same distribution. The reader who is familiar with the central limit theorem from Chapter 5 for the sum of one-dimensional random variables will not be surprised to learn that the random vector $(S_{n1}, S_{n2}) = (X_1 + \cdots + X_n, Y_1 + \cdots + Y_n)$ satisfies the conditions of the two-dimensional version of the central limit theorem. In general form, the two-dimensional version of the central limit theorem reads as:

$$\lim_{n\to\infty} P\left(\frac{S_{n1} - n\mu_1}{\sigma_1\sqrt{n}} \leq x, \frac{S_{n2} - n\mu_2}{\sigma_2\sqrt{n}} \leq y\right)$$
$$= \frac{1}{2\pi\sqrt{(1-\rho^2)}} \int_{-\infty}^{x} \int_{-\infty}^{y} e^{-\frac{1}{2}(v^2 - 2\rho v w + w^2)/(1-\rho^2)}\, dv\, dw,$$

where

$$\mu_1 = E(X_i),\ \mu_2 = E(Y_i),\ \sigma_1^2 = \sigma^2(X_i),\ \sigma_2^2 = \sigma^2(Y_i)$$

and

$$\rho = \frac{\operatorname{cov}(X_i, Y_i)}{\sigma_1\sigma_2}.$$

In the particular case of the drunkard's walk, we have

$$\mu_1 = \mu_2 = 0,\ \sigma_1 = \sigma_2 = \frac{1}{\sqrt{2}} \qquad \text{and} \qquad \rho = 0.$$

The derivation of this result is simple and instructive. The random variable Θ has the uniform density function $f(\theta) = \frac{1}{2\pi}$ for $0 < \theta < 2\pi$. Applying the substitution rule gives

$$\mu_1 = E[\cos(\Theta)] = \int_0^{2\pi} \cos(\theta) f(\theta)\, d\theta = \frac{1}{2\pi} \int_0^{2\pi} \cos(\theta)\, d\theta = 0,$$

where the last equality uses the fact that $\cos(x + \pi) = -\cos(x)$ for $0 \leq x \leq \pi$. In the same way, $\mu_2 = 0$. Using the formula $\sigma^2(X) = E(X^2) - [E(X)]^2$ with $X = \cos(\Theta)$, we find

$$\sigma_1^2 = E\left[\cos^2(\Theta)\right] = \int_0^{2\pi} \cos^2(\theta) f(\theta)\, d\theta = \frac{1}{2\pi} \int_0^{2\pi} \cos^2(\theta)\, d\theta.$$

In the same way, $\sigma_2^2 = \frac{1}{2\pi} \int_0^{2\pi} \sin^2(\theta)\, d\theta$. Invoking the celebrated formula $\cos^2(\theta) + \sin^2(\theta) = 1$ from goniometry, we obtain $\sigma_1^2 + \sigma_2^2 = 1$. Hence, for reasons of symmetry, $\sigma_1^2 = \sigma_2^2 = \frac{1}{2}$. Finally,

$$\text{cov}(X_1, Y_1) = E\left[(\cos(\Theta) - 0)(\sin(\Theta) - 0)\right] = \frac{1}{2\pi} \int_0^{2\pi} \cos(\theta)\sin(\theta)\, d\theta.$$

This integral is equal to zero since $\cos(x + \frac{\pi}{2})\sin(x + \frac{\pi}{2}) = -\cos(x)\sin(x)$ for each of the ranges $0 \leq x \leq \frac{\pi}{2}$ and $\pi \leq x \leq \frac{3\pi}{2}$. This shows that $\rho = 0$, as was to be proved.

Next we can formulate two interesting results using the two-dimensional central limit theorem. The first result states that

$$P\left(S_{n1} \leq x, S_{n2} \leq y\right) \approx \frac{1}{\pi n} \int_{-\infty}^{x} \int_{-\infty}^{y} e^{-(t^2+u^2)/n}\, dt\, du,$$

for n large. In other words, the position of the particle after n steps has approximately the bivariate normal density function

$$\phi_n(x, y) = \frac{1}{\pi n} e^{-(x^2+y^2)/n}$$

when n is large. That is, the probability of finding the particle after n steps in a small rectangle with sides Δa and Δb around the point (a, b) is approximately equal to $\phi_n(a, b)\Delta a \Delta b$ for n large. In Figure 12.1, we display the bivariate normal density function $\phi_n(x, y)$ for $n = 25$. The correlation coefficient of the bivariate normal density $\phi_n(x, y)$ is zero. Hence, in accordance with our intuition, the coordinates of the position of the particle after many steps are practically independent of each other.

The second result states that

$$E(D_n) \approx \frac{1}{2}\sqrt{\pi n}$$

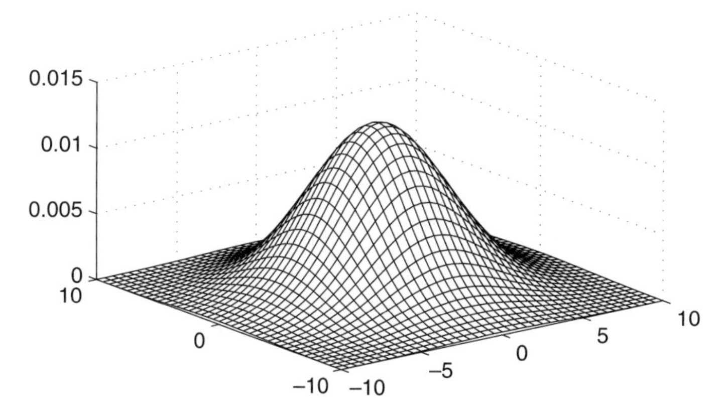

Figure 12.1 The density of the particle's position after 25 steps.

for n large, where the random variable D_n is defined by

$D_n =$ the distance from the origin to the position of the particle after n steps.

The proof of these results goes as follows. Rewrite $P(S_{n1} \leq x, S_{n2} \leq y)$ as

$$P\left(\frac{S_{n1} - n\mu_1}{\sigma_1\sqrt{n}} \leq \frac{x - n\mu_1}{\sigma_1\sqrt{n}}, \frac{S_{n2} - n\mu_2}{\sigma_2\sqrt{n}} \leq \frac{y - n\mu_2}{\sigma_2\sqrt{n}}\right).$$

Substituting the values of μ_1, μ_2, σ_1, σ_2, and ρ, it next follows from the two-dimensional central limit theorem that

$$P(S_{n1} \leq x, S_{n2} \leq y) \approx \frac{1}{2\pi}\int_{-\infty}^{x/\sqrt{n/2}}\int_{-\infty}^{y/\sqrt{n/2}} e^{-\frac{1}{2}(v^2+w^2)}\,dv\,dw$$

for n large. By the change of variables $t = v\sqrt{n/2}$ and $u = w\sqrt{n/2}$, the first result is obtained. To find the approximation formula for $E(D_n)$, note that

$$D_n = \sqrt{S_{n1}^2 + S_{n2}^2}.$$

An application of Rule 11.2 yields that

$$E(D_n) \approx \frac{1}{\pi n}\int_{-\infty}^{\infty}\int_{-\infty}^{\infty} \sqrt{x^2 + y^2}\, e^{-(x^2+y^2)/n}\,dx\,dy$$

for n large. To evaluate this integral, we use several results from advanced calculus. By a change to polar coordinates $x = r\cos(\theta)$ and $y = r\sin(\theta)$ with

$dx\,dy = r dr\,d\theta$ and using the identity $\cos^2(\theta) + \sin^2(\theta) = 1$, we find

$$\int_{-\infty}^{\infty}\int_{-\infty}^{\infty} \sqrt{x^2 + y^2}\, e^{-(x^2+y^2)/n}\, dx\, dy$$

$$= \int_{0}^{\infty}\int_{0}^{2\pi} \sqrt{r^2\cos^2(\theta) + r^2\sin^2(\theta)}\, e^{-(r^2\cos^2(\theta)+r^2\sin^2(\theta))/n}\, r\, dr\, d\theta$$

$$= \int_{0}^{\infty}\int_{0}^{2\pi} r^2 e^{-r^2/n}\, dr\, d\theta = 2\pi \int_{0}^{\infty} r^2 e^{-r^2/n}\, dr.$$

Obviously,

$$\int_{0}^{\infty} r^2 e^{-r^2/n}\, dr = \frac{1}{2}\int_{-\infty}^{\infty} r^2 e^{-r^2/n}\, dr$$

$$= \frac{\sqrt{n/2}\sqrt{2\pi}}{2} \frac{1}{\sqrt{n/2}\,\sqrt{2\pi}} \int_{-\infty}^{\infty} r^2 e^{\ r^2/n}\, dr$$

$$= \frac{\sqrt{n\pi}}{2}\frac{1}{2}n,$$

using the fact that an $N(0, n/2)$ distributed variable U satisfies $E(U^2) = \sigma^2(U) = \frac{1}{2}n$. Putting the pieces together, we get $E(D_n) \approx \frac{1}{2}\sqrt{\pi n}$. This is an excellent approximate for $n = 10$ onwards.

Problem 12.4 The rectangular distance from the origin to the position of the particle after n steps is defined by $R_n = |S_{n1}| + |S_{n2}|$. Verify that $E(R_n) \approx \sqrt{\frac{4n}{\pi}}$ for n large.

Problem 12.5 Two particles carry out a drunkard's walk on the two-dimensional plane, independently of each other. Both particles start at the origin (0, 0). One particle performs n steps and the other m steps. Can you give an intuitive explanation why the expected distance between the final positions of the two particles is equal to $\frac{1}{2}\sqrt{\pi}\sqrt{n+m}$?

12.1.2 Drunkard's walk in dimension three

When the drunkard's walk occurs in three-dimensional space, it can be shown that the joint probability density function of the (x, y, z)-coordinates of the position of the particle after n steps is approximately given by the trivariate normal probability density function

$$\frac{1}{(2\pi n/3)^{3/2}} e^{-\frac{3}{2}(x^2+y^2+z^2)/n}$$

for n large. The expected value of the distance between the origin and the position of the particle after n steps is approximately equal to $\sqrt{\frac{8n}{3\pi}}$ for n large. An application of this formula in physics can be found in Section 2.4.

12.2 Multivariate normal distribution

The multivariate normal distribution is a very useful probability model to describe dependencies between two or more random variables. In finance, the multivariate normal distribution is frequently used to model the joint distribution of the returns in a portfolio of assets.

First we give a general definition of the multivariate normal distribution.

Definition 12.1 *A d-dimensional random vector $(S_1, S_2, \ldots, S_d)$ is said to be multivariate normal distributed if for any d-tuple of real numbers $\alpha_1, \ldots, \alpha_d$ the one-dimensional random variable*

$$\alpha_1 S_1 + \alpha_2 S_2 + \cdots + \alpha_d S_d$$

has a (univariate) normal distribution.

In particular, each of the individual random variables $S_1, \ldots, S_d$ is normally distributed. Let us define the vector $\mu = (\mu_i), i = 1, \ldots, d$ of expected values and the matrix $\mathbf{C} = (\sigma_{ij}), i, j = 1, \ldots, d$ of covariances by

$$\mu_i = E(X_i) \qquad \text{and} \qquad \sigma_{ij} = \text{cov}(X_i, X_j).$$

Note that $\sigma_{ii} = \text{var}(X_i)$. The multivariate normal distribution is called *nonsingular* if the determinant of the covariance matrix $\mathbf{C}$ is nonzero; otherwise, the distribution is called singular. By a basic result from linear algebra, a singular covariance matrix $\mathbf{C}$ means that the probability mass of the multivariate normal distribution is concentrated on a subspace with a dimension lower than d. In applications, the covariance matrix of the multivariate normal distribution is often singular. In the example of the drunkard's walk on the two-dimensional plane, however, the approximate multivariate normal distribution of the position of the particle after n steps has the nonsingular covariance matrix

$$\begin{pmatrix} \frac{1}{2}n & 0 \\ 0 & \frac{1}{2}n \end{pmatrix}.$$

A very useful result for practical applications is the fact that the multivariate normal distribution is *uniquely determined* by the vector of expected values and the covariance matrix. The proof of this general result is rather complicated and is not given. Also it can be shown that each component of the random vector $(S_1, \ldots, S_d)$ can be expressed as a linear combination of d independent standard normal random variables. That is,

$$S_i = \mu_i + \sum_{j=1}^{d} a_{ij} Z_j \quad \text{for} \quad i = 1, \ldots, d,$$

where $Z_1, \ldots, Z_n$ are independent random variables each having the standard normal distribution. Readers who are familiar with linear algebra will recognize that the matrix $\mathbf{A} = (a_{ij})$ satisfies $\mathbf{C} = \mathbf{A}\mathbf{A}^T$ and can be computed from

$$\mathbf{A} = \mathbf{U}\mathbf{D}^{1/2}.$$

The diagonal matrix $\mathbf{D}^{1/2}$ has the square roots of the eigenvalues of the covariance matrix $\mathbf{C}$ on its diagonal (these eigenvalues are real and non-negative, since the covariance matrix $\mathbf{C}$ is symmetric and positive semi-definite). The orthogonal matrix $\mathbf{U}$ has the normalized eigenvectors of the matrix $\mathbf{C}$ as column vectors (Cholesky decomposition is a convenient method to compute the matrix $\mathbf{A}$ when $\mathbf{C}$ is nonsingular). The decomposition result for the vector $(S_1, \ldots, S_d)$ is particularly useful when simulating random observations from the multivariate normal distribution. In Section 11.3, we explained how to simulate from the one-dimensional standard normal distribution.

Remark 12.1 The result that the S_i are distributed as $\mu_i + \sum_{j=1}^{d} a_{ij} Z_j$ has a useful corollary. By taking the inproduct of the vector $(S_1 - \mu_1, \ldots, S_d - \mu_d)$ with itself, it is a matter of basic linear algebra to prove that $\sum_{j=1}^{d}(S_j - \mu_j)^2$ is distributed as $\sum_{j=1}^{d} \lambda_j {Z_j}^2$. This is a useful result for establishing the chi-square test in Section 12.4. If the eigenvalues λ_k of the covariance matrix $\mathbf{C}$ are 0 or 1, then the random variable $\sum_{j=1}^{d} \lambda_j Z_j^2$ has a chi-square distribution.

Remark 12.2 If the covariance matrix $\mathbf{C}$ of the multivariate normal distribution is nonsingular, it is possible to give an explicit expression for the corresponding multivariate probability density function. To do so, let us define the matrix $\mathbf{Q} = (q_{ij})$ by

$$q_{ij} = \frac{\sigma_{ij}}{\sigma_i \sigma_j} \quad \text{for} \quad i, j = 1, \ldots, d,$$

where σ_ℓ is a shorthand for $\sqrt{\sigma_{\ell\ell}}$. Denote by γ_{ij} the (i, j)th element of the inverse matrix $\mathbf{Q}^{-1}$ and let the polynomial $Q(x_1, \ldots, x_d)$ denote

$$Q(x_1, \ldots, x_d) = \sum_{i=1}^{d} \sum_{j=1}^{d} \gamma_{ij} x_i x_j.$$

Then the standardized vector $\left(\frac{S_1 - \mu_1}{\sigma_1}, \ldots, \frac{S_d - \mu_d}{\sigma_d}\right)$ has the standard multivariate normal probability density function

$$\frac{1}{(2\pi)^{d/2}\sqrt{\det(\mathbf{Q})}} e^{-\frac{1}{2}Q(x_1,\ldots,x_d)}.$$

This multidimensional density function reduces to the standard bivariate normal probability density function from Section 12.1 when $d = 2$.

12.3 Multidimensional central limit theorem

The central limit theorem is the queen of all theorems in probability theory. The one-dimensional version is extensively discussed in Chapter 5. The analysis of the drunkard's walk on the two-dimensional plane used the two-dimensional version. The multidimensional version of the central limit theorem is as follows. Suppose that

$$\mathbf{X}_1 = (X_{11}, \ldots, X_{1d}), \mathbf{X}_2 = (X_{21}, \ldots, X_{2d}), \ldots, \mathbf{X}_n = (X_{n1}, \ldots, X_{nd})$$

are independent random vectors of dimension d. The random vector $\mathbf{X}_k$ has the one-dimensional random variable X_{kj} as its jth component. The random vectors $\mathbf{X}_1, \ldots, \mathbf{X}_n$ are said to be independent if

$$P(\mathbf{X}_1 \in A_1, \ldots, \mathbf{X}_n \in A_n) = P(\mathbf{X}_1 \in A_1) \cdots P(\mathbf{X}_n \in A_n)$$

for any subsets $A_1, \ldots, A_n$ of the d-dimensional Euclidean space. Note that, for fixed k, the random variables $X_{k1}, \ldots, X_{kd}$ need not be independent. Also assume that $\mathbf{X}_1, \ldots, \mathbf{X}_n$ have the same individual distributions, that is, $P(\mathbf{X}_1 \in A) = \cdots = P(\mathbf{X}_n \in A)$ for any subset A of the d-dimensional space. Under this assumption, let

$$\mu_j = E(X_{1j}) \qquad \text{and} \qquad \sigma_{ij} = \operatorname{cov}(X_{1i}, X_{1j})$$

for $i, j = 1, \ldots, d$, assuming that the expectations exist. For $j = 1, \ldots, d$, we now define the random variable S_{nj} by

$$S_{nj} = X_{1j} + X_{2j} + \cdots + X_{nj}.$$

Multi-dimensional central limit theorem. *For n large, the random vector $\mathbf{S}_n = (S_{n1}, S_{n2}, \ldots, S_{nd})$ has approximately a multivariate normal distribution. The vector μ of expected values and the covariance matrix $\mathbf{C}$ are given by*

$$\mu = (n\mu_1, \ldots, n\mu_s) \qquad \textit{and} \qquad \mathbf{C} = (n\sigma_{ij})$$

when the random vectors $\mathbf{X}_k$ are identically distributed.

In the next section we discuss two applications of the multidimensional central limit theorem. In the first application, we will use the fact that the assumption of identically distributed random vectors $\mathbf{X}_k$ may be weakened in the multi-dimensional central limit theorem.

Problem 12.6 The annual rates of return on the three stocks A, B, and C have a trivariate normal distribution. The rate of return on stock A has expected value 7.5% and standard deviation 7%, the rate of return on stock B has expected value 10% and standard deviation 12%, and the rate of return on stock C has expected value 20% and standard deviation 18%. The correlation coefficient of the rates of return on stocks A and B is 0.7, the correlation coefficient is -0.5 for the stocks A and C, and the correlation coefficient is -0.3 for stocks B and C. An investor has \$100,000 in cash. Any cash that is not invested in the three stocks will be put in a riskless asset that offers an annual interest rate of 5%.

(a) Suppose the investor puts \$20,000 in stock A, \$20,000 in stock B, \$40,000 in stock C, and \$20,000 in the riskless asset. What are the expected value and the standard deviation of the portfolio's value next year?
(b) Can you find a portfolio whose risk is smaller than the risk of the portfolio from Question (a) but whose expected return is not less than that of the portfolio from Question (a)?
(c) For the investment plan from Question (a), use computer simulation to find the probability that the portfolio's value next year will be less than \$112,500 and the probability that the portfolio's value next year will be more than \$125,000.

Problem 12.7 The random vector (X_1, X_2, X_3) has a trivariate normal distribution. What is the joint distribution of X_1 and X_2?

12.3.1 Predicting election results

The multivariate normal distribution is also applicable to the problem of predicting election results. In Section 3.6, we discuss a polling method whereby a respondent is not asked to choose a favorite party, but instead is asked to indicate how likely the respondent is to vote for each party. Consider the situation in which there are three parties A, B, and C and n representative voters are interviewed. A probability distribution (p_{iA}, p_{iB}, p_{iC}) with $p_{iA} + p_{iB} + p_{iC} = 1$ describes the voting behavior of respondent i for $i = 1, \ldots, n$. That is, p_{iP} is the probability that respondent i will vote for party P on election day. Let the random variable S_{nA} be the number of respondents of the n interviewed voters who actually vote for party A on election day. The random variables S_{nB} and S_{nC} are defined in a similar manner. The vector $\mathbf{S}_n = (S_{nA}, S_{nB}, S_{nC})$ can be written as the sum of n random vectors $\mathbf{X}_1 = (X_{1A}, X_{1B}, X_{1C}), \ldots, \mathbf{X}_n = (X_{nA}, X_{nB}, X_{nC})$, where the random variable X_{iP} is defined by

$$X_{iP} = \begin{cases} 1 & \text{if respondent } i \text{ votes for party } P \\ 0 & \text{otherwise.} \end{cases}$$

The random vector $\mathbf{X}_i = (X_{iA}, X_{iB}, X_{iC})$ describes the voting behavior of respondent i. The simplifying assumption is made that the random vectors $\mathbf{X}_1, \ldots, \mathbf{X}_n$ are independent. These random vectors do not have the same individual distributions. However, under the crucial assumption of independence, the multidimensional central limit theorem can be shown to remain valid and thus the random vector (S_{nA}, S_{nB}, S_{nC}) has approximately a multivariate normal distribution for n large. This multivariate normal distribution has

$$\mu = \left(\sum_{i=1}^{n} p_{iA}, \sum_{i=1}^{n} p_{iB}, \sum_{i=1}^{n} p_{iC} \right)$$

as vector of expected values and

$$\mathbf{C} = \begin{pmatrix} \sum_{i=1}^{n} p_{iA}(1-p_{iA}) & -\sum_{i=1}^{n} p_{iA}p_{iB} & -\sum_{i=1}^{n} p_{iA}p_{iC} \\ -\sum_{i=1}^{n} p_{iA}p_{iB} & \sum_{i=1}^{n} p_{iB}(1-p_{iB}) & -\sum_{i=1}^{n} p_{iB}p_{iC} \\ -\sum_{i=1}^{n} p_{iA}p_{iC} & -\sum_{i=1}^{n} p_{iB}p_{iC} & \sum_{i=1}^{n} p_{iC}(1-p_{iC}) \end{pmatrix}$$

as covariance matrix (this matrix is singular, since for each row the sum of the elements is zero). The result for the vector μ of expected values is obvious, but a few words of explanation are in order for the covariance matrix $\mathbf{C}$. By the independence of $X_{1A}, \ldots, X_{nA}$,

$$\sigma^2(S_{nA}) = \sigma^2\left(\sum_{i=1}^{n} X_{iA}\right) = \sum_{i=1}^{n} \sigma^2(X_{iA}) = \sum_{i=1}^{n} p_{iA}(1-p_{iA}).$$

Also, noting that X_{iA} and X_{jB} are independent for $j \neq i$ and noting that X_{iA} and X_{iB} cannot be both positive, it follows that $\operatorname{cov}(S_{nA}, S_{nB})$ is given by

$$E\left[\left(\sum_{i=1}^{n} X_{iA}\right)\left(\sum_{j=1}^{n} X_{jB}\right)\right] - E\left(\sum_{i=1}^{n} X_{iA}\right) E\left(\sum_{j=1}^{n} X_{jB}\right)$$

$$= \sum_{i=1}^{n} \sum_{j \neq i} E(X_{iA}X_{jB}) - \left(\sum_{i=1}^{n} p_{iA}\right)\left(\sum_{j=1}^{n} p_{jB}\right)$$

$$= \sum_{i=1}^{n} \sum_{j \neq i} p_{iA}p_{jB} - \sum_{i=1}^{n} \sum_{j=1}^{n} p_{iA}p_{jB} = -\sum_{i=1}^{n} p_{iA}p_{iB}.$$

Similarly, the other terms in matrix $\mathbf{C}$ are explained.

It is standard fare in statistics to simulate random observations from the multivariate normal distribution. This means that computer simulation provides a fast and convenient tool to estimate probabilities of interest such as the probability that party A will receive the most votes or the probability that the two parties A and C will receive more than half of the votes.

12.3.2 Numerical illustration

Suppose that a representative group of $n = 1{,}000$ voters is polled. The probabilities assigned by each of the 1,000 voters to parties A, B, and C are summarized in Table 12.1: the vote of each of 230 persons will go to parties A, B, and C with probabilities 0.80, 0.20, and 0, the vote of each of 140 persons will go to parties A, B, and C with probabilities 0.65, 0.35, and 0, and so on. Each person votes independently. Let the random variable S_A be defined as

$$S_A = \text{the number of votes on party } A \text{ when the 1,000 voters actually vote on election day.}$$

Similarly, the random variables S_B and S_C are defined. How do we calculate probabilities such as the probability that party A will become the largest party and the probability that parties A and C together will get the majority of the votes? These probabilities are given by $P(S_A > S_B,\ S_A > S_C)$ and $P(S_A + S_C > S_B)$. Simulating from the trivariate normal approximation for the random vector (S_A, S_B, S_C) provides a simple and fast method to get approximate values for these probabilities. The random vector (S_A, S_B, S_C) has approximately a trivariate normal distribution with

$$\mu = (454, 485, 61)$$

Table 12.1 *Voting probabilities*

No. of voters	(p_{iA}, p_{iB}, p_{iC})
230	(0.20, 0.80, 0)
140	(0.65, 0.35, 0)
60	(0.70, 0.30, 0)
120	(0.45, 0.55, 0)
70	(0.90, 0.10, 0)
40	(0.95, 0, 0.05)
130	(0.60, 0.35, 0.05)
210	(0.20, 0.55, 0.25)

as vector of expected values and the matrix

$$\mathbf{C} = \begin{pmatrix} 183.95 & -167.65 & -16.30 \\ -167.65 & 198.80 & -31.15 \\ -16.30 & -31.15 & 47.45 \end{pmatrix}$$

as covariance matrix. In order to simulate random observations from this trivariate normal distribution, the eigenvalues λ_1, λ_2, λ_3 and the corresponding normalized eigenvectors $\mathbf{u}_1, \mathbf{u}_2, \mathbf{u}_3$ of matrix $\mathbf{C}$ must be first calculated. Using standard software, we find

$$\lambda_1 = 70.6016,\ \mathbf{u}_1 = \begin{pmatrix} 0.4393 \\ 0.3763 \\ -0.8157 \end{pmatrix}, \quad \lambda_2 = 359.5984,\ \mathbf{u}_2 = \begin{pmatrix} -0.6882 \\ 0.7246 \\ -0.0364 \end{pmatrix},$$

$$\lambda_3 = 0,\ \mathbf{u}_3 = \begin{pmatrix} 0.5774 \\ 0.5774 \\ 0.5774 \end{pmatrix}.$$

The diagonal matrix $\mathbf{D}^{1/2}$ has $\sqrt{\lambda_1}$, $\sqrt{\lambda_2}$, and $\sqrt{\lambda_3}$ on its diagonal and the orthogonal matrix $\mathbf{U}$ has $\mathbf{u}_1$, $\mathbf{u}_2$, and $\mathbf{u}_3$ as column vectors. The matrix product $\mathbf{U}\mathbf{D}^{1/2}$ gives the desired decomposition matrix

$$\mathbf{A} = \begin{pmatrix} 3.6916 & -13.0508 & 0 \\ 3.1622 & 13.7405 & 0 \\ -6.8538 & -0.6897 & 0 \end{pmatrix}.$$

Thus, the random vector (S_A, S_B, S_C) can approximately be represented as

$$\begin{aligned} S_A &= 454 + 3.6916Z_1 - 13.0508Z_2 \\ S_B &= 485 + 3.1622Z_1 + 13.7405Z_2 \\ S_C &= 61 - 6.8538Z_1 - 0.6897Z_2, \end{aligned}$$

where Z_1 and Z_2 are independent random variables each having the standard normal distribution. Note that the condition $S_A + S_B + S_C = 1{,}000$ is preserved in this decomposition. Using the decomposition, it is standard fare to simulate random observations from the trivariate normal approximation to (S_A, S_B, S_C). A simulation study with 100,000 random observations leads to the following estimates

$$P(\text{party A becomes the largest party}) = 0.123(\pm 0.002)$$

$$P(\text{party B becomes the largest party}) = 0.877(\pm 0.002)$$

$$P(\text{parties A and C get the majority of the votes}) = 0.855(\pm 0.002),$$

where the numbers between the parentheses indicate the 95% confidence intervals. How accurate is the model underlying these predictions? They are based on an approximatively multivariate normal distribution. To find out how good this approximative model works, we use the bootstrap method to simulate directly from the data in Table 12.1 (see Section 3.6 for more details on this powerful method). Performing 100,000 simulation runs, we obtain the values 0.120 ($\pm$ 0.002), 0.872 ($\pm$ 0.002) and 0.851 ($\pm$ 0.002) for the three probabilities above. The approximative values of the multivariate normal model are very close to the exact values of the bootstrap method. This justifies the use of this model which is computationally less demanding than the bootstrap method.

12.3.3 Lotto r/s

In the Lotto 6/45, six balls are drawn out of a drum with 45 balls numbered from 1 to 45. More generally, in the Lotto r/s, r balls are drawn from a drum with s balls. For the Lotto r/s, define the random variable S_{nj} by

$$S_{nj} = \text{the number of times ball number } j \text{ is drawn in } n \text{ drawings}$$

for $j = 1, \ldots, s$. Letting

$$X_{kj} = \begin{cases} 1 & \text{if ball number } j \text{ is drawn at the } k\text{th drawing} \\ 0 & \text{otherwise,} \end{cases}$$

we can represent S_{nj} in the form

$$S_{nj} = X_{1j} + X_{2j} + \cdots + X_{nj}.$$

Thus, by the multidimensional central limit theorem, the random vector $\mathbf{S}_n = (S_{n1}, \ldots, S_{ns})$ approximately has a multivariate normal distribution. The quantities $\mu_j = E(X_{1j})$ and $\sigma_{ij} = \text{cov}(X_{1i}, X_{1j})$ are given by

$$\mu_j = \frac{r}{s}, \ \sigma_{jj} = \frac{r}{s}\left(1 - \frac{r}{s}\right) \quad \text{and} \quad \sigma_{ij} = -\frac{r(s-r)}{s^2(s-1)} \quad \text{for} \quad i \neq j.$$

It is left to the reader to verify this with the help of

$$P(X_{1j} = 1) = \frac{r}{s}, \ P(X_{1j} = 0) = 1 - \frac{r}{s} \quad \text{for} \quad \text{all } j$$

and

$$P(X_{1i} = 1, X_{1j} = 1) = \frac{r}{s} \times \frac{r-1}{s-1} \quad \text{for} \quad \text{all } i, j \text{ with } i \neq j.$$

The covariance matrix $\mathbf{C} = (n\sigma_{ij})$ is singular. The reason is that the sum of the elements of each row is zero. The matrix $\mathbf{C}$ has rank $s - 1$.

An interesting random walk is the stochastic process that describes how the random variable $\max_{1 \leq j \leq s} S_{nj} - \min_{1 \leq j \leq s} S_{nj}$ behaves as a function of n. This random variable gives the difference between the number of occurrences of the most-drawn-ball number and that of the least-drawn-ball number in the first n drawings. Simulation experiments reveal that the sample paths of the random walk exhibit the tendency to increase proportionally to $\sqrt{n}$ as n gets larger. The central limit theorem is at work here. In particular, it can be proved that a constant c exists such that

$$E\left[\max_{1\leq j\leq s} S_{nj} - \min_{1\leq j\leq s} S_{nj}\right] \approx c\sqrt{n}$$

for n large. Using computer simulation, we find the value $c = 1.52$ for the Lotto 6/45 and the value $c = 1.48$ for the Lotto 6/49.

Problem 12.8 For the Lotto 6/45, simulate from the multivariate normal distribution in order to find approximately the probability

$$P\left(\max_{1\leq j\leq s} S_{nj} - \min_{1\leq j\leq s} S_{nj} > 1.5\sqrt{n}\right) \quad \text{for} \quad n = 100, 200, \text{and } 500.$$

12.4 The chi-square test

The chi-square (χ^2) test is tailored to measure how well an assumed distribution fits given data when the data are the result of independent repetitions of an experiment with a finite number of possible outcomes. Let's consider an experiment with d possible outcomes $j = 1, \ldots, d$, where the outcome j occurs with probability p_j for $j = 1, \ldots, d$. Suppose that n independent repetitions of the experiment are done. Define the random variable X_{kj} by

$$X_{kj} = \begin{cases} 1 & \text{if the outcome of the } k\text{th experiment is } j \\ 0 & \text{otherwise.} \end{cases}$$

Then, the random vectors $\mathbf{X}_1 = (X_{11}, \ldots, X_{1d}), \ldots, \mathbf{X}_\mathrm{n} = (X_{n1}, \ldots, X_{nd})$ are independent and identically distributed. Let the random variable N_j represent the number of times that the outcome j appears in the n repetitions of the experiment. That is,

$$N_j = X_{1j} + \cdots + X_{nj} \quad \text{for} \quad j = 1, \ldots, d.$$

A convenient measure of the distance between the random variables N_j and their expected values np_j is the weighted sum of squares

$$\sum_{j=1}^{d} w_j (N_j - np_j)^2$$

for appropriately chosen weights $w_1, \ldots, w_d$. How do we choose the constants w_j? Naturally, we want to make the distribution of the weighted sum of squares as simple as possible. This is achieved by choosing $w_j = (np_j)^{-1}$. For large n, the test statistic

$$D = \sum_{j=1}^{d} \frac{(N_j - np_j)^2}{np_j}$$

has approximately a chi-square distribution with $d - 1$ degrees of freedom (one degree of freedom is lost because of the linear relationship $\sum_{j=1}^{d} N_j = n$). We briefly outline the proof of this very useful result that goes back to Karl Pearson (1857–1936), one of the founders of modern statistics. Using the multi-dimensional central limit theorem, it can be shown that, for large n, the random vector

$$\left(\frac{N_1 - np_1}{\sqrt{np_1}}, \ldots, \frac{N_d - np_d}{\sqrt{np_d}} \right)$$

has approximately a multivariate normal distribution with the zero vector as its vector of expected values and the matrix $\mathbf{C} = \mathbf{I} - \sqrt{\mathbf{p}}\sqrt{\mathbf{p}}^T$ as its covariance matrix, where $\mathbf{I}$ is the identity matrix and the column vector $\sqrt{\mathbf{p}}$ has $\sqrt{p_j}$ as its jth component. Using the fact that $\sum_{j=1}^{d} p_j = 1$, the reader familiar with linear algebra may easily verify that one of the eigenvalues of the matrix $\mathbf{C}$ is equal to zero, and all other $d - 1$ eigenvalues are equal to 1. Thus, by appealing to a result stated in Remark 12.1 in Section 12.2, the random variable $\sum_{j=1}^{d} \left(\frac{N_j - np_j}{\sqrt{np_j}} \right)^2$ is approximately distributed as the sum of the squares of $d - 1$ independent $N(0, 1)$ random variables and thus has an approximate chi-square distribution with $d - 1$ degrees of freedom.

To get an idea as to how well the chi-square approximation performs, consider Question 9 from Chapter 1 again. The problem deals with an experiment having the six possible outcomes $1, \ldots, 6$ where the probabilities are hypothesized to be $\frac{1}{6}, \ldots, \frac{1}{6}$. In 1,200 rolls of a fair die the outcomes 1, 2, 3, 4, 5 and 6 occurred 196, 202, 199, 198, 202 and 203 times. In this case the test statistic D takes on the value

$$\frac{4^2 + 2^2 + 1^2 + 2^2 + 2^2 + 3^2}{200} = 0.19.$$

We immediately notice that the value 0.19 lies far below the expected value 5 of the χ_5^2-distribution. Since the frequencies of the outcomes are very close to the expected values, the suspicion is that the data are fabricated. Therefore, we determine the probability that the test statistic $D = \sum_{j=1}^{6} (N_j - 200)^2/200$ is smaller than or equal to 0.19, where N_j is the number of occurrences of outcome

j in 1,200 rolls of a fair die. The chi-square approximation of this probability is equal to $P(\chi_5^2 \leq 0.19) = 0.00078$. This approximation is very close to the simulated value $P(D \leq 0.19) = 0.00083$ obtained from four million simulation runs of 1,200 rolls of a fair die. The very small value of this probability indicates that the data are most likely fabricated. The finding that $P(\chi_5^2 \leq 0.19)$ is an excellent approximation to the exact value of $P(D \leq 0.19)$ confirms the widely used rule of thumb that the chi-square approximation can be applied when $np_j \geq 5$ for all j.

Problem 12.9 In a classical experiment, Gregor Mendel observed the shape and color of peas that resulted from certain crossbreedings. A sample of 556 peas was studied with the result that 315 produced round yellow, 108 produced round green, 101 produced wrinkled yellow, and 32 produced wrinkled green. According to Mendelian theory, the frequencies should be in the ratio 9:3:3:1. What do you conclude from a chi-square test?

Problem 12.10 Use a chi-square test to investigate the quality of the random-number generator on your computer. Generate 10,000 random numbers and count the frequencies of the numbers falling in each of the intervals $\left(\frac{i-1}{10}, \frac{i}{10}\right)$ for $i = 1, \ldots, 10$.

Problem 12.11 In the Dutch Lotto six so-called *main numbers* and one so-called *bonus number* are drawn from the numbers $1, \ldots, 45$ and in addition one color is drawn from six differing colors. For each ticket you are asked to mark six distinct numbers and one color. You win the jackpot (first prize) by matching the six main numbers and the color; the second prize by matching the six main numbers, but not the color; the third prize by matching five main numbers, the color, and the bonus number; the fourth prize by matching five main numbers and the bonus number, but not the color; the fifth prize by matching five main numbers and the color, but not the bonus number; and the sixth prize by matching only five main numbers. A total of 98,364,597 tickets filled in during a half-year period resulted in 2 winners of the jackpot, 6 winners of the second prize, 9 winners of the third prize, 35 winners of the fourth prize, 411 winners of the fifth prize, and 2,374 winners of the sixth prize. Use a chi-square test to find out whether or not the tickets were randomly filled in.

Problem 12.12 A study of D. Kadell and D. Ylvisaker entitled "Lotto play: the good, the fair and the truly awful," *Chance Magazine*, 1991, Vol. 4, pp. 22–25, analyses the behavior of players in the lotto. They took 111, 221, 666 tickets that were manually filled in for a specific draw of the California Lotto 6/53 and counted how many combinations were filled in exactly k times for $k = 0, 1, \ldots, 20$. Letting N_k denote the number of combinations filled in k

times, the following results were found:

k	N_k	k	N_k
0	288,590	11	217,903
1	1,213,688	12	126,952
2	2,579,112	13	77,409
3	3,702,310	14	50,098
4	4,052,043	15	33,699
5	3,622,666	16	23,779
6	2,768,134	17	17,483
7	1,876,056	18	13,146
8	1,161,423	19	10,158
9	677,368	20	7,969
10	384,186	>20	53,308

Use a chi-square test to find out whether the picks chosen by the players are random or not.

13
Conditional distributions

In Chapter 8, conditional probabilities are introduced by conditioning upon the occurrence of an event B of nonzero probability. In applications, this event B is often of the form $Y = b$ for a discrete random variable Y. However, when the random variable Y is continuous, the condition $Y = b$ has probability zero for any number b. The purpose of this chapter is to develop techniques for handling a condition provided by the observed value of a continuous random variable. We will see that the conditional probability density function of X given $Y = b$ for continuous random variables is analogous to the conditional probability mass function of X given $Y = b$ for discrete random variables. The conditional distribution of X given $Y = b$ enables us to define the natural concept of conditional expectation of X given $Y = b$. This concept allows for an intuitive understanding and is of utmost importance. In statistical applications, it is often more convenient to work with conditional expectations instead of the correlation coefficient when measuring the strength of the relationship between two dependent random variables. In applied probability problems, the computation of the expected value of a random variable X is often greatly simplified by conditioning on an appropriately chosen random variable Y. Learning the value of Y provides additional information about the random variable X and for that reason the computation of the conditional expectation of X given $Y = b$ is often simple.

13.1 Conditional probability densities

Suppose that the random variables X and Y are defined on the same sample space Ω with probability measure P. A basic question for dependent random variables X and Y is: if the observed value of Y is y, what distribution now describes the distribution of X? To answer this question, we first consider the case of discrete random variables X and Y with joint probability mass function $p(x, y) = P(X = x, Y = y)$. The *conditional probability mass function* of X given that $Y = b$ is denoted and defined by

$$P(X = x \mid Y = b) = \frac{P(X = x, Y = b)}{P(Y = b)}$$

for any fixed b with $P(Y = b) > 0$. This definition is just $P(A \mid B) = \frac{P(AB)}{P(B)}$ written in terms of random variables, where $A = \{\omega : X(\omega) = x\}$ and $B = \{\omega : Y(\omega) = b\}$ with ω denoting an element of the sample space. The notation $p_X(x \mid b)$ is often used for the conditional mass function $P(X = x \mid Y = b)$. Writing

$$P(X = a, Y = b) = P(X = a \mid Y = b)P(Y = b)$$

and using the fact that $\sum_b P(X = a, Y = b) = P(X = a)$, we have the useful relation

$$P(X = a) = \sum_b P(X = a \mid Y = b)P(Y = b),$$

in agreement with the law of conditional probabilities from Section 8.1.2.

Example 13.1 Let's return to Example 11.1. In this example, two fair dice are rolled. The random variable X denotes the smallest of the outcomes of the two rolls, and the random variable Y denotes the sum of the outcomes of the two rolls. The joint probability mass function $p(x, y) = P(X = x, Y = y)$ of X and Y is given in Table 11.1. The conditional mass functions follow directly from this table. For example, the conditional mass function $p_X(x \mid 7) = P(X = x \mid Y = 7)$ is given by

$$p_X(1 \mid 7) = \frac{2/36}{6/36} = \frac{1}{3},\ p_X(2 \mid 7) = \frac{2/36}{6/36} = \frac{1}{3},\ p_X(3 \mid 7) = \frac{2/36}{6/36} = \frac{1}{3},$$

$$p_X(x \mid 7) = 0 \quad \text{for } x = 4, 5, 6.$$

This conditional distribution is a discrete uniform distribution on $\{1, 2, 3\}$. We also give the conditional mass function $p_Y(y \mid 3) = P(Y = y \mid X = 3)$:

$$p_Y(6 \mid 3) = \frac{1/36}{7/36} = \frac{1}{7},\ p_Y(7 \mid 3) = p_Y(8 \mid 3) = p_Y(9 \mid 3) = \frac{2/36}{7/36} = \frac{2}{7}$$

$$p_Y(y \mid 3) = 0 \quad \text{for } y = 2, 3, 4, 5, 10, 11, 12.$$

What is the continuous analog of the conditional probability mass function when X and Y are continuous random variables with a joint probability density function $f(x, y)$? In this situation, we have the complication that $P(Y = y) = 0$ for each real number y. Nevertheless, this situation also allows for a natural definition of the concept of conditional distribution. Toward this end, we need the probabilistic interpretations of the joint density function $f(x, y)$ and the marginal densities $f_X(x)$ and $f_Y(y)$ of the random variables X and Y. For small values of $\Delta a > 0$ and $\Delta b > 0$,

$$P\left(a - \frac{1}{2}\Delta a \leq X \leq a + \frac{1}{2}\Delta a \mid b - \frac{1}{2}\Delta b \leq Y \leq b + \frac{1}{2}\Delta b\right)$$

$$= \frac{P\left(a - \frac{1}{2}\Delta a \leq X \leq a + \frac{1}{2}\Delta a, b - \frac{1}{2}\Delta b \leq Y \leq b + \frac{1}{2}\Delta b\right)}{P(b - \frac{1}{2}\Delta b \leq Y \leq b + \frac{1}{2}\Delta b)}$$

$$\approx \frac{f(a, b)\Delta a \Delta b}{f_Y(b)\Delta b} = \frac{f(a, b)}{f_Y(b)}\Delta a$$

provided that (a, b) is a continuity point of $f(x, y)$ and $f_Y(b) > 0$. This leads to the following definition.

Definition 13.1 *If X and Y are continuous random variables with joint probability density function $f(x, y)$ and $f_Y(y)$ is the marginal density function of Y, then the conditional probability density function of X given that $Y = b$ is defined by*

$$f_X(x \mid b) = \frac{f(x, b)}{f_Y(b)}, \qquad -\infty < x < \infty$$

for any fixed b with $f_Y(b) > 0$.

A probabilistic interpretation can be given to $f_X(a \mid b)$: given that the observed value of Y is b, the probability of the other random variable X taking on a value in a small interval of length Δa around point a is approximately equal to $f_X(a \mid b)\Delta a$. The conditional probability that the random variable X takes on a value smaller than or equal to x is denoted by $P(X \leq x \mid Y = b)$ and is defined by

$$P(X \leq x \mid Y = b) = \int_{-\infty}^{x} f_X(u \mid b)\, du.$$

Before discussing implications of this definition, we illustrate the concept of conditional probability density function with two examples.

Example 13.2 A point (X, Y) is chosen at random inside the unit circle. In Example 11.4, we determined the joint density function $f(x, y)$ of X and Y together with the marginal density function $f_Y(y)$ of Y. This gives for any fixed

b with $-1 < b < 1$,

$$f_X(x \mid b) = \begin{cases} \frac{1}{2\sqrt{1-b^2}} & \text{for} \quad -\sqrt{1-b^2} < x < \sqrt{1-b^2} \\ 0 & \text{otherwise.} \end{cases}$$

In other words, the conditional distribution of X given that $Y = b$ is the uniform distribution on the interval $(-\sqrt{1-b^2}, \sqrt{1-b^2})$. The same distribution as that of the x-coordinate of a randomly chosen point of the horizontal chord through the point $(0, b)$. This chord has length $2\sqrt{1-b^2}$, by Pythagoras.

Example 13.3 Suppose that the random variables X and Y have a bivariate normal distribution with parameters $(\mu_1, \mu_2, \sigma_1^2, \sigma_2^2, \rho)$. The joint density function $f(x,y)$ is specified in Section 12.1. Also in this section, we find that the marginal probability densities $f_X(x)$ and $f_Y(y)$ of X and Y are given by the $N(\mu_1, \sigma_1^2)$ density and the $N(\mu_2, \sigma_2^2)$ density. Substituting the expressions for these densities in the formulas for the conditional densities, we find after simple algebra that the conditional probability density $f_X(x \mid b)$ of X given that $Y = b$ is the

$$N\left[\mu_1 + \rho\frac{\sigma_1}{\sigma_2}(b - \mu_2), \sigma_1^2(1 - \rho^2)\right]$$

density and the conditional probability density $f_Y(y \mid a)$ of Y given that $X = a$ is the

$$N\left[\mu_2 + \rho\frac{\sigma_2}{\sigma_1}(a - \mu_1), \sigma_2^2(1 - \rho^2)\right]$$

density. Thus, the expected values of the conditional densities are linear functions of the conditional variable, and the conditional variances are constants.

The relation $f_X(x \mid y) = \frac{f(x,y)}{f_Y(y)}$ can be written in the more insightful form

$$f(x, y) = f_X(x \mid y) f_Y(y),$$

in analogy with $P(AB) = P(A \mid B)P(B)$. This representation of $f(x,y)$ may be helpful in simulating a random observation from the joint probability distribution of X and Y. First, a random observation for Y is generated from the density function $f_Y(y)$. If the value of this observation is y, a random observation for X is generated from the conditional density function $f_X(x \mid y)$. For example the results of Examples 11.4 and 13.2 show that a random point (X, Y) in the unit circle can be simulated by generating first a random observation Y from the density function $\frac{2}{\pi}\sqrt{1 - y^2}$ on $(-1, 1)$ and next a random observation

X from the uniform density on $(-\sqrt{1-y^2}, \sqrt{1-y^2})$. A generally applicable method to generate a random observation from any given univariate density function is the acceptance-rejection method. This method is discussed in any textbook on simulation.

Example 13.4 A very tasty looking toadstool growing in the forest is nevertheless so poisonous that it is fatal to squirrels that consume more than half of it. Squirrel 1, however, does partake of it, and later on squirrel 2 does the same. What is the probability that both squirrels survive? Assume that the first squirrel consumes a uniformly distributed amount of the toadstool, the second squirrel a uniformly distributed amount of the remaining part of the toadstool.

To answer the question, let the random variable X represent the proportion of the toadstool consumed by squirrel 1 and let Y be the proportion of the toadstool consumed by squirrel 2. Using the uniformity assumption, it follows that $f_X(x) = 1$ for all $0 < x < 1$ and $f_Y(y \mid x) = \frac{1}{1-x}$ for $0 < y < 1 - x$. Applying the representation $f(x, y) = f_X(x) f_Y(y \mid x)$ leads to

$$f(x, y) = \frac{1}{1-x} \quad \text{for} \quad \text{all } x, y > 0 \qquad \text{with} \qquad x + y < 1.$$

The probability of both squirrels surviving is equal to

$$P\left(X \leq \frac{1}{2}, Y \leq \frac{1}{2}\right) = \int_0^{\frac{1}{2}} \int_0^{\frac{1}{2}} f(x,y)\, dx\, dy = \int_0^{\frac{1}{2}} \frac{dx}{1-x} \int_0^{\frac{1}{2}} dy$$

$$= \frac{1}{2} \int_{\frac{1}{2}}^{1} \frac{du}{u} = \frac{1}{2} \ln(2) = 0.3466.$$

13.2 Law of conditional probabilities

For discrete random variables X and Y, the unconditional probability $P(X = a)$ can be calculated from

$$P(X = a) = \sum_b P(X = a \mid Y = b) P(Y = b).$$

This *law of conditional probabilities* is a special case of Rule 8.1 in Chapter 8. In the situation of continuous random variables X and Y, the continuous analog of the law of conditional probabilities is:

Rule 13.1 *If the random variables X and Y are continuously distributed with a joint density function $f(x, y)$ and $f_Y(y)$ is the marginal density function of*

Y, then

$$P(X \leq a) = \int_{-\infty}^{\infty} P(X \leq a \mid Y = y) f_Y(y)\, dy.$$

This statement is a direct consequence of the definition of the conditional probability $P(X \leq a \mid Y = y) = \int_{-\infty}^{a} f_X(x \mid y)\, dx$. Thus,

$$\int_{-\infty}^{\infty} P(X \leq a \mid Y = y) f_Y(y)\, dy = \int_{-\infty}^{\infty} \left[\int_{-\infty}^{a} \frac{f(x, y)}{f_Y(y)}\, dx \right] f_Y(y)\, dy$$

$$= \int_{-\infty}^{\infty} \left[\int_{-\infty}^{a} f(x, y)\, dx \right] dy = \int_{-\infty}^{a} \left[\int_{-\infty}^{\infty} f(x, y)\, dy \right] dx$$

$$= \int_{-\infty}^{a} f_X(x)\, dx = P(X \leq a).$$

The importance of the continuous analog of the law of conditional probabilities can be hardly overestimated. In applications, the conditional probability $P(X \leq a \mid Y = y)$ is often calculated without explicitly using the joint distribution of X and Y, but through a direct physical interpretation of the conditional probability in the context of the concrete application. To illustrate this, let's return to Example 13.4 and calculate the probability that squirrel 2 will survive. This probability can be obtained as

$$P\left(Y \leq \frac{1}{2}\right) = \int_{0}^{1} P\left(Y \leq \frac{1}{2} \,\middle|\, X = x\right) f_X(x)\, dx =$$

$$= \int_{0}^{\frac{1}{2}} \frac{0.5}{1 - x}\, dx + \int_{\frac{1}{2}}^{1} 1\, dx = \frac{1}{2} \ln(2) + 0.5 = 0.8466.$$

In the following example, the law of conditional probabilities is used for the situation of a discrete random variable X and a continuous random variable Y. A precise definition of $P(X \leq a \mid Y = y)$ for this situation requires some technical machinery and will not be given. However, in the context of the concrete problem, it is immediately obvious what is meant with the conditional probability.

Example 13.5 Every morning at exactly the same time, Mr. Johnson rides the metro to work. He waits for the metro at the same place in the metro station. Every time the metro arrives at the station, the exact spot where it comes to a stop is a surprise. From experience, Mr. Johnson knows that the distance between him and the nearest metro door once the metro has stopped is uniformly distributed between 0 and 2 meters. Mr. Johnson is able to find a place to sit with probability $1 - \sqrt{\frac{1}{2} y}$ if the nearest door is y meters from where he is standing.

On any given morning, what is the probability that Mr. Johnson will succeed in finding a place to sit down?

The probability is not $1 - \sqrt{\frac{1}{2} \times 1} = 0.293$ as some people believe (they substitute the expected value of the distance to the nearest door for y into the formula $1 - \sqrt{\frac{1}{2}y}$). The correct value can be obtained as follows. Define the random variable X as equal to 1 when Mr. Johnson finds a seat and 0 otherwise. Now define the random variable Y as the distance from Mr. Johnson's waiting place to the nearest metro door. It will be intuitively clear that

$$P(X = 1 \mid Y = y) = 1 - \sqrt{\frac{1}{2}y}.$$

The random variable Y has the probability density function $f_Y(y) = \frac{1}{2}$ for $0 < y < 2$. Hence, the unconditional probability that Mr. Johnson will succeed in finding a place to sit down on any given morning is equal to

$$P(X = 1) = \int_0^2 \left(1 - \sqrt{\frac{1}{2}y}\right) \frac{1}{2}\, dy = \frac{1}{3}.$$

Problem 13.1 The length of time required to unload a ship has an $N(\mu, \sigma^2)$ distribution. The crane to unload the ship has just been overhauled and the time it will operate until the next breakdown has an exponential distribution with an expected value of $1/\lambda$. What is the probability of no breakdown during the unloading of the ship?

Problem 13.2 Customers arrive at a service facility according to a Poisson process with an average of λ customers per unit time. The service facility can handle only one customer at a time. The service time of any customer has an exponential density with parameter μ. A newly arriving customer finds no other customers present upon arrival. What is the probability distribution of the number of customers present upon service completion of this customer?

Problem 13.3 Use twice the law of conditional probabilities to calculate the probability that the quadratic equation $Ax^2 + Bx + C$ will have two real roots when A, B, and C are independent samples from the uniform distribution on $(0, 1)$.

Problem 13.4 You leave work at random times between 5.45 pm and 6.00 pm to take the bus home. Bus numbers 1 and 3 bring you home. You take the first bus that arrives. Bus number 1 arrives exactly every 15 minutes starting from the hour, whereas bus number 3 arrives according to a Poisson process with the same average frequency as bus number 1. What is the probability that you take bus number 1 home on any given day? Use the law of conditional probabilities to verify that this probability is equal to $1 - \frac{1}{e}$. Can you give an intuitive explanation why the probability is larger than $\frac{1}{2}$?

13.3 Law of conditional expectations

In the case that the random variables X and Y have a discrete distribution, the *conditional expectation* of X given that $Y = b$ is defined by

$$E(X \mid Y = b) = \sum_x x P(X = x \mid Y = b)$$

for each b with $P(Y = b) > 0$ (assuming that the sum is well-defined). In the case that X and Y are continuously distributed with joint probability density function $f(x, y)$, the *conditional expectation* of X given that $Y = b$ is defined by

$$E(X \mid Y = b) = \int_{-\infty}^{\infty} x f_X(x \mid b)\, dx$$

for each b with $f_Y(b) > 0$ (assuming that the integral is well-defined).

Just as the law of conditional probabilities directly follows from the definition of the conditional distribution of X given that $Y = y$, the law of conditional expectations is a direct consequence of the definition of $E(X \mid Y = y)$. In the discrete case the *law of conditional expectations* reads as

$$E(X) = \sum_y E(X \mid Y = y) P(Y = y),$$

while for the continuous case the law reads as

$$E(X) = \int_{-\infty}^{\infty} E(X \mid Y = y) f_Y(y)\, dy.$$

In words, the law of conditional expectations expresses that the expected value of what would be the expected value of X after observing another random variable Y gives the unconditional expected value of X.

Remark 13.1 For two dependent random variables X and Y, let $m(x) = E(Y \mid X = x)$. The curve of the function $y = m(x)$ is called the *regression curve* of Y on X. It is a better measure for the dependence between X and Y than the correlation coefficient (recall that dependence does not necessarily imply a nonzero correlation coefficient). In statistical applications, it is often the case that we can observe the random variable X but we want to know the dependent random variable Y. The function value $y = m(x)$ can be used as a prediction of the value of the random variable Y given the observation x of the random variable X. The function $m(x) = E(Y \mid X = x)$ is an optimal prediction function in the sense that this function minimizes

$$E[(Y - g(X))^2]$$

over all functions $g(x)$. We only sketch the proof of this result. For any random variable U, the minimum of $E[(U - c)^2]$ over all constants c is achieved for the constant $c = E(U)$, as follows by differentiating $E[(U - c)^2] = E(U^2) - 2cE(U) + c^2$ with respect to c. Using the law of conditional expectations, $E[(Y - g(X))^2]$ can be expressed as

$$E[(Y - g(X))^2] = \int_{-\infty}^{\infty} E[(Y - g(X))^2 \mid X = x] f_X(x) dx.$$

For every x, the inner side of the integral is minimized by $g(x) = E(Y \mid X = x)$, yielding that $m(X)$ is the minimum mean squared error predictor of Y from X. By the law of conditional expectation, the statistic $m(X)$ has the same expected value as Y. But the predictor $m(X)$ has the nice feature that its variance is usually smaller than var(Y) itself. An intuitive explanation of this fact is that the conditional distributions of Y given the values of X involve more information than the distribution of Y alone. For the case that X and Y have a bivariate normal distribution, it follows from the results in Example 13.3 that the optimal prediction function $m(x)$ coincides with the best linear prediction function discussed in Section 11.4. The best linear prediction function uses only the expected values, the variances, and the correlation coefficient of the random variables X and Y.

In applied probability problems, the law of conditional expectations is a very useful result to calculate unconditional expectations. As illustration, we give two examples. The first example concerns the calculation of the expected value of a mixed random variable.

Example 13.6 Someone purchases a liability insurance policy. The probability that a claim will be made on the policy is 0.1. In case of a claim, the size of the claim has an exponential distribution with an expected value of \$1,000,000. The maximum insurance policy payout is \$2,000,000. What is the expected value of the insurance payout? The insurance payout is a mixed random variable: it takes on one of the discrete values 0 and 2×10^6 or a value in the continuous interval $(0, 2 \times 10^6)$. Its expected value is calculated through a two-stage process. First, we condition on the outcome of the random variable I, where $I = 0$ if no claim is made and $I = 1$ otherwise. The insurance payout is 0 if I takes on the value 0, and otherwise the insurance payout is distributed as $\max(2 \times 10^6, D)$, where the random variable D has an exponential distribution with parameter

$\lambda = 1/10^6$. Thus, by conditioning,

$$E(\text{insurance payout}) = 0.9 \times 0 + 0.1 \times E[\max(2 \times 10^6, D)].$$

Using the substitution rule, it follows that

$$E[\max(2 \times 10^6, D)] = \int_0^\infty \max(2 \times 10^6, x)\, \lambda e^{-\lambda x}\, dx$$
$$= \int_0^{2\times 10^6} x \lambda e^{-\lambda x}\, dx + \int_{2\times 10^6}^\infty (2 \times 10^6)\, \lambda e^{-\lambda x}\, dx.$$

It is left to the reader to verify the calculations leading to

$$E[\max(2 \times 10^6, D)] = 10^6(1 - e^{-2}) = 864{,}665 \text{ dollars}.$$

Hence, we can conclude that $E(\text{insurance payout}) = \$86{,}466.50$.

Example 13.7 Suppose that a ship transporting grain is being unloaded in Gotham city harbor. The amount of time required to unload the ship is uniformly distributed between a and b hours as long as the ship is unloaded without interruption. The crane used to unload the grain, however, is unreliable and apt to breakdown. Breakdowns of the crane occur according to a Poisson process with an average of λ breakdowns per hour. Each time the crane breaks down, a fixed time of d hours is required to fix it. What is the expected time needed to complete the unloading of the ship?

Let the random variable T denote the time elapsed between the start and the completion of the unloading of the ship. The hour is taken as time unit. Denote by the continuous random variable U the pure unloading time of the ship (not including delays due to breakdowns). The random variable U has the uniform probability density function $f(u) = 1/(b - a)$ for $a < u < b$. We first calculate the conditional expected value of T given that $U = u$. Once $E(T \mid U = u)$ has been calculated, the expected value $E(T)$ follows from

$$E(T) = \int_a^b E(T \mid U = u) f(u)\, du.$$

In order to calculate $E(T \mid U = u)$, let the discrete random variable N_u denote the number of interruptions during the pure unloading time U given that $U = u$. The random variable N_u has a Poisson distribution with expected value λu (see

Section 4.2.4). By conditioning on N_u, we find

$$E(T \mid U = u) = \sum_{n=0}^{\infty} E(T \mid U = u, N_u = n) P(N_u = n)$$

$$= \sum_{n=0}^{\infty} (u + nd) P(N_u = n)$$

$$= u \sum_{n=0}^{\infty} P(N_u = n) + d \sum_{n=0}^{\infty} n P(N_u = n),$$

yielding $E(T \mid U = u) = u + d\lambda u$. Hence,

$$E(T) = \int_a^b (u + d\lambda u) f(u)\, du = (1 + \lambda d) \int_a^b u f(u)\, du$$

$$= (1 + \lambda d) E(U).$$

Since $E(U) = (a + b)/2$, the expected value of the time until completion of the unloading of the ship is equal to $(1 + \lambda d)(a + b)/2$.

Problem 13.5 Consider the casino game Red Dog from Problem 3.27 again. Suppose that the initial stake of the player is \$10. What are the expected values of the total amount staked and the payout in any given play? Use the law of conditional expectations to find these expected values.

Problem 13.6 A replenishment order is placed to raise the stock level of a given product. The current stock level is s units. The lead time of the replenishment order is a continuous random variable having an exponential distribution with a mean of $1/\mu$ days. Customer demand for the product occurs according to a Poisson process with an average demand of λ units per day. Each customer asks for one unit of the product. What is the probability of a shortage occurring during the replenishment lead time and what is the expected value of the total shortage?

Problem 13.7 A fair coin is tossed no more than n times, where n is fixed in advance. You stop the coin-toss experiment as soon as the proportion of heads exceeds $\frac{1}{2}$ or as soon as n tosses are done, whichever occurs first. Use the law of conditional expectations to calculate, for $n = 5, 10, 25,$ and 50, the expected value of the proportion of heads at the moment the coin-toss experiment is stopped. *Hint*: define the random variable $X_k(i)$ as the proportion of heads upon stopping given that k tosses are still possible and heads turned up i times so far. Set up a recursion equation for $E[X_k(i)]$.

Problem 13.8 Let's return to Problem 13.4. Use the law of conditional expectations to verify that the expected value of your waiting time until the next bus arrival is equal to $\frac{15}{e}$.

Problem 13.9 In a television game show, you may choose to roll any number of dice. Your score is zero if at least one of the dice comes up with the face value one, and otherwise your score is the sum of the face values shown by the dice. Use the law of conditional expectations to calculate the expected value and the standard deviation of your score if you choose to roll five dice. How do you calculate the probability of getting a score of 20 or more points?

Problem 13.10 What is the expected value of the number of draws needed to obtain each of the numbers $1, \ldots, 45$ at least once in the Lotto $6/45$? *Hint*: use the same idea as in Problem 13.7 to set up a recursion.

14
Generating functions

Generating functions were introduced by the Swiss genius Leonhard Euler (1707–1783) in the eighteenth century to facilitate calculations in counting problems. However, this important concept is also extremely useful in applied probability, as was first demonstrated by the work of Abraham de Moivre (1667–1754) who discovered the technique of generating functions independently of Euler. In modern probability theory, generating functions are an indispensable tool in combination with methods from numerical analysis. The purpose of this chapter is to give the basic properties of generating functions and to show the utility of this concept. First, the generating function is defined for a discrete random variable on nonnegative integers. Next, we consider the more general moment-generating function, which is defined for any random variable. The (moment) generating function is a powerful tool for both theoretical and computational purposes. In particular, it can be used to prove the central limit theorem. A sketch of the proof will be given.

14.1 Generating functions

We first introduce the concept of generating function for a discrete random variable X whose possible values belong the set of nonnegative integers.

Definition 14.1 If X is a nonnegative, integer-valued random variable, then the generating function of X is defined by

$$G_X(z) = \sum_{k=0}^{\infty} z^k P(X = k), \qquad |z| \leq 1.$$

The power series $G_X(z)$ is absolutely convergent for any $|z| \leq 1$ (why?). For any z, we can interpret $G_X(z)$ as

$$G_X(z) = E\left(z^X\right),$$

as follows by applying Rule 9.2. The probability mass function of X is uniquely determined by the generating function of X. To see this, use the fact that the derivative of an infinite series is obtained by differentiating the series term by term. Thus,

$$\frac{d^r}{dz^r} G_X(z) = \sum_{k=r}^{\infty} k(k-1)\cdots(k-r+1) z^{k-r} P(X=k), \qquad r = 1, 2, \ldots.$$

In particular, by taking $z = 0$,

$$P(X = r) = \frac{1}{r!} \frac{d^r}{dz^r} G_X(z)|_{z=0}, \qquad r = 1, 2, \ldots .$$

This proves that the generating function uniquely determines the probability mass function. This basic result explains the importance of the generating function. In many applications, it is relatively easy to obtain the generating function of a random variable X even when the probability mass function is not explicitly given. An example will be given below. Once we know the generating function of a random variable X, it is a simple matter to obtain the factorial moments of the random variable X. The rth factorial moment of the random variable X is defined by $E[X(X-1)\cdots(X-r+1)]$ for $r = 1, 2, \ldots$. In particular, the first factorial moment of X is the expected value of X. The variance of X is determined by the first and the second factorial moment of X. Putting $z = 1$ in the above expression for the rth derivative of $G_X(z)$, we obtain

$$E\left[X(X-1)\cdots(X-r+1)\right] = \frac{d^r}{dz^r} G_X(z)|_{z=1}, \qquad r = 1, 2, \ldots .$$

In particular,

$$E(X) = G'_X(1) \qquad \text{and} \qquad E\left(X^2\right) = G''_X(1) + G'_X(1).$$

Example 14.1 Suppose that the random variable X has a Poisson distribution with expected value λ. Then,

$$\sum_{k=0}^{\infty} z^k e^{-\lambda} \frac{\lambda^k}{k!} = e^{-\lambda} \sum_{k=0}^{\infty} \frac{(\lambda z)^k}{k!} = e^{-\lambda} e^{\lambda z},$$

using the series expansion $e^x = \sum_{n=0}^{\infty} x^n/n!$. Hence,

$$G_X(z) = e^{-\lambda(1-z)}, \qquad |z| \leq 1.$$

Differentiating $G_X(z)$ gives $G'_X(1) = \lambda$ and $G''_X(1) = \lambda^2$. Hence, $E(X) = \lambda$ and $E(X^2) = \lambda^2 + \lambda$. This implies that both the expected value and the variance of a Poisson distributed random variable with parameter λ are given by λ, in agreement with earlier results in Example 9.2.

14.1.1 Convolution rule

The importance of the concept of generating function comes up especially when calculating the probability mass function of a sum of independent random variables that are nonnegative and integer-valued.

Rule 14.1 *Let X and Y be two nonnegative, integer-valued random variables. If the random variables X and Y are independent, then*

$$G_{X+Y}(z) = G_X(z)G_Y(z), \qquad |z| \leq 1.$$

Rule 14.1 is known as the *convolution rule* for generating functions and can be directly extended to the case of a finite sum of independent random variables. The proof is simple. If X and Y are independent, then the random variables $U = z^X$ and $V = z^Y$ are independent for any fixed z (see Rule 9.3). Also, by Rule 9.5, $E(UV) = E(U)E(V)$ for independent U and V. Thus,

$$E\left(z^{X+Y}\right) = E\left(z^X z^Y\right) = E\left(z^X\right)E\left(z^Y\right),$$

proving that $G_{X+Y}(z) = G_X(z)G_Y(z)$. The converse of the statement in Rule 14.1 is, in general, not true. The random variables X and Y are not necessarily independent if $G_{X+Y}(z) = G_X(z)G_Y(z)$. It is left to the reader to verify that a counterexample is provided by the random vector (X, Y) that takes on the values (1, 1), (2, 2) and (3, 3) each with probability $\frac{1}{9}$ and the values (1, 2), (2, 3) and (3, 1) each with probability $\frac{2}{9}$. This counterexample was communicated to me by Fred Steutel.

Example 14.2 Suppose that X and Y are independent random variables that are Poisson distributed with respective parameters λ and μ. What is the probability mass function of $X + Y$?

Using the result from Example 14.1, we find

$$G_{X+Y}(z) = e^{-\lambda(1-z)}e^{-\mu(1-z)} = e^{-(\lambda+\mu)(1-z)}, \qquad |z| \leq 1.$$

Since a Poisson-distributed random variable with parameter $\lambda + \mu$ has the generating function $e^{-(\lambda+\mu)(1-z)}$ and the generating function $G_{X+Y}(z)$ uniquely determines the probability mass function of $X + Y$, it follows that $X + Y$ has a Poisson distribution with parameter $\lambda + \mu$.

Problem 14.1 Suppose that the random variable X has a binomial distribution with parameters n and p. Use the fact that X can be represented as the sum of n independent Bernoulli variables to derive the generating function of X.

Problem 14.2 Independent trials of a Bernouilli experiment with success probability p are done until a success occurs for the rth time. Determine the generating function of the probability distribution of the number of trials required. *Remark*: this generating function is the generating function of the so-called *negative binomial distribution*.

Problem 14.3 Suppose that you draw a number at random from the unit interval r times. A draw is called a "record draw" when the resulting number is larger than the previously drawn numbers. Determine the generating function of the number of record draws. What are the expected value and variance of the number of record draws?

In many applications, it is possible to derive an explicit expression for the generating function of a random variable X whose probability mass function is not readily available and has a complicated form. Is this explicit expression for the generating function of practical use apart from calculating the moments of X? The answer is yes! If an explicit expression for the generating function of the random variable X is available, then the numerical values of the (unknown) probability mass function of X can be calculated by appealing to the discrete Fast Fourier Transform method from numerical analysis (this algorithm functions in the seemingly mystical realm of complex numbers, which world nonetheless is of great real-world significance). An explanation of how this extremely powerful method works is beyond the scope of this book.

Example 14.3 In the coupon collector's problem from Section 3.2, we calculated the expected value of the random variable X representing the number of bags of chips that must be purchased in order to get a complete set of n distinct flippos. How do we calculate the probability distribution of the random variable X? This can be done with the help of the generating function of X. The random variable X can be written as

$$X = Y_1 + Y_2 + \cdots + Y_n,$$

where the random variable Y_i denotes the number of bags of chips in order to go from $i - 1$ to i different flippos. The random variables $Y_1, \ldots, Y_n$ are independent and the random variable Y_i has a geometric distribution with parameter $p_i = 1 - \frac{(i-1)}{n}$ for $i = 1, \ldots, n$. A random variable Y is said to have a geometric distribution with parameter p if $P(Y = k) = (1 - p)^{k-1} p$ for $k = 1, 2, \ldots$. The random variable Y satisfies

$$\sum_{k=0}^{\infty} P(Y = k) z^k = \sum_{k=1}^{\infty} (1 - p)^{k-1} p z^k = pz \sum_{k=1}^{\infty} \Big((1 - p) z \Big)^{k-1}$$

$$= pz \sum_{j=0}^{\infty} \Big((1 - p) z \Big)^{j} = \frac{pz}{1 - (1 - p) z},$$

using the fact that the geometric series $\sum_{j=0}^{\infty} x^j$ equals $\frac{1}{1-x}$ for $|x| < 1$ (see the Appendix). Since $G_X(z) = G_{Y_1}(z) G_{Y_2}(z) \cdots G_{Y_n}(z)$ by the independence of $Y_1, \ldots, Y_n$, it follows that

$$G_X(z) = \frac{p_1 p_2 \cdots p_n z^n}{(1 - z + p_1 z)(1 - z + p_2 z) \cdots (1 - z + p_n z)}.$$

The coupon collector's problem with $n = 365$ flippos enables us to calculate how many persons are needed to have a group of persons in which all 365 possible birthdays (excluding February 29) are represented with a probability of at least 50%. Using the discrete Fast Fourier Transform method we can calculate that the group should consist of 2,287 randomly picked persons.

14.1.2 Branching processes and generating functions

The family name is inherited by sons only. Take a father who has one or more sons. In turn, each of his sons will have a random number of sons, each son of the second generation will have a random number of sons, and so forth. What is the probability that the family name will ultimately die out? The process describing the survival of family names is an example of a so-called branching process. Branching processes arise naturally in many situations. In physics, the model of branching processes can be used to study neutron chain reaction. A chance collision of a nucleus with a neutron yields a random number of new neutrons. Each of these secondary neutrons may hit some other nuclei, producing more additional neutrons, and so forth. In genetics, the model can be used to estimate the probability of long-term survival of genes that are subject to mutation. All of these examples possess the following structure. There is a population of individuals able to produce offspring of the same kind. Each individual will, by the end of its lifetime, have produced j new offspring with probability p_j for $j = 0, 1, \ldots$. All offspring behave independently. The number of individuals initially present, denoted by X_0, is called the size of the 0th generation. All offspring of the 0th generation constitute the first generation, and their number is denoted by X_1. In general, let X_n denote the size of the nth generation. We are interested in the probability that the population will eventually die out. To avoid uninteresting cases, it is assumed that $0 < p_0 < 1$. In order to find the extinction probability, it is no restriction to assume that $X_0 = 1$ (why?). Define the probability u_n by

$$u_n = P(X_n = 0).$$

Obviously, $u_0 = 0$ and $u_1 = p_0$. Noting that $X_n = 0$ implies $X_{n+1} = 0$, it follows that $u_{n+1} \geq u_n$ for all n. Since u_n is a nondecreasing sequence of numbers, $\lim_{n\to\infty} u_n$ exists. Denote this limit by u_∞. The probability u_∞ is the desired extinction probability. This requires some explanation. The probability that extinction will ever occur is defined as $P(X_n = 0 \text{ for some } n \geq 1)$. However, $\lim_{n\to\infty} P(X_n = 0) = P(X_n = 0 \text{ for some } n \geq 1)$, using the fact that $\lim_{n\to\infty} P(A_n) = P(\bigcup_{n=1}^{\infty} A_n)$ for any nondecreasing sequence of events A_n. The probability u_∞ can be computed by using the generating function $P(z) = \sum_{j=0}^{\infty} p_j z^j$ of the offspring distribution p_j. To do so, we first argue that

$$u_n = \sum_{k=0}^{\infty} (u_{n-1})^k p_k \quad \text{for} \quad n = 2, 3, \ldots .$$

This relation can be explained using the law of conditional probabilities. Fix $n \geq 2$. Now, condition on $X_1 = k$ and use the fact that the k subpopulations generated by the distinct offspring of the original parent behave independently and follow the same distributional law. The probability that any particular one of them will die out in $n-1$ generations is u_{n-1} by definition. Thus, the probability that all k subpopulations die out in $n-1$ generations is equal to $P(X_n = 0 | X_1 = k) = (u_{n-1})^k$ for $k \geq 1$. This relation is also true for $k = 0$, since $X_1 = 0$ implies that $X_n = 0$ for all $n \geq 2$. The equation for u_n next follows using the fact that

$$P(X_n = 0) = \sum_{k=0}^{\infty} P(X_n = 0 | X_1 = k) p_k,$$

by the law of conditional probabilities.

Using the definition of the generating function $P(z) = \sum_{k=0}^{\infty} p_k z^k$, the recursion equation for u_n can be rewritten as

$$u_n = P(u_{n-1}) \quad \text{for} \quad n = 2, 3, \ldots .$$

Next, by letting $n \to \infty$ in both sides of this equation and using a continuity argument, it can be shown that the desired probability u_∞ satisfies the equation

$$u = P(u).$$

This equation may have more than one solution. However, it can be shown that u_∞ is the smallest positive root of the equation $u = P(u)$. It may happen that $u_\infty = 1$, that is, the population is sure to die out ultimately. The case of $u_\infty = 1$ can only happen if the expected value of the offspring distribution p_j is smaller than or equal to 1. The proof of this fact is omitted. As illustration, consider the numerical example with $p_0 = 0.25$, $p_1 = 0.25$ and $p_2 = 0.5$. The equation $u = P(u)$ then becomes the quadratic equation $u = \frac{1}{4} + \frac{1}{4}u + \frac{1}{2}u^2$. This equation has roots $u = 1$ and $u = \frac{1}{2}$. The smallest root gives the extinction probability $u_\infty = \frac{1}{2}$.

Problem 14.4 Every adult male in a certain society is married. Twenty percent of the married couples have no children. The other 80% have two or three children with respective probabilities $\frac{1}{3}$ and $\frac{2}{3}$, each child being equally likely to be a boy or a girl. What is the probability that the male line of a father with one son will eventually die out?

Problem 14.5 A population of bacteria begins with a single individual. In each generation, each individual dies with probability $\frac{1}{3}$ or splits in two with probability $\frac{2}{3}$.

(a) What is the probability that the population will die out by generation 3?
(b) What is the probability that the population will die out eventually?
(c) What are the probabilities in (a) and (b) if the initial population consists of two individuals?

14.2 Moment-generating functions

How do we generalize the concept of generating function when the random variable is not integer-valued and nonnegative? The idea is to work with $E(e^{tX})$ instead of $E(z^X)$. Since e^{tX} is a nonnegative random variable, $E(e^{tX})$ is defined for any value of t. However, it may happen that $E(e^{tX}) = \infty$ for some values of t. For any nonnegative random variable X, we have that $E(e^{tX}) < \infty$ for any $t \leq 0$ (why?), but $E(e^{tX})$ need not be finite when $t > 0$. To illustrate this, suppose that the nonnegative random variable X has the one-sided Cauchy density function $f(x) = (2/\pi)/(1+x^2)$ for $x > 0$. Then, $E(e^{tX}) = \int_0^\infty e^{tx} f(x)\,dx = \infty$ for any $t > 0$, since $e^{tx} \geq 1 + tx$ and $\int_0^\infty \frac{x}{1+x^2}dx = \infty$. In case the random variable X can take on both positive and negative values, then it may happen that $E(e^{tX}) = \infty$ for all $t \neq 0$. An example is provided by the random variable X having the two-sided Cauchy density function $f(x) = (1/\pi)/(1+x^2)$ for $-\infty < x < \infty$. Fortunately, most random variables

X of practical interest have the property that $E(e^{tX}) < \infty$ for all t in a neighborhood of 0.

Definition 14.2 *A random variable X is said to have a moment-generating function if $E(e^{tX}) < \infty$ for all t in an interval of the form $-\delta < t < \delta$ for some $\delta > 0$. For those t with $E(e^{tX}) < \infty$ the moment-generating function of X is defined and denoted by*

$$M_X(t) = E(e^{tX}).$$

If the random variable X has a probability density function $f(x)$, then

$$M_X(t) = \int_{-\infty}^{\infty} e^{tx} f(x) dx.$$

As an illustration, consider the case of an exponentially distributed random variable X. The density function $f(x)$ of X is equal to $\lambda e^{-\lambda x}$ for $x > 0$ and 0 otherwise. Then, $M_X(t) = \lambda \int_0^{\infty} e^{(t-\lambda)x}\, dx$. This integral is finite only if $t - \lambda < 0$. Thus, $M_X(t)$ is defined only for $t < \lambda$ and is then given by $M_X(t) = \lambda/(\lambda - t)$.

The explanation of the name moment-generating function is as follows. If the moment-generating function $M_X(t)$ of the random variable X exists, then it can be shown that

$$M_X(t) = 1 + tE(X) + t^2 \frac{E(X^2)}{2!} + t^3 \frac{E(X^3)}{3!} + \cdots$$

for $-\delta < t < \delta$. Heuristically, this result can be seen by using the expansion $E(e^{tX}) = E(\sum_{n=0}^{\infty} t^n \frac{X^n}{n!})$ and interchanging the order of expectation and summation. Conversely, the moments $E(X^r)$ for $r = 1, 2, \ldots$ can be obtained from the moment-generating function $M_X(t)$, when $E(e^{tX})$ exists in a neighborhood of $t = 0$. Assuming that X has a probability density function $f(x)$, it follows from advanced calculus that

$$\frac{d^r}{dt^r} \int_{-\infty}^{\infty} e^{tx} f(x)\, dx = \int_{-\infty}^{\infty} x^r e^{tx} f(x)\, dx$$

for $-\delta < t < \delta$. Taking $t = 0$, we obtain

$$E(X^r) = \frac{d^r}{dt^r} M_X(t)|_{t=0}, \qquad r = 1, 2, \ldots .$$

In particular,

$$E(X) = M_X'(0) \qquad \text{and} \qquad E(X^2) = M_X''(0).$$

A moment-generating function determines not only the moments of a random variable X, but also determines uniquely the probability distribution of X. The following uniqueness theorem holds for the moment-generating function.

Rule 14.2 *If the moment-generating functions $M_X(t)$ and $M_Y(t)$ of the random variables X and Y exist and $M_X(t) = M_Y(t)$ for all t satisfying $-\delta < t < \delta$ for some $\delta > 0$, then the random variables X and Y are identically distributed.*
The proof of this rule is beyond the scope of this book. Also, we have the following very useful rule.

Rule 14.3 *Let X and Y be two random variables with generating functions $M_X(t)$ and $M_Y(t)$. If the random variables X and Y are independent, then*

$$M_{X+Y}(t) = M_X(t)M_Y(t)$$

for all t in a neighborhood of $t = 0$.

The proof is easy. If X and Y are independent, then the random variables e^{tX} and e^{tY} are independent for any fixed t. Since $E(UV) = E(U)E(V)$ for independent random variables U and V, it follows that

$$E\left[e^{t(X+Y)}\right] = E\left[e^{tX}e^{tY}\right] = E(e^{tX})E(e^{tY}).$$

Example 14.4 Suppose that the random variable X has an $N(\mu, \sigma^2)$ distribution. Then,

$$M_X(t) = e^{\mu t + \frac{1}{2}\sigma^2 t^2}, \qquad -\infty < t < \infty.$$

The derivation is as follows. Let $Z = (X - \mu)/\sigma$. Then, Z has the $N(0, 1)$ distribution and

$$\begin{aligned} M_Z(t) &= \frac{1}{\sqrt{2\pi}} \int_{-\infty}^{\infty} e^{tx} e^{-\frac{1}{2}x^2} dx = e^{\frac{1}{2}t^2} \frac{1}{\sqrt{2\pi}} \int_{-\infty}^{\infty} e^{-\frac{1}{2}(x-t)^2} dx \\ &= e^{\frac{1}{2}t^2}, \end{aligned}$$

where the last equality uses the fact that for fixed t the function $\frac{1}{\sqrt{2\pi}} e^{-\frac{1}{2}(x-t)^2}$ is the probability density function of an $N(t, 1)$ distribution. This implies that the integral of this function over the interval $(-\infty, \infty)$ equals 1. The desired expression for $M_X(t)$ next follows from

$$E\left(e^{tX}\right) = E\left(e^{t(\mu+\sigma Z)}\right) = e^{t\mu} E\left(e^{t\sigma Z}\right) = e^{t\mu} e^{\frac{1}{2}\sigma^2 t^2}.$$

The first and the second derivatives of $M_X(t)$ at the point $t = 0$ are given by

$$M_X'(0) = \mu \qquad \text{and} \qquad M_X''(0) = \mu^2 + \sigma^2,$$

showing that the expected value and variance of an $N(\mu, \sigma^2)$ distributed random variable are equal to μ and σ^2.

Remark 14.1 The moment-generating function $M_X(t)$ of the normal distribution enables us also to derive the expected value and the variance of the lognormal distribution. If X is $N(\mu, \sigma^2)$ distributed, then $Y = e^X$ has a lognormal distribution with parameters μ and σ. Taking $t = 1$ in the moment-generating function $M_X(t) = E(e^{tX})$, we obtain $E(Y)$. Also, by $e^{2X} = Y^2$, we obtain $E(Y^2)$ by putting $t = 2$ in $M_X(t) = E(e^{tX})$.

Using the result of Example 14.4, we easily verify that a linear combination of independent normal variates has, again, a normal distribution.

Example 14.5 Suppose that the random variables $X_1, \ldots, X_n$ are independent and normally distributed, where X_i has an $N(\mu_i, \sigma_i^2)$ distribution for $i = 1, \ldots, n$. Let $a_1, \ldots, a_n$ be any constants not all equal to zero. What is the probability distribution of the random variable

$$U = a_1 X_1 + \cdots + a_n X_n?$$

The answer is that U has an $N(\mu, \sigma^2)$ distribution with

$$\mu = a_1 \mu_1 + \cdots + a_n \mu_n \qquad \text{and} \qquad \sigma^2 = a_1^2 \sigma_1^2 + \cdots + a_n^2 \sigma_n^2.$$

It suffices to prove this result for $n = 2$. Next, the general result follows by induction. Using Rule 14.3 and the result from Example 14.4, we find

$$\begin{aligned} E\left[e^{t(a_1 X_1 + a_2 X_2)}\right] &= E\left(e^{ta_1 X_1}\right) E\left(e^{ta_2 X_2}\right) \\ &= e^{\mu_1 a_1 t + \frac{1}{2}\sigma_1^2 (a_1 t)^2} e^{\mu_2 a_2 t + \frac{1}{2}\sigma_2^2 (a_2 t)^2} \\ &= e^{(a_1 \mu_1 + a_2 \mu_2) t + \frac{1}{2}(a_1^2 \sigma_1^2 + a_2^2 \sigma_2^2) t^2}, \end{aligned}$$

proving the desired result with an appeal to the uniqueness Rule 14.2.

The above example shows that the class of normal distributions is closed. A similar result holds for the class of gamma distributions.

Example 14.6 Suppose that $X_1, \ldots, X_n$ are independent random variables, each having a gamma distribution with the same scale parameter β. Let α_i be the shape parameter of the gamma distribution of the random variable X_i for $i = 1, \ldots, n$. Then, the sum $X_1 + \cdots + X_n$ is gamma distributed with shape parameter $\alpha = \alpha_1 + \cdots + \alpha_n$ and scale parameter $\lambda = \beta$. To prove this useful result, we first verify that

$$M_X(t) = \left(\frac{\lambda}{\lambda - t}\right)^\alpha, \qquad t < \lambda$$

when the random variable X is gamma distributed with shape parameter α and scale parameter λ. To do so, fix t with $t < \lambda$ and note that

$$M_X(t) = \int_0^\infty e^{tx} \frac{\lambda^\alpha}{\Gamma(\alpha)} x^{\alpha-1} e^{-\lambda x}\, dx = \int_0^\infty \frac{\lambda^\alpha}{\Gamma(\alpha)} x^{\alpha-1} e^{-(\lambda-t)x}\, dx$$

$$= \left(\frac{\lambda}{\lambda-t}\right)^\alpha \int_0^\infty \frac{(\lambda-t)^\alpha}{\Gamma(\alpha)} x^{\alpha-1} e^{-(\lambda-t)x}\, dx = \left(\frac{\lambda}{\lambda-t}\right)^\alpha,$$

where the last equality uses the fact that $(\lambda - t)^\alpha x^{\alpha-1} e^{-(\lambda-t)x} / \Gamma(\alpha)$ is a gamma density for any fixed t with $t < \lambda$ and thus integrates to 1. Next, we apply Rule 14.3 to the sum $X_1 + \cdots + X_n$. This gives

$$M_{X_1+\cdots+X_n}(t) = \left(\frac{\lambda}{\lambda-t}\right)^{\alpha_1} \cdots \left(\frac{\lambda}{\lambda-t}\right)^{\alpha_n} = \left(\frac{\lambda}{\lambda-t}\right)^{\alpha_1+\cdots+\alpha_n}.$$

The desired result that the sum $X_1 + \cdots + X_n$ is gamma distributed with shape parameter $\alpha = \alpha_1 + \cdots + \alpha_n$ and scale parameter $\lambda = \beta$ next follows by appealing to the uniqueness Rule 14.2. Taking $\alpha_1 = \cdots = \alpha_n = 1$ and noting that the gamma distribution with shape parameter 1 reduces to the exponential distribution, we obtain as corollary that a gamma-distributed random variable with shape parameter n and scale parameter λ can be decomposed as the sum of n independent random variables each having an exponential distribution with the same scale parameter λ. This is an extremely useful result in probability applications.

In Chapter 9, we stated the result that a chi-squared distributed random variable with n degrees of freedom has a gamma density function with shape parameter $\frac{1}{2}n$ and scale parameter 1. Using the moment-generating function approach, this result is easily verified. Let $Z_1, \ldots, Z_n$ be independent random variables each having a standard normal distribution. The chi-squared distributed random variable U is defined by $U = Z_1^2 + \cdots + Z_n^2$. Letting Z be an $N(0, 1)$ random variable, it follows that

$$E(e^{tZ^2}) = \frac{1}{\sqrt{2\pi}} \int_{-\infty}^\infty e^{tx^2} e^{-\frac{1}{2}x^2}\, dx = \frac{1}{\sqrt{2\pi}} \int_{-\infty}^\infty e^{-\frac{1}{2}(1-2t)x^2}\, dx$$

$$= \frac{1}{\sqrt{1-2t}} \frac{1}{(1/\sqrt{1-2t})\sqrt{2\pi}} \int_{-\infty}^\infty e^{-\frac{1}{2}x^2/(1/\sqrt{1-2t})^2}\, dx$$

$$= \frac{1}{\sqrt{1-2t}}, \quad t < \frac{1}{2}.$$

Next, by applying Rule 14.3,

$$M_U(t) = \frac{1}{\sqrt{1-2t}} \cdots \frac{1}{\sqrt{1-2t}} = \frac{1}{(1-2t)^{n/2}}, \qquad t < \frac{1}{2}.$$

Comparing this expression with the expression for the moment-generating function of a gamma-distributed random variable in Example 14.6 and using the uniqueness Rule 14.2, it follows that the chi-squared distributed random variable U has a gamma density with shape parameter $\frac{1}{2}n$ and scale parameter 1.

Problem 14.6 Determine the moment-generating function of the random variable X having as density function the so-called *Laplace* density function $f(x) = \frac{1}{2}ae^{-a|x|}$ for $-\infty < x < \infty$, where a is a positive constant. Use the moment-generating function of X to find $E(X)$ and $\text{var}(X)$.

Problem 14.7 The moment-generating function of two jointly distributed random variables X and Y is defined by $M_{X,Y}(t,u) = E(e^{tX+uY})$, provided that this integral is finite for all (t,u) in a neighborhood of $(0,0)$. A basic result is that the random variables X and Y are independent if and only if $M_{X,Y}(t,u) = M_X(t)M_Y(u)$ for all t, u. Use this result to prove that random variables X and Y having a bivariate normal distribution are independent if and only if the correlation coefficient of X and Y is zero.

Problem 14.8 An insurance company receives any year a random number of claims. The yearly number of claims has a Poisson distribution with expected value μ. The amounts of the individual claims are independent random variables having a gamma density with parameters α and λ. The sizes of the individual claims are also independent of the number of claims. What is the moment-generating function of the total amount claimed in any given year?

Chernoff bound

Let X be a random variable for which the moment generating function $M_X(t) = E(e^{tX})$ is defined for all $t > 0$. The so-called *Chernoff bound* states that

$$P(X \geq c) \leq \min_{t>0} \left[e^{-ct} M_X(t) \right] \quad \text{for} \quad \text{any } c.$$

This is a very useful bound for tail probabilities.

The proof of the Chernoff bound is very simple. The bound follows directly from *Markov's inequality*, which states that

$$P(U \geq a) \leq \frac{1}{a} E(U) \quad \text{for} \quad \text{any } a > 0$$

when U is a nonnegative random variable. Apply Markov's inequality with

$U = e^{tX}$ and $a = e^{ct} > 0$ and use the fact that

$$P(X \geq c) = P(tX \geq tc) = P(e^{tX} \geq e^{tc}) \quad \text{for} \quad \text{any } t > 0.$$

This gives $P(X \geq c) \leq e^{-ct} M_X(t)$ for any $t > 0$, implying the desired result. For its part, Markov's inequality is simply proved. Define the indicator variable I as equal to 1 if $U \geq a$ and 0 otherwise. Then, by $U \geq aI$ and $E(I) = P(U \geq a)$, it follows that $E(U) \geq aP(U \geq a)$. Incidentally, Chebyshev's inequality from Section 5.2 is also an immediate consequence of Markov's inequality (take $U = (X - \mu)^2$). The Chernoff bound is more powerful than Chebyshev's inequality.

Problem 14.9 Prove that $P(X \leq c) \leq \inf_{t<0} \left[e^{-tc} M_X(t)\right]$ for any constant c, assuming that $M_X(t)$ exists for all $t < 0$.

Problem 14.10 Let the random variable X be the number of successes in n independent Bernoulli trials with success probability p. Choose any $\delta > 0$ such that $(1 + \delta)p < 1$. Use the Chernoff bound to verify that

$$P\big(X \geq (1 + \delta)np\big) \leq \left[\left(\frac{p}{a}\right)^a \left(\frac{1-p}{1-a}\right)^{1-a}\right]^n,$$

where $a = (1 + \delta)p$. (*Remark*: the upper bound can be shown to be smaller than or equal to $e^{-2p^2\delta^2 n}$.)

14.2.1 The central limit theorem revisited

We cannot end this book without offering at least a glimpse of the steps involved in the proof of the central limit theorem, which plays such a prominent role in probability theory. The mathematical formulation of the central limit theorem is as follows. Suppose that $X_1, X_2, \ldots$ are independent and identically distributed random variables with expected value μ and standard deviation σ. Then,

$$\lim_{n\to\infty} P\left(\frac{X_1 + \cdots + X_n - n\mu}{\sigma\sqrt{n}} \leq x\right) = \frac{1}{\sqrt{2\pi}} \int_{-\infty}^{x} e^{-\frac{1}{2}y^2}\, dy \quad \text{for} \quad \text{all } x.$$

We make this result plausible for the case that the moment-generating function of the X_i exists and is finite for all t in some neighborhood of $t = 0$. To do so, consider the standardized variables

$$U_i = \frac{X_i - \mu}{\sigma}, \qquad i = 1, 2, \ldots.$$

Then, $E(U_i) = 0$ and $\sigma(U_i) = 1$. Letting

$$Z_n = \frac{U_1 + \cdots + U_n}{\sqrt{n}},$$

we have $Z_n = (X_1 + \cdots + X_n - n\mu)/\sigma\sqrt{n}$. Denoting by $M_{Z_n}(t) = E\left(e^{tZ_n}\right)$ the moment-generating function of Z_n, it will be proved in a moment that

$$\lim_{n\to\infty} M_{Z_n}(t) = e^{\frac{1}{2}t^2}$$

for all t in a neighborhood of $t = 0$. In other words,

$$\lim_{n\to\infty} M_{Z_n}(t) = E\left(e^{tZ}\right)$$

when Z is a standard normal random variable. From this result, we can conclude that

$$\lim_{n\to\infty} P(Z_n \leq x) = P(Z \leq x) \quad \text{for} \quad \text{all } x,$$

using a deep continuity theorem for moment-generating functions. This theorem linking the convergence of moment-generating functions to convergence of probability distribution functions must be taken for granted by the reader.

To verify that the moment-generating function of Z_n converges to the moment-generating function of the standard normal random variable, let $M_U(t)$ be the moment-generating function of the U_i. Using the assumption that $U_1, \ldots,$ U_n are independent and identically distributed, it follows that

$$\begin{aligned} E\left(e^{tZ_n}\right) &= E\left(e^{t(U_1+\cdots+U_n)/\sqrt{n}}\right) \\ &= E\left(e^{(t/\sqrt{n})U_1}\right)\cdots E\left(e^{(t/\sqrt{n})U_n}\right) \end{aligned}$$

and so

$$M_{Z_n}(t) = \left[M_U(t/\sqrt{n})\right]^n, \qquad n = 1, 2, \ldots.$$

Since $M_U(t) = 1 + t\,E(U_1) + \frac{t^2}{2!}\,E(U_1^2) + \frac{t^3}{3!}\,E(U_1^3) + \cdots$ in some neighborhood of $t = 0$ and using the fact that $E(U_1) = 0$ and $\sigma(U_1) = 1$, it follows that

$$M_U(t) = 1 + \frac{1}{2}t^2 + \epsilon(t)$$

in a neighborhood of $t = 0$, where $\epsilon(t)$ tends faster to zero than t^2 as $t \to 0$. That is,

$$\lim_{t\to 0} \frac{\epsilon(t)}{t^2} = 0.$$

Now, fix t and let $\epsilon_n = \epsilon(t/\sqrt{n})$. Then,

$$M_{Z_n}(t) = \left(1 + \frac{1}{2}\frac{t^2}{n} + \frac{n\epsilon_n}{n}\right)^n, \qquad n = 1, 2, \ldots .$$

Since $\lim_{u \to 0} \epsilon(u)/u^2 = 0$, we have that $\lim_{n\to\infty} n\epsilon_n = 0$. Using the fact that $\lim_{n\to\infty}(1 + \frac{a}{n})^n = e^a$ for any constant a, it is now a matter of standard manipulation in analysis to conclude that

$$\lim_{n\to\infty} M_{Z_n}(t) = e^{\frac{1}{2}t^2},$$

as was to be proved.

Appendix

Counting methods and e^x

This appendix first gives some background material on counting methods. Many probability problems require counting techniques. In particular, these techniques are extremely useful for computing probabilities in a chance experiment in which all possible outcomes are equally likely. In such experiments, one needs effective methods to count the number of outcomes in any specific event. In counting problems, it is important to know whether the order in which the elements are counted is relevant or not. After the discussion on counting methods, the Appendix summarizes a number of properties of the famous number e and the exponential function e^x both playing an important role in probability.

Permutations

How many different ways can you arrange a number of different objects such as letters or numbers? For example, what is the number of different ways that the three letters A, B, and C can be arranged? By writing out all the possibilities ABC, ACB, BAC, BCA, CAB, and CBA, you can see that the total number is 6. This brute-force method of writing down all the possibilities and counting them is naturally not practical when the number of possibilities gets large, for example the number of different ways to arrange the 26 letters of the alphabet. You can also determine that the three letters A, B, and C can be written down in 6 different ways by reasoning as follows. For the first position, there are 3 available letters to choose from, for the second position there are 2 letters over to choose from, and only one letter for the third position. Therefore, the total number of possibilities is $3 \times 2 \times 1 = 6$. The general rule should now be evident. Suppose that you have n distinguishable objects. How many ordered arrangements of these objects are possible? Any ordered sequence of the objects is called a *permutation*. Reasoning similar to that described shows that there are n ways for choosing the first object, leaving $n-1$ choices for the second object, etc. Therefore, the total number of ways to order n distinguishable objects is $n \times (n-1) \times \cdots \times 2 \times 1$. A convenient shorthand for this product is $n!$ (pronounce: n factorial). Thus, for any positive integer n,

$$n! = 1 \times 2 \times \cdots \times (n-1) \times n.$$

A convenient convention is $0! = 1$. Summarizing,

> **the total number of ordered sequences (permutations) of n distinguishable objects is $n!$**

Example A.1 A scene from the movie "The Quick and the Dirty" depicts a Russian roulette type of duel. Six identical shot glasses of whiskey are set on the bar, one of which is laced with deadly strychnine. The bad guy and the good guy must drink in turns. The bad guy offers \$1000 to the good guy, if the latter will go first. Is this an offer that should not be refused?

A handy way to think of the problem is as follows. Number the six glasses from 1 to 6 and assume that the glasses are arranged in a random order after strychnine has been put in one of the glasses. There are 6! possible arrangements of the six glasses. If the glass containing strychnine is in the first position, there remain 5! possible arrangements for the other five glasses. Thus, the probability that the glass in the first position contains strychnine is equal to $5!/6! = 1/6$. By the same reasoning, the glass in each of the other five positions contains strychnine with a probability of $1/6$, before any glass is drunk. It is a fair game. Each of the two "duelists" will drink the deadly glass with a probability of $(3 \times 5!)/6! = 1/2$. The good guy will do well to accept the offer of the bad guy. After the good guy has drunk the first glass and survived it, the probability that the bad guy will get the glass with strychnine becomes $(3 \times 4!)/5! = 3/5$.

Example A.2 Eight important heads of state, including the U.S. President and the British Premier, are present at a summit conference. For the perfunctory group photo, the eight dignitaries are lined up randomly next to one other. What is the probability that the U.S. President and the British Premier will stand next to each other?

Number the eight heads of state as $1, \ldots, 8$, where the number 1 is assigned to the U.S. President and number 2 to the British Premier. The eight statesmen are put in a random order in a row. There are 8! possible arrangements. If the positions of the U.S. President and the British Premier are fixed, there remain 6! possible arrangements for the other six statesmen. The U.S. President and the British Premier stand next to each other if they take up the positions i and $i+1$ for some i with $1 \leq i \leq 7$. In case these two statesmen take up the positions i and $i+1$, there are 2! possibilities for the order among them. Thus, there are $6! \times 7 \times 2!$ arrangements in which the U.S. President and the British Premier stand next to each other, and so the sought probability equals $(6! \times 7 \times 2!)/8! = 1/4$.

Combinations

How many different juries of three persons can be formed from five persons A, B, C, D and E? By direct enumeration, you see that the answer is 10: $\{A, B, C\}$, $\{A,B,D\}$, $\{A,B,E\}$, $\{A,C,D\}$, $\{A,C,E\}$, $\{A,D,E\}$, $\{B,C,D\}$, $\{B,C,E\}$, $\{B,D,E\}$, $\{C,D,E\}$. In this problem, the order in which the jury members are chosen is not relevant. The answer 10 juries could also have been obtained by a basic principle of counting. First, count how many juries of three persons are possible when attention is paid to the order. Then determine how often each group of three persons has been counted. Thus, the reasoning is as follows. There are 5 ways to select the first jury member, 4 ways to then select the next member, and 3 ways to select the final member. This would give $5 \times 4 \times 3$ ways of forming the jury when the order in which the members are chosen would be relevant. However, this order makes no difference. For example, for the jury consisting of the persons A, B, and C, it is not relevant which of the 3! ordered sequences ABC, ACB, BAC, BCA, CAB, CBA has lead to the jury. Hence the total number of ways a jury of 3 persons can be formed from a group of 5 persons is equal to $\frac{5\times4\times3}{3!}$. This expression can be rewritten as

$$\frac{5 \times 4 \times 3 \times 2 \times 1}{3! \times 2!} = \frac{5!}{3! \times 2!}.$$

In general, you can calculate that the total number of possible ways to choose a jury of k persons out of a group of n persons is equal to

$$\frac{n \times (n-1) \times \cdots \times (n-k+1)}{k!}$$
$$= \frac{n \times (n-1) \times \cdots \times (n-k+1) \times (n-k) \times \cdots \times 1}{k! \times (n-k)!}$$
$$= \frac{n!}{k! \times (n-k)!}.$$

For nonnegative integers n and k with $k \leq n$, we define

$$\binom{n}{k} = \frac{n!}{k! \times (n-k)!}.$$

The quantity $\binom{n}{k}$ (pronounce: n over k) has the interpretation:

$\binom{n}{k}$ is the total number of ways to choose k different objects out of n distinguishable objects, paying no attention to their order.

The numbers $\binom{n}{k}$ are referred to as the *binomial coefficients*. The binomial coefficients arise in numerous counting problems.

Example A.3 Is the probability of winning the jackpot with a single ticket in Lotto 6/45 larger than the probability of getting 22 heads in a row when tossing a fair coin 22 times?

In Lotto 6/45, six different numbers are drawn out of the numbers $1, \ldots, 45$. The total number of ways the winning six numbers can be drawn is equal to $\binom{45}{6}$. Hence, the probability of hitting the jackpot with a single ticket is

$$\frac{1}{\binom{45}{6}} = 1.23 \times 10^{-7}.$$

This probability is smaller than the probability $\left(\frac{1}{2}\right)^{22} = 2.38 \times 10^{-7}$ of getting 22 heads in a row.

Example A.4 In the Powerball lottery, five distinct white balls are drawn out of a drum with 53 white balls, and one red ball is drawn from a drum with 42 red balls. The white balls are numbered $1, \ldots, 53$ and the red balls are numbered $1, \ldots, 42$. You have filled in a single ticket with five different numbers for the white balls and one number for the red ball (the Powerball number). What is the probability that you match only the Powerball number?

There are $42 \times \binom{53}{5}$ ways to choose your six numbers. Your five white numbers must come from the 48 white numbers not drawn by the lottery. This can happen in $\binom{48}{5}$ ways. There is only one way to match the Powerball number. Hence, the probability that you match the red Powerball alone is

$$\frac{1 \times \binom{48}{5}}{42 \times \binom{53}{5}} = 0.0142.$$

Example A.5 What is the probability that a bridge player's hand of 13 cards contains exactly k aces for $k = 0, 1, 2, 3, 4$?

There are $\binom{4}{k}$ ways to choose k aces from the four aces and $\binom{48}{13-k}$ ways to choose the other $13 - k$ cards from the remaining 48 cards. Hence, the desired probability is

$$\frac{\binom{4}{k}\binom{48}{13-k}}{\binom{52}{13}}.$$

This probability has the values 0.3038, 0.4388, 0.2135, 0.0412, and 0.0026 for $k = 0, 1, 2, 3$, and 4.

Example A.6 The following question is posed in the sock problem from Chapter 1. What are the probabilities of seven and four matching pairs of socks remaining when six socks are lost during the washing of ten different pairs of socks?

There are $\binom{20}{6}$ possible ways to choose six socks out of ten pairs of socks. You are left with seven complete pairs of socks only if both socks of three pairs are missing. This can happen in $\binom{10}{3}$ ways. Hence, the probability that you are left with seven complete pairs of socks is equal to

$$\frac{\binom{10}{3}}{\binom{20}{6}} = 0.0031.$$

You are left with four matching pairs of socks only if exactly one sock of each of six pairs is missing. These six pairs can be chosen in $\binom{10}{6}$ ways. There are two possibilities to choose one sock from a given pair. This means that there are $\binom{10}{6}2^6$ ways to choose six socks so that four matching pairs of socks are left. Hence, the probability of four matching pairs of socks remaining is equal to

$$\frac{\binom{10}{6}2^6}{\binom{20}{6}} = 0.3467.$$

It is remarkable that the probability of the worst case of four matching pairs of socks remaining is more than hundred times as large as the probability of the best case of seven matching pairs of socks remaining. When things go wrong, they really go wrong.

Exponential function

The history of the number e begins with the discovery of logarithms by John Napier in 1614. At this time in history, international trade was experiencing a period of strong growth, and, as a result, there was much attention given to the concept of compound interest. At that time, it was already noticed that $(1+\frac{1}{n})^n$ tends to a certain limit if n is allowed to increase without bound:

$$\lim_{n\to\infty}\left(1+\frac{1}{n}\right)^n = e,$$

where e is the famous number $e = 2.71828\ldots$.[1] The *exponential function* is defined by e^x, where the variable x runs through the real numbers. A fundamental property of e^x is that this function has itself as derivative. That is,

$$\frac{de^x}{dx} = e^x.$$

This property is easy to explain. Consider the function $f(x) = a^x$ for some constant $a > 0$. It then follows from $f(x+h) - f(x) = a^{x+h} - a^x = a^x(a^h - 1)$ that

$$\lim_{h\to 0}\frac{f(x+h)-f(x)}{h} = cf(x)$$

for the constant $c = \lim_{h\to 0}(a^h - 1)/h$. The proof is omitted that this limit always exists. Next, one might wonder for what value of a the constant $c = 1$ so that $f'(x) = f(x)$. Noting that the condition $(a^h - 1)/h = 1$ can be written as $a = (1+h)^{1/h}$, it can easily be shown that $\lim_{h\to 0}(a^h - 1)/h = 1$ boils down to $a = \lim_{h\to 0}(1+h)^{1/h}$, yielding $a = e$.

How do we calculate the function e^x? The generally valid relation

$$\lim_{n\to\infty}\left(1+\frac{x}{n}\right)^n = e^x \quad \text{for} \quad \text{each real number } x$$

is not useful for this purpose. The calculation of e^x is based on the power series expansion

$$e^x = 1 + x + \frac{x^2}{2!} + \frac{x^3}{3!} + \cdots.$$

In a compact notation,

$$e^x = \sum_{n=0}^{\infty}\frac{x^n}{n!} \quad \text{for} \quad \text{each real number } x.$$

The proof of this power series expansion requires Taylor's theorem from calculus. The fact that e^x has itself as derivative is crucial in the proof. Note that term-by-term differentiation of the series $1 + x + \frac{x^2}{2!} + \cdots$ leads to the same series, in agreement with the fact that e^x has itself as derivative. The series expansion of e^x shows that $e^x \approx 1 + x$ for x close to 0. In other words,

$$1 - e^{-\lambda} \approx \lambda \quad \text{for} \quad \lambda \text{ close to } 0.$$

This approximation formula is very useful in probability theory.

[1] A wonderful account of the number e and its history can be found in E. Maor, *e:The Story of a Number*, Princeton University Press, Princeton, NJ, 1994.

Geometric series

For any nonnegative integer n,

$$\sum_{k=0}^{n} x^k = \frac{1 - x^{n+1}}{1 - x} \quad \text{for} \quad \text{each real number } x \neq 1.$$

This useful result is a direct consequence of

$$\begin{aligned}(1 - x) \sum_{k=0}^{n} x^k &= \sum_{k=0}^{n} x^k - \sum_{k=0}^{n} x^{k+1} \\ &= (1 + x + \cdots + x^n) - \left(x + x^2 + \cdots + x^n + x^{n+1}\right) \\ &= 1 - x^{n+1}.\end{aligned}$$

The term x^{n+1} converges to 0 for $n \to \infty$ if $|x| < 1$. This leads to the important result

$$\sum_{k=0}^{\infty} x^k = \frac{1}{1 - x} \quad \text{for} \quad \text{each real number } x \text{ with } |x| < 1.$$

This series is called the *geometric series* and is frequently encountered in probability problems. The series $\sum_{k=1}^{\infty} kx^{k-1}$ may be obtained by differentiating the geometric series $\sum_{k=0}^{\infty} x^k$ term by term and using the fact that the derivative of $1/(1 - x)$ is given by $1/(1 - x)^2$. The operation of term-by-term differentiation is justified by a general theorem for the differentiation of power series and leads to the result

$$\sum_{k=1}^{\infty} kx^{k-1} = \frac{1}{(1 - x)^2} \quad \text{for} \quad \text{each real number } x \text{ with } |x| < 1.$$

Recommended readings

There are many fine books on probability theory available. The following more applied books are recommended for further reading.

1. W. Feller, *Introduction to Probability Theory and Its Applications*, Vol. I, third edition, Wiley, New York, 1968.

   This classic in the field of probability theory is still up-to-date and offers a rich assortment of material. Intended for the somewhat advanced reader.

2. S. M. Ross, *Introduction to Probability Models*, eighth edition, Academic Press, New York, 2002.

   A delightfully readable book that makes a good companion to Feller (noted above). Provides a clear introduction to many advanced topics in applied probability.

3. H. C. Tijms, *A First Course in Stochastic Models*, Wiley, Chichester, UK, 2003.

   This is an advanced textbook on stochastic processes and gives particular attention to applications and solution tools in computational probability.

Answers to odd-numbered problems

Chapter 2

2.1 The answer is yes. Use the sample space for this conclusion.

2.3 Take a sample space Ω with the eight elements KKK, KKM, KMK, MKK, MMM, MMK, MKM, and KMM. Assign a probability of $\frac{1}{8}$ to each element of Ω. The desired probability is $\frac{6}{8}$.

2.5 Take $\Omega = \{(i_1, i_2, i_3, i_4) | i_k = 0, 1 \text{ for } k = 1, \ldots, 4\}$ as sample space and assign a probability of $\frac{1}{16}$ to each element of Ω. The probability of three puppies of one gender and one of the other is $\frac{8}{16}$. The probability of two puppies of each gender is $\frac{6}{16}$.

2.7 Take the set of all 10! permutations of the integers $1, \ldots, 10$ as sample space. The number of permutations having the winning number in any given position i is 9! for each $i = 1, \ldots, 10$. In both cases, your probability of winning is $9!/10! = 1/10$.

2.9 Take $\Omega = \{(i, j) | i, j = 1, \ldots, 6\}$ as sample space and assign a probability of $\frac{1}{36}$ to each element of Ω. The expected payoff is $2 \times \frac{15}{36} + 0 \times \frac{21}{36} = \frac{30}{36}$ dollars for both bets.

2.11 Take $\Omega = \{(i, j, k, l) | i, j, k, l = 1, \ldots, 6\}$ as sample space. The expected payoff is $100 \times \frac{6}{1{,}296} + 10 \times \frac{54}{1{,}296} = \frac{95}{108}$ dollars.

2.13 Invest 35.5% of your bankroll in the risky project each time. The effective rate of return is 6.6%.

2.15 $\Omega = \{(i, j) | i, j = 1, \ldots, 100\}$ as sample space. There are 5,000 elements (i, j) satisfying $i + j \leq 50$ or $100 < i + j \leq 150$. The wager is fair.

2.17 This problem is a variant on the daughter-son problem from Chapter 1. Your probability of winning is $\frac{1}{3}$. The bet is not fair.

2.19 The probability is 0.875.

2.21 The probability is 0.627.

2.23 The expected values in parts **(a)**, **(b)**, **(c)**, and **(d)** are 0.333, 0.521, 0.905, and 0.365, respectively.

2.25 The expected values of your loss and the total amount you bet are \$0.942 and \$34.86.

2.27 The probability is 0.529.

2.29 The probability of the bank winning is 0.81, and the average number of points collected by the bank is 9.4.

2.31 The probabilities are 0.750 and 0.689 for $G = 50$ and are 0.698 and 0.705 for $G = 55$.
2.33 The expected payoff is 0.60 dollars and the probability of getting 25 or more points is 0.693.
2.35 The probability is 0.257.
2.37 The game is not fair. The expected value of the number of tosses required is 14.
2.39 The probabilities are 0.396 and 0.335 for $n = 25$ and $n = 100$.
2.41 The probabilities are 0.587, 0.312, 0.083, and 0.015, respectively.
2.43 The probability is 0.60.
2.45 The probabilities are 0.329 and 0.536.

Chapter 3

3.1 Yes.
3.3 No, the probability is $\frac{1}{10{,}000}$.
3.5 The probability is 0.01.
3.7 The proposition is unfavorable for the friends who stay behind. Their leaving friend wins on average $1 \times \frac{65}{81} - 4 \times \frac{16}{81} = \frac{1}{81}$ drink per round.
3.9 Your probability of winning is $1 - (100 \times 99 \times \cdots \times 86)/100^{15} = 0.6687$.
3.11 The probabilities in parts (a) and (b) are $1 - (25 \times 24 \times \cdots \times 19)/25^7 = 0.6031$ and $1 - \frac{24^6}{25^6} = 0.2172$.
3.13 Substituting $n = 78{,}000$ and $c = \binom{52}{13}$ into the formula in part **(b)** of Problem 3.12 gives the probability 0.0048.
3.15 Substituting $n = 500$ and $c = 2{,}400{,}000$ into the formula in part **(b)** of Problem 3.12 gives the probability 0.051.
3.17 The probabilities in parts **(a)** and **(b)** are 0.764 and 0.006.
3.19 **(a)** $\left(\frac{20}{22}\right)^{20} = 0.1486$, **(b)** $22 \times \left(\frac{20}{22}\right)^{20} = 3.27$, and **(c)** 0.043.
3.21 Let A be the event that the sports car is won and let B_i represent the event that the contestant selects i keys for $i = 0, 1, 2$. Then, $P(A) = \sum_{i=0}^{2} P(A \mid B_i)P(B_i)$. It holds that $P(A \mid B_0) = 0$, $P(A \mid B_1) = \frac{2}{5}$ and $P(A \mid B_2) = 1 - \frac{3}{10} = \frac{7}{10}$. This leads to $P(A) = 0 \times \frac{1}{4} + \frac{2}{5} \times \frac{1}{2} + \frac{7}{10} \times \frac{1}{4} = \frac{3}{8}$.
3.23 The main point will be equal to m with probability $r_m = p_m / \sum_{k=5}^{9} p_k$ for $m = 5, 6, \ldots, 9$, where the p_i are the same as the probabilities p_i in the Craps example. Using this result and the law of conditional probabilities, we obtain that the probability of the player winning on the main point is $p_5 r_5 + (p_6 + p_7) r_6 + (p_7 + p_{11}) r_7 + (p_8 + p_{12}) r_8 + p_9 r_9 = 0.1910$. Similarly, the conditional probability of the player winning on a chance point given that the main point is m equals $\sum_{i \notin A_m} p_i \frac{p_i}{p_i + p_m}$, where $A_5 = \{5, 11, 12\}$, $A_6 = \{6, 7, 11\}$, $A_7 = \{7, 11, 12\}$, $A_8 = \{8, 11, 12\}$, and $A_9 = \{9, 11, 12\}$. Using the law of conditional probabilities, we arrive at the conclusion that the probability of the player winning on a chance point is 0.3318. The house percentage is 5%.
3.27 The house percentage is 7.45%.
3.29 The probabilities are 0.0515, 0.1084, 0.1592, and 0.1790.
3.31 The recursion is $a_k = \frac{1}{2} a_{k-1} + \frac{1}{4} a_{k-2}$ for $k \geq 2$ with the boundary conditions $a_0 = a_1 = 1$. This leads to $a_5 = 0.4063$, $a_{10} = 0.1406$, $a_{25} = 5.85 \times 10^{-3}$, and $a_{50} = 2.93 \times 10^{-5}$.

3.33 The probability is 0.0009. The expected value is $\frac{1}{2} \times \binom{n}{2}$.

3.35 Let the random variable N_i denote the number of times that number i will be drawn in the next 250 draws of Lotto 6/45. Using computer simulation, we find that $P\left(\sum_{i=1}^{45} |N_i - 33.333| > 202\right) = 0.333$.

Chapter 4

4.1 Poisson distribution.

4.3 Apply the binomial distribution with $n = 1500$ and $p = \frac{1}{1000}$. The desired probability is 0.7770.

4.5 Apply the binomial distribution with $n = 125$ and $p = \left(\frac{1}{2}\right)^7$. The desired probabilities are 0.625 an 0.075.

4.7 Ten dice.

4.9 The probability is $0.45 \times \sum_{n=4}^{7} \binom{n-1}{3}(0.45)^3(0.55)^{n-1-3} = 0.3917$.

4.11 The expected payoff is $5 \times 0.1820 = 0.91$ per dollar staked.

4.13 Let E be the expected payoff for any newly purchased ticket. The equation $E = 50{,}000 \times (3.5 \times 10^{-5}) + 500 \times 0.00168 + 50 \times 0.03025 + 0.2420 \times E$ gives $E = 5.41$ dollars. The house percentage is 45.9%.

4.15 Apply the multinomial distribution with $n = 5$, $p_1 = \frac{1}{6}$, $p_2 = \frac{1}{6}$, and $p_3 = \frac{2}{3}$. Your probability of winning is 0.3279. The game is unfavorable for the player.

4.17 Use the multinomial distribution to find the house percentage of 12.3%

4.19 The first method. The corresponding binomial probabilities are 0.634 and 0.630.

4.21 The binomial distribution with parameters $n = 4 \times 6^{r-1}$ and $p = \frac{1}{6^r}$ converges to a Poisson distribution with expected value $\frac{2}{3}$ as $r \to \infty$.

4.23 The number of winners is approximately Poisson distributed with an expected value of $\lambda = 200 \times \frac{25}{2{,}500{,}000}$. The monthly amount the corporation will have to give away is zero with probability 0.9980 and \$25,000 with probability 0.002.

4.25 This is a birthday problem with a group of 100,000 persons and $\binom{45}{6} = 8{,}145{,}060$ birthdays. Letting $\lambda = \binom{100{,}000}{8}\frac{1}{(8{,}145{,}060)^7}$, the Poisson approximation for the desired probability is $1 - e^{-\lambda} = 1.04 \times 10^{-13}$.

4.27 Using the approach from Section 4.2.3 for the birthday problem, a trial is associated with each sample of three persons from the group of 25 persons. This leads to the Poisson approximations $1 - e^{-\lambda_0} = 0.0171$ and $1 - e^{-\lambda_1} = 0.1138$, where $\lambda_0 = \binom{25}{3} \times \left(\frac{1}{365}\right)^2$ and $\lambda_1 = \binom{25}{3} \times \left(7 \times \left(\frac{1}{365}\right)^2\right)$. The simulated values of the desired probabilities are 0.0164 and 0.1030. An alternative approximation approach is to associate a trial with each of the 365 days or to each of the 365 combinations of two consecutive days. In the first problem, the trial associated with day i is called succesful if three or more persons have their birthdays on day i. In the second problem, the trial associated with the days i and $i + 1$ is called succesful if three or more persons have their birthdays on one of these two days. This leads to the Poisson approximations $1 - e^{-\eta_0} = 0.0164$ and $1 - e^{-\eta_1} = 0.1186$, where $\eta_0 = 365 \times 0.002739$ and $\eta_1 = 365 \times 0.0054795$.

4.29 Letting $\lambda = \binom{25}{2} \times \left(\frac{14}{365} \times \frac{1}{365}\right)$, a Poisson approximation for the desired probability is $1 - e^{-\lambda} = 0.0310$. The simulated value is 0.0299.

4.31 Letting $\lambda_0 = 44 \times \left[\binom{43}{4} / \binom{45}{6}\right]$ and $\lambda_1 = 43 \times \left[\binom{42}{3} / \binom{45}{6}\right]$, the Poisson approximations are 0.487 and 0.059. The simulated values are 0.529 and 0.056.

4.33 A remarkably accurate approximation is given by the Poisson distribution with expected value $\lambda = 8 \times \frac{1}{15} = \frac{8}{15}$; see also the answer to Problem 2.41. In the Poisson approximation approach, the ith trial concerns your prediction of the ith match. The success probability of each trial is given by $p = \frac{8 \times 2 \times 14!}{16!} = \frac{1}{15}$.

4.35 Letting $\lambda = 365 \times \left[1 - \left(\frac{354}{365}\right)^{75} - 75 \times \frac{1}{365} \times \left(\frac{364}{365}\right)^{74}\right] = 6.6603$, a Poisson approximation for the probability of having at least 7 days on which two or more employees have birthdays is $1 - \sum_{k=0}^{6} e^{-\lambda} \lambda^k / k! = 0.499$. The simulated value is 0.516.

4.37 The probability is $1 - e^{-\alpha} \times e^{-\alpha} = 1 - e^{-2\alpha}$. Use the property that the Poisson process has independent increments.

4.39 The simulated value is 0.203.

4.41 The number of illegal parking customers is binomially distributed with parameters $n = 75$ and $p = 5/(45 + 5) = \frac{1}{10}$. The desired probability is 0.0068.

4.43 The probability is $\binom{50}{5}\binom{50}{5} / \binom{100}{10} = 0.2593$.

4.45 The hypergeometric model with $R = W = 25$ and $n = 25$ is applicable under the hypothesis that the psychologist blindly guesses which 25 persons are left-handed. Then, the probability of identifying correctly 18 or more of the 25 left-handers is 2.1×10^{-3}. This small probability provides evidence against the hypothesis.

4.47 Apply the hypergeometric model with $R = 122$, $W = 244$, and $n = 31$. The desired probability is 0.0083.

4.49 The probabilities are 7.15×10^{-9}, 6.44×10^{-8}, 4.29×10^{-7}, 1.80×10^{-5}, 4.51×10^{-5}, 9.23×10^{-4}, 0.00123, and 0.01667. The probability of not winning the jackpot in the coming m years is $1 - e^{-52 \times 12 \times m \times p}$ with $p = \frac{1}{10} \times \frac{1}{\binom{49}{6}}$. Putting this probability equal to 0.5 yields $m = 155{,}334$ years.

Chapter 5

5.1 Statement **(b)**.

5.3 $\Phi\left(\frac{550-799.5}{121.4}\right) = 0.3250$.

5.5 $1 - \Phi\left(\frac{20}{16}\right) = 0.1056$.

5.7 If Y is distributed as $2X$, then $\sigma(Y) = 2\sigma(X)$.

5.9 **(a)** The correlation coefficient is -1. **(b)** Invest $\frac{1}{2}$ of your capital in stock A and $\frac{1}{2}$ in stock B. The expected value of the rate of return is 7% and the standard deviation is zero. In other words, the portfolio has a guaranteed rate of return of 7%.

5.11 For the case of $p = 0.5$ and $f = 0.2$ the simulated probability mass function is given by (0.182, 0.093, 0.047, 0.029, 0.022, 0.016, 0.014, 0.012, 0.011, 0.011, 0.010, 0.011, 0.011, 0.012, 0.012, 0.014, 0.013, 0.015, 0.029, 0.001, 0.433).

5.13 For the case of $p = 0.8$ and $f = 0.1$, the simulated values of the expected value and the standard deviation of the investor's capital after 20 years are about \$270,000 and \$71,000. For the case of $p = 0.5$ and $f = 0.2$, the simulated values are about \$430,000 and \$2,150,000.

5.15 This value converges to $\frac{1}{2}$, since $P(X \geq \mu) = \frac{1}{2}$ for any $N(\mu, \sigma^2)$ random variable X.

5.17 The Poisson model is applicable. A Poisson distribution with mean 81 can be approximated by an $N(81, 81)$ distribution. The observed value of 117 lies 4 standard deviations above the expected value of 81. This is difficult to explain as a chance variation.

5.19 The observed value of 70 lies 3 standard deviations below the expected value of 70. This is difficult to explain as a chance variation.

5.21 This can hardly be explained as a chance variation. In 1,000 rolls of a fair die, the average number of points per roll is approximately $N(\mu, \sigma^2)$ distributed with $\mu = 3.5$ and $\sigma = \frac{1.708}{\sqrt{1000}} = 0.054$. The reported value of 3.25 lies 4.63 standard deviations above the expected value of 3.5.

5.23 The outcome can hardly be explained as a chance variation. It lies 3.93 standard deviations above the expected value of 17.83.

5.25 The probability is about $1 - \Phi(2.8) = 0.093$.

5.27 The 95% confidence interval is $0.52 \pm 1.96\frac{\sqrt{0.52 \times 0.48}}{\sqrt{400}} = 0.52 \pm 0.049$. The enlarged sample size must be about 2,400 students.

5.29 Under the hypothesis that the aspirin has no effect, the observed value of 104 for the aspirin group lies $\left(\frac{313}{2} - 104\right) / \left(\frac{1}{2}\sqrt{313}\right) = 5.93$ standard deviations below the expected value of 156.5. This is strong evidence against the null hypothesis.

5.31 Under the hypothesis that the generator produces true random numbers, the number of runs is distributed as $1 + R$, where R has a binomial distribution with parameters $n = 99{,}999$ and $p = \frac{1}{2}$. The observed value of 49,487 runs lies 3.25 standard deviations below the expected value. This is a strong indication that the new random number generator is a bad one.

5.33 The recursion is $u_k(i) = \frac{1}{6}u_{k-1}(i+1) + \frac{5}{6}u_k(0)$ for $i = 0, 1, 2$ and $k = 1, \ldots, 25$, where the boundary conditions are $u_0(i) = 0$ and $u_k(3) = 1$. The desired probability is $u_{25}(0) = 0.087$.

Chapter 6

6.1 Disagree.

6.3 A chance tree leads to the probability $\frac{1}{5}$.

6.5 A chance tree leads to the probability $4(0.2 \times 0.5) = 0.4$.

6.7 A chance tree leads to the probability $\frac{0.375}{0.375+0.075} = 0.8333$.

6.9 A chance tree gives $P(\text{not drunk}|\text{positive}) = \frac{0.0475}{0.0475+0.045} = 0.5135$.

6.11 A chance tree gives $P(\text{white cab}|\text{white cab seen}) = \frac{0.12}{0.12+0.17} = 0.4138$.

6.13 A chance tree leads to the probability $\frac{1/3}{1/3+1/6} = \frac{2}{3}$.

6.15 The probability is $\frac{1}{1+9.9} = 0.092$.

6.17 Pick three marbles out of the vase. Guess the dominant color among these three marbles. Under this strategy you win \$8,500, with the probability 0.7407.

Chapter 7

7.1 The sample space is $\Omega = \{(x, y)|0 \le x \le a, 0 \le y \le a\}$. The probability $P(A) =$ (area of $A)/a^2$ is assigned to each subset A of Ω. The desired probability is $(a - d)^2/a^2$.

7.3 Let $B_n = \cup_{k=n}^{\infty} A_k$ for $n \geq 1$, then $B_1, B_2, \ldots$ is a nonincreasing sequence of sets. Note that $\omega \in C$ if and only if $\omega \in B_n$ for all $n \geq 1$. This implies that set C equals the intersection of all sets B_n. Using the continuity of probabilities, $P(C) = \lim_{n\to\infty} P(B_n)$. This gives $P(C) = \lim_{n\to\infty} P(\cup_{k=n}^{\infty} A_k) \leq \lim_{n\to\infty} \sum_{k=n}^{\infty} P(A_k)$. The latter limit is zero, since $\sum_{k=1}^{\infty} P(A_k) < \infty$.

7.5 The desired probability is equal to $\sum_{k=0}^{\infty} \left(\frac{7}{10}\right)^{2k} \frac{3}{10} = 0.5882$.

7.7 The desired probability is $1 - P(A \cup B) = 1 - 0.7 - 0.5 + 0.3 = 0.1$.

7.9 Let $A = \{3k|k = 1, \ldots, 333\}$, $B = \{5k|k = 1, \ldots, 200\}$, and $C = \{7k|k = 1, \ldots, 142\}$. The first probability is $P(A \cup B) = \frac{333}{1000} + \frac{200}{1000} - \frac{66}{1000} = 0.467$ and the second probability is $P(A \cup B \cup C) = \frac{333}{1000} + \frac{200}{1000} + \frac{142}{1000} - \frac{66}{1000} - \frac{47}{1000} - \frac{28}{1000} + \frac{9}{1000} = 0.543$.

7.11 The inclusion-exclusion formula leads to the probability 0.051.

Chapter 8

8.1 Take as sample space the set of four pairs (G, G), (G, F), (F, G), and (F, F), where G stands for a "correct prediction" and F stands for a "false prediction," and the first and second components of each pair refer to the predictions of weather station 1 and weather station 2. The probabilities $0.9 \times 0.8 = 0.72$, $0.9 \times 0.2 = 0.18$, $0.1 \times 0.8 = 0.08$, and $0.1 \times 0.2 = 0.02$ are assigned to the elements (G, G), (G, F), (F, G), and (F, F). The desired probability is $P(\{(G, F)\} | \{(G, F), (F, G)\}) = 0.18/0.26 = 0.692$.

8.3 The probabilities are $\frac{0.4388}{0.6962} = 0.630$ and $\frac{0.1097}{0.25} = 0.439$.

8.5 The desired probability is $\frac{2}{7} \times \frac{2}{6} \times \frac{2}{5} \times \frac{2}{4} \times \frac{2}{3} = 0.0127$. *Remark*: using clever counting arguments it can be reasoned that the desired probability is $2^6/7!$; however, the derivation using conditional probabilities is simpler.

8.7 Condition on the number of matches among the one million tickets. As shown in Example 7.7, the probability of j matches is $e^{-1}/j!$ for $j \geq 0$. The desired probability is $1 - \sum_j \left[\binom{1{,}000{,}000-j}{500{,}000} \Big/ \binom{1{,}000{,}000}{500{,}000}\right] \frac{e^{-1}}{j!} = 0.3935$.

8.9 The desired probability p satisfies $\frac{p}{1-p} = \frac{1/5}{4/5} \frac{(0.75)^3}{(0.5)^3} = 0.84375$, which gives $p = 0.4576$.

Chapter 9

9.1 The expected values are $\frac{10}{3}$ and 2.75.

9.3 Your stake should be

$$2\left[1 + \frac{1}{2} + \cdots + \frac{1}{m+1} - 1\right] \approx 2\left[\ln(m+1) + \gamma - 1 + \frac{1}{2(m+1)}\right],$$

where $\gamma = 0.57722\ldots$ is Euler's constant.

9.5 Let X_i be equal to 1 if there is a birthday on day i and 0 otherwise. For each i, $P(X_i = 0) = \left(\frac{364}{365}\right)^{100}$ and $P(X_i = 1) = 1 - P(X_i = 0)$. The expected number of distinct birthdays is $365 \times \left(1 - \left(\frac{364}{365}\right)^{100}\right) = 87.6$.

9.7 Let P be the net profit in any given month. Then, $E(P) = 1522.50$ dollars and $\sigma(P) = 536.77$ dollars.

9.9 The random variables X and Y are dependent (e.g. $P(X = 2, Y = 1)$ is not equal to $P(X = 2)P(Y = 1)$). The values of $E(XY)$ and $E(X)E(Y)$ are given by $\frac{1232}{36}$ and $7 \times \frac{161}{36}$.

Chapter 10

10.1 The random variable V satisfies $P(V \leq v) = P(X \leq v/(1+v)) = \frac{v}{1+v}$ for $v \geq 0$. Its density function is equal to $\frac{1}{(1+v)^2}$ for $v > 0$ and 0 otherwise. The random variable W satisfies $P(W \leq w) = 1 - \sqrt{1-4w}$ for $0 \leq w \leq \frac{1}{4}$ and its density function is equal to $2(1-4w)^{-1/2}$ for $0 < w < \frac{1}{4}$ and 0 otherwise.

10.3 Let X be the distance of the point from the origin. Then, $P(X \leq a) = \frac{1}{4}\pi a^2$ for $0 \leq a \leq 1$ and $P(X \leq a) = \frac{1}{4}\pi a^2 - 2\int_1^a \sqrt{a^2 - x^2}\,dx = \frac{1}{4}\pi a^2 - a^2\{\arcsin(1) - \arcsin(\frac{1}{a})\} + \sqrt{a^2 - 1}$ for $1 < a \leq \sqrt{2}$. The density function $f(x)$ of X satisfies $f(x) = \frac{1}{2}\pi x$ for $0 < x \leq 1$ and $f(x) = \frac{1}{2}\pi x - 2x\{\arcsin(1) - \arcsin(\frac{1}{x})\}$ for $1 < x < \sqrt{2}$. Numerical integration leads to $E(X) = \int_0^{\sqrt{2}} xf(x)\,dx = 0.765$.

10.5 If you choose point c, your expected loss is $E\left[(X-c)^2\right] = E(X^2) - 2cE(X) + c^2$. This expression is minimal for $c = E(X)$. The minimal value is the variance of X.

10.7 Let the random variable X represent the total demand during the lead time of the replenishment order. Define the function $g(x)$ by $g(x) = x - s$ for $x > s$ and $g(x) = 0$ for $0 \leq x \leq s$. Simple calculations lead to $E[g(X)] = e^{-\lambda s}/\lambda$ and $E[(g(X))^2] = 2e^{-\lambda s}/\lambda^2$. Hence, the expected value and standard deviation of the shortage are given by $\frac{1}{\lambda}e^{-\lambda s}$ and $\frac{1}{\lambda}\left[e^{-\lambda s}(2 - e^{-\lambda s})\right]^{1/2}$.

10.9 The random variable X is distributed as $\min(U, 1-U)$, where U has the uniform density on (0,1). **(a)** The substitution rule leads to $E(X) = \frac{1}{4}$, $E[X/(1-X)] = 2\ln(2) - 1$ and $E[(1-X)/X] = \infty$. Also, $\text{var}(X) = \frac{1}{48}$ and $\text{var}[X/(1-X)] = 0.078$, **(b)** The random variable X satisfies $P(X \leq x) = 2x$ for $0 \leq x \leq \frac{1}{2}$, and its density function equals 2 for $0 < x < \frac{1}{2}$ and 0 otherwise. Since $P(X \leq x) = 2x$, the random variables $V = X/(1-X)$ and $W = (1-X)/X$ satisfy $P(V \leq v) = 2v/(1+v)$ for $0 \leq v \leq 1$ and $P(W \leq w) = 1 - 2/(1+w)$ for $w \geq 1$. The density function of V equals $2/(1+v)^2$ for $0 < v < 1$ and 0 otherwise. The density function of W equals $2/(1+w)^2$ for $w \geq 1$ and 0 otherwise.

10.11 If X gamma(α, λ) distributed, then

$$E(X) = \int_0^\infty x\frac{\lambda^\alpha}{\Gamma(\alpha)}x^{\alpha-1}e^{-\lambda x}\,dx = \frac{\Gamma(\alpha+1)}{\lambda\Gamma(\alpha)}\int_0^\infty \frac{\lambda^{\alpha+1}}{\Gamma(\alpha+1)}x^\alpha e^{-\lambda x}\,dx,$$

yielding that $E(X) = \frac{\Gamma(\alpha+1)}{\lambda\Gamma(\alpha)} = \frac{\alpha\Gamma(\alpha)}{\lambda\Gamma(\alpha)} = \frac{\alpha}{\lambda}$. Similarly, $E(X^2) = \frac{\Gamma(\alpha+2)}{\lambda^2\Gamma(\alpha)} = \frac{(\alpha+1)\alpha}{\lambda^2}$. Use the change of variable $z = (\lambda x)^\alpha$ to obtain $E(X^k) = \frac{\Gamma(1+k/\alpha)}{\lambda^k}$ for a Weibull-distributed random variable X.

10.13 $$P(X \leq x) = \frac{1}{\sigma\sqrt{2\pi}}\int_{-\infty}^{x} e^{-\frac{1}{2}w^2/\sigma^2}\,dw - \frac{1}{\sigma\sqrt{2\pi}}\int_{-\infty}^{-x} e^{-\frac{1}{2}w^2/\sigma^2}\,dw \quad \text{for } x \geq 0.$$

Taking the derivative shows that the density function of X is equal to $\frac{2}{\sigma\sqrt{2\pi}}e^{-\frac{1}{2}x^2}$ for $x > 0$ and 0 otherwise.

10.15 **(a)** $P(I(U) \leq x) = P(U \leq F(x)) = F(x)$ for all x. **(b)** Apply twice the substitution rule.

Chapter 11

11.1 Let X denote the low points rolled and Y the high points rolled. Then, $P(X = i, Y = i) = \frac{1}{36}$ for $1 \leq i \leq 6$ and $P(X = i, Y = j) = \frac{2}{36}$ for $1 \leq i < j \leq 6$.

11.3 The area of the triangle is $\frac{1}{2}$. The joint density function $f(x, y)$ of X and Y equals 2 for (x, y) inside the triangle and 0 otherwise. The random variable $V = X + Y$ satisfies $P(V \leq v) = 2\int_0^v dx \int_0^{v-x} dy$ and so $P(V \leq v) = v^2$ for $0 \leq v \leq 1$. The density function of V is $2v$ for $0 < v < 1$ and 0 otherwise. The random variable $W = \max(X, Y)$ satisfies

$$P(W \leq w) = P(X \leq w, Y \leq w) = 2\int_0^w dx \int_0^{\min(1-x,w)} dy,$$

yielding that $P(W \leq w) = 2w^2$ for $0 \leq w \leq \frac{1}{2}$ and $P(W \leq w) = 4w - 2w^2 - 1$ for $\frac{1}{2} \leq w \leq 1$. The density function of W equals $4w$ for $0 < w < \frac{1}{2}$, $4 - 4w$ for $\frac{1}{2} \leq w < 1$ and 0 otherwise.

11.5 **(a)** The joint density function $f(x, y)$ of X and Y satisfies $f(x, y) = f_X(x)f_Y(y)$ and is equal to 1 for all $0 < x, y < 1$ and 0 otherwise. The random variable $Z = X/Y$ satisfies $P(Z \leq z) = \int_0^1 dy \int_0^{\min(1,zy)} dx$, which yields the desired result.

(b) The probability that the first significant digit of Z equals 1 is given by

$$\sum_{n=0}^{\infty} P(10^n \leq Z < 2 \times 10^n) + \sum_{n=1}^{\infty} P(10^{-n} \leq Z < 2 \times 10^{-n})$$
$$= \frac{5}{18} + \frac{1}{18} = \frac{1}{3}.$$

In general, $P(\text{first significant digit of } Z \text{ is } k) = \frac{10}{18} \times \frac{1}{k(k+1)} + \frac{1}{18}$ for $k = 1, \ldots, 9$.

(c) It holds that $P(V \leq v) = \int_0^1 dx \int_0^{\min(1,v/x)} dy$, which yields $P(V \leq v) = v - v\ln(v)$ for $0 < v \leq 1$. Hence, $P(\text{first significant digit of } V \text{ is } k) = \sum_{n=1}^{\infty} \left[\frac{1}{10^n} + \frac{k}{10^n}\ln\left(\frac{k}{10^n}\right) - \frac{(k+1)}{10^n}\ln\left(\frac{k+1}{10^n}\right)\right]$.

(d) The joint density function of $Z = X/Y$ and U equals $f_Z(z)f_U(u)$. The density function $f_Z(z)$ equals $\frac{1}{2}$ for $0 < z \leq 1$ and $\frac{1}{2z^2}$ for $z > 1$. The density function $f_U(u)$ equals 1 for $0 < u \leq 1$ and 0 otherwise. The random variable $W = ZU$ satisfies

$$P(W \leq w) = \int_0^{\infty} f_Z(z) \int_0^{\min(w/z,1)} f_U(u)\,du.$$

This leads to $P(W \leq w) = \frac{3}{4}w - \frac{1}{2}w\ln(w)$ for $0 < w \leq 1$ and $P(W \leq w) = 1 - \frac{1}{4w}$ for $w > 1$. The density function of W equals $\frac{1}{4} - \frac{1}{2}\ln(w)$ for $0 < w \leq 1$ and $\frac{1}{4w^2}$ for $w > 1$.

11.7 The joint density function $f(x, y)$ of X and Y is equal to $4/\sqrt{3}$ for points (x, y) inside the triangle and 0 otherwise. The marginal density function $f_X(x)$ is equal to $\int_0^{x\sqrt{3}} f(x, y)\,dy = 4x$ for $0 < x < \frac{1}{2}$, $\int_0^{(1-x)\sqrt{3}} f(x, y)\,dy = 4(1 - x)$ for $\frac{1}{2} < x < 1$ and 0 otherwise. The marginal density function $f_Y(y)$ is equal to $\int_{y/\sqrt{3}}^{1-y/\sqrt{3}} f(x, y)\,dx = 4/\sqrt{3} - 8y/3$ for $0 < y < \frac{1}{2}\sqrt{3}$ and 0 otherwise.

11.9 The inverse functions are given by $x = \frac{v}{\sqrt{v^2+w^2}} e^{-\frac{1}{4}(v^2+w^2)}$ and $y = \frac{w}{\sqrt{v^2+w^2}} e^{-\frac{1}{4}(v^2+w^2)}$, and the Jacobian is $\frac{1}{2}e^{-\frac{1}{2}(v^2+w^2)}$.

11.11 $E(\text{annual rainfall}) = 799.5$ mm and $\sigma(\text{annual rainfall}) = 121.4$ mm.

11.13 $\rho(X, Y) = \frac{441/36-(91/36)(161/36)}{(1.40408)^2} = 0.479$.

Chapter 12

12.1 It suffices to prove the result for the standard bivariate normal distribution. Also it is no restriction to take $b > 0$. Let $W = aX + bY$. Differentiating

$$P(W \le w) = \frac{1}{2\pi\sqrt{1-\rho^2}} \int_{-\infty}^{\infty} dx[\int_{-\infty}^{(w-ax)/b} e^{-\frac{1}{2}(x^2-2\rho xy+y^2)/(1-\rho^2)}\,dy]$$

yields that the density function of W is given by

$$f_W(w) = \frac{(1/b)}{2\pi\sqrt{1-\rho^2}} \int_{-\infty}^{\infty} e^{-\frac{1}{2}[x^2-2\rho x(w-ax)/b+(w-ax)^2/b^2]/(1-\rho^2)}\,dx.$$

This expression for $f_W(w)$ can be reduced to $(\eta\sqrt{2\pi})^{-1}\exp(-\frac{1}{2}w^2/\eta^2)$ with $\eta = \sqrt{a^2 + b^2 + 2ab\rho}$.

12.3 $P(Z \le z) = \int_0^{\infty} dy \int_{-\infty}^{yz} f(x, y)\,dx + \int_{-\infty}^{0} dy \int_{yz}^{\infty} f(x, y)\,dx$. Differentiation leads to $f_Z(z) = \int_0^{\infty} yf(yz, y)\,dy - \int_{-\infty}^{0} yf(yz, y)\,dy$. Hence, $f_Z(z) = \int_{-\infty}^{\infty} |y|\, f(yz, y)\,dy$. Inserting the standard bivariate normal density for $f(x, y)$ and using the result of Problem 10.13, the desired result follows.

12.5 Go through the path of length n in opposite direction and next continue this path with m steps.

12.7 The vector (X_1, X_2) has a bivariate normal distribution. Use the fact that $aX_1 + bX_2$ is normally distributed for all constants a and b not both equal to zero.

12.9 The observed value of test statistic D is 0.470. The probability $P(\chi_3^2 > 0.470) = 0.925$. The agreement with the theory is very good.

12.11 The value of the test statistic D is 20.848. The probability $P(\chi_5^2 > 20.848) = 8.65 \times 10^{-4}$. This is a strong indication that the tickets are not randomly filled in.

Chapter 13

13.1 Condition on the unloading time. The probability of no breakdown is given by $\int_{-\infty}^{\infty} e^{-\lambda y} \frac{1}{\sigma\sqrt{2\pi}} e^{-\frac{1}{2}(y-\mu)^2/\sigma^2}\,dy = e^{-\mu\lambda+\frac{1}{2}\sigma^2\lambda^2}$.

13.3 The desired probability $P(B^2 \geq 4AC)$ can be calculated as

$$\int_0^1 P\left(AC \leq \frac{b^2}{4}\right) db = \int_0^1 db \left[\int_0^1 P\left(C \leq \frac{b^2}{4a}\right) da\right]$$

$$= \int_0^1 db \left[\int_0^{\frac{b^2}{4}} 1\, da + \int_{\frac{b^2}{4}}^1 \frac{b^2}{4a}\, da\right] = \int_0^1 db \left[\frac{b^2}{4} - \frac{b^2}{4}\ln\left(\frac{b^2}{4}\right)\right].$$

This leads to $P\left(B^2 \geq 4AC\right) = \frac{5}{36} + \frac{1}{6}\ln(2) = 0.2544$.

13.5 Condition on the spread. The probability of a spread of i points is given by $\alpha_i = [(12 - i) \times 4 \times 4 \times 2]/(52 \times 51)$ for $i = 0, 1, \ldots, 11$. Define the constants $\gamma_1 = 6$, $\gamma_2 = 5$, $\gamma_3 = 3$ and $\gamma_i = 2$ for $i \geq 4$. Then,

$$E(\text{stake}) = 10 + \sum_{i=7}^{11} \alpha_i \times 10 = \$11.81$$

$$E(\text{payoff}) = \sum_{i=1}^{6} \alpha_i \times \frac{4i}{50} \times \gamma_i \times 10 + \sum_{i=7}^{11} \alpha_i \times \frac{4i}{50} \times \gamma_i \times 20$$

$$+\alpha_0 \times 10 + 13 \times \frac{4}{52} \times \frac{3}{51} \times \frac{2}{50} \times 120 = \$10.93.$$

The house percentage is 7.45%.

13.7 For fixed n, let $u_k(i) = E\left[X_k(i)\right]$. The goal is to find $u_n(0)$. Apply the recursion $u_k(i) = \frac{1}{2}u_{k-1}(i+1) + \frac{1}{2}u_{k-1}(i)$ for i satisfying $\frac{i}{n-k} \leq \frac{1}{2}$, and use the boundary conditions $u_0(i) = \frac{i}{n}$ and $u_k(i) = \frac{i}{n-k}$ for $i > \frac{1}{2}(n-k)$ and $1 \leq k \leq n$. The desired value $u_n(0)$ has the values 0.7083, 0.7437, 0.7675 and 0.7761 for $n = 5$, 10, 25 and 50 (*Remark*: $u_n(0)$ tends to $\frac{\pi}{4}$ as n increases without bound).

13.9 Let the random variable S represent your score. Define the random variable Y as equal to 0 if none of the five dice shows the face value one and 1 otherwise. It follows from $E(S) = E(S \mid Y = 0)P(Y = 0) + E(S \mid Y = 1)P(Y = 1)$ that $E(S)$ is equal to

$$5\left(\frac{2+3+\cdots+6}{5}\right) \times \left(\frac{5}{6}\right)^5 + 0 \times \left(1 - \left(\frac{5}{6}\right)^5\right).$$

This gives $E(S) = 8.038$. Similarly, if follows from $E(S^2) = E(S^2 \mid Y = 0)P(Y = 0) + E(S^2 \mid Y = 1)P(Y = 1)$ that $E(S^2)$ is equal to

$$\left[5\left(\frac{2^2+3^2+\cdots+6^2}{5}\right) + 20\left(\frac{2+3+\cdots+6}{5}\right)^2\right] \times \left(\frac{5}{6}\right)^5.$$

This gives $E(S^2) = 164.77$ and so $\sigma(S) = 10.01$. To find the probability $P(S \geq 20)$, apply repeatedly the convolution formula in order to calculate the probability mass function of $U_1 + \cdots + U_5$, where $U_1, \ldots, U_5$ are independent random variables each having the discrete uniform distribution on $2, 3, \ldots, 6$. Using the fact that $P(S \geq 20) = \left(\frac{5}{6}\right)^5 \sum_{j=20}^{30} P(U_1 + \cdots + U_5 = j)$, this leads to $P(S \geq 20) = 0.2254$.

Chapter 14

14.1 $G_X(z) = (1 - p + pz)^n$.

14.3 The number of record draws is distributed as $R = X_1 + \cdots + X_r$, where X_i equals 1 if the ith draw is a record draw and 0 otherwise. For each i, $P(X_i = 1) = \frac{1}{i}$ and $P(X_i = 0) = 1 - \frac{1}{i}$. The random variables $X_1, \ldots, X_r$ are independent (the proof of this fact is not trivial). This leads to $G_R(z) = z(\frac{1}{2} + \frac{1}{2}z) \cdots (1 - \frac{1}{r} + \frac{1}{r}z)$. The expected value and variance of R are given by $\sum_{i=1}^{r} 1/i$ and $\sum_{i=1}^{r}(i-1)/i^2$.

14.5 The generating function of the offspring distribution is $P(u) = \frac{1}{3} + \frac{2}{3}u^2$. **(a)** To find u_3, iterate $u_n = P(u_{n-1})$ starting with $u_0 = 0$. This gives $u_1 = P(0) = \frac{1}{3}$, $u_2 = P(\frac{1}{3}) = \frac{1}{3} + \frac{2}{3}\left(\frac{1}{3}\right)^2 = \frac{11}{27}$, and $u_3 = P(\frac{11}{27}) = \frac{1}{3} + \frac{2}{3}\left(\frac{11}{27}\right)^2 = 0.4440$. **(b)** The equation $u = \frac{1}{3} + \frac{2}{3}u^2$ has roots $u = 1$ and $u = \frac{1}{2}$. The probability $u_\infty = \frac{1}{2}$. **(c)** The probabilities are $u_3^2 = 0.1971$ and $u_\infty^2 = 0.25$.

14.7 Put $\mu = tE(X) + uE(Y)$ and $\sigma^2 = t^2\sigma^2(X) + u^2\sigma^2(Y) + 2tu\,\text{cov}(X, Y)$ for fixed t, u. Then $M_{X,Y}(t, u) = \exp(\mu + \frac{1}{2}\sigma^2)$. The function $M_{X,Y}(t, u)$ can be written as the product of $M_X(t) = \exp\left(tE(X) + \frac{1}{2}t^2\sigma^2(X)\right)$ and $M_Y(u) = \exp\left(uE(Y) + \frac{1}{2}u^2\sigma^2(Y)\right)$ if and only if $\text{cov}(X, Y) = 0$.

14.9 If $t < 0$, then $P(X \leq c) = P(tX \geq tc) = P(e^{tX} \geq e^{tc})$. Next apply Markov's inequality.

Bibliography

Aldous, D.J. and P. Diaconis. "Shuffling cards and stopping times." *The American Mathematical Monthly* 93 (1986): 333–348.

Bennett, D.J. *Randomness*. Cambridge, MA: Harvard University Press, 1999.

Berry, D.A. *Statistics: A Bayesian Perspective*. Belmont, CA: Duxbury Press, 1996.

Diaconis, P. and F. Mosteller. "Methods for studying coincidences." *Journal of the American Statistical Association* 84 (1989): 853–861.

Gigerenzer, G. *Calculated Risks*. New York, NY: Simon & Schuster, 2002.

Hanley, J.A. "Jumping to coincidences: Defying odds in the realm of the preposterous." *Journal of the American Statistical Association* 46 (1992): 197–202.

Harmer, G.P. and D. Abbott. "Losing strategies can win by Parrondo's paradox." *Nature* 402 (1999, 23/30 December): 864.

Henze, N. and H. Riedwyl. *How to Win More*. Natics, MA: A.K. Peters, 1998.

Hill, T.P. "The difficulty of faking data." *Chance Magazine* 12 (1999): 27–31.

Kadell, D. and D. Ylvisaker. "Lotto play: The good, the fair and the truly awful." *Chance Magazine* 4 (1991): 22–25.

Kahneman, D., P. Slovic and A. Tversky. *Judgment under Uncertainty: Heuristics and Bias*. Cambridge, MA: Cambridge University Press, 1982.

Maor, E. *e: The Story of a Number*. Princeton, NJ: Princeton University Press, 1994.

Matthews, R. "Ladies in waiting." *New Scientist* 167 (July 29, 2000): 40.

McKean, K. "Decisions, decisions, ..." *Discover* (June 1985): 22–31.

Merz, J.F. and J.P. Caulkins. "Propensity to abuse-propensity to murder?" *Chance Magazine* 8 (1995): 14.

Paulos, J.A. *Innumeracy: Mathematical Illiteracy and its Consequences*. New York, NY: Vintage Books, 1988.

Savage, S. "The flaw of averages." *San Jose Mercury News* (October 8, 2000).

Thaler, R.H. *The Winner's Curse, Paradoxes and Anomalies in Economic Life*. Princeton, NJ: Princeton University Press, 1992.

vos Savant, M. *The Power of Logical Thinking*. New York, NY: St. Martin's Press, 1997.

vos Savant, M. "Ask Marilyn." *Parade* (February 7, 1999).

Weiss, G. "Random walks and their applications." *American Scientist* 71 (January–February, 1983): 65–70.

Index